BRADY

Fire Department Hydraulics

Eugene Mahoney

Upper Saddle River, New Jersey 07458

Library of Congress Cataloging-in-Publication Data

Mahoney, Eugene
   Fire department hydraulics / Eugene Mahoney.
      p. cm.
Includes bibliographical references and index.
   ISBN 0-13-111309-7
   1. Fire extinction—Water supply.   I. Title.
   TH9311 .M33 2003
   628.9′252—dc21

                         2003008474

Publisher: *Julie Levin Alexander*
Publisher's Assistant: *Regina Bruno*
Senior Acquisitions Editor: *Katrin Beacom*
Assistant Editor: *Kierra Kashickey*
Senior Marketing Manager: *Katrin Beacom*
Product Information Manager: *Rachele Strober*
Director of Production and Manufacturing: *Bruce Johnson*
Managing Production Editor: *Patrick Walsh*
Manufacturing Buyer: *Pat Brown*
Production Liaison: *Julie Li*
Production Editor: *Karen Ettinger, The GTS Companies/York, PA Campus*
Creative Director: *Cheryl Asherman*
Cover Design Coordinator: *Christopher Weigand*
Cover Designer: *Christopher Weigand*
Compositor: *The GTS Companies/York, PA Campus*
Printer/Binder: *Courier Westford*
Cover Printer: *Phoenix Color Corporation*

Cover photo courtesy of Task Force Tips.

Credits and acknowledgments borrowed from other sources and reproduced, with
permission, in this textbook appear on appropriate page within text.

Pearson Education LTD.
Pearson Education Singapore, Pte. Ltd
Pearson Education, Canada, Ltd
Pearson Education–Japan

Pearson Education Australia PTY, Limited
Pearson Education North Asia Ltd
Pearson Educación de Mexico, S.A. de C.V.
Pearson Education Malaysia, Pte. Ltd

10 9 8 7 6 5 4 3 2 1
ISBN 0-13-111309-7

The Fireman's Prayer

When I am called to duty, God, whenever flames may rage;
Give me the strength to save some life, whatever be its age.
Help me embrace a little child before it's too late
Or save an older person from the horror of that fate.
Enable me to be alert and hear the weakest shout,
And quickly and efficiently to put the fire out.
I want to fill my calling and give the best in me,
To guard my every neighbor and protect their property.
And if, according to Your will, I lose my life;
Please bless with Your protective hand
my children and my wife.

(Author Unknown)

Contents

Preface

Hydraulics is the branch of physics having to do with the mechanical properties of water and other liquids and the application of these properties in engineering. Fire department hydraulics essentially involves the application of water and other liquids in the many aspects of fire protection, particularly fire fighting. People who study fire department hydraulics are generally divided into two groups: those interested in the engineering aspects of fire protection and those responsible for producing adequate hose streams on the fire ground. The first group is most interested in the "why" and the second group in the "how" of hydraulics.

This book intends to meet the needs of both groups by presenting material in such a way that the text can be used either in a formalized classroom situation or for self-study. Review and test questions are included at the end of each chapter so that students can check their comprehension of the subject matter. When a reader can answer the questions and work the problems without difficulty, he or she can feel that they have an adequate foundation for competing in promotional examinations in the area of hydraulics.

Over the years I have asked many students and scores of experienced firefighters what they thought was needed in a book on fire department hydraulics. The answers kept repeating themselves: lots of examples, abundant illustrations, and step-by-step solutions to problems. The format of this book reflects the requests of those people.

I have also sought the advice of my many associates over the years, seeking opinions about course content and sequence of presentation. Their contributions were invaluable.

Perhaps the greatest contributor to this manual is the Los Angeles City Fire Department for providing me with the fire-fighting experience that is essential to the understanding of the application of fire department hydraulic principles. It all comes together, of course, on the fire ground.

I wish to thank the following individuals and organizations who so willingly contributed material or advice for use in this edition of the book: Deputy Chief Jim Beery, Portland, Oregon, Fire Department; Battalion Chief Billy Goldfedder, Loveland–Symmes, Ohio, Fire Department; Captain Steven Gobel, Henderson, Nevada, Fire Department; Lieutenant George Fulcher, Columbus, Ohio, Division of Fire; Executive Officer Jeffrey T. Lindsey, M. Ed. Estero, Florida, Fire Rescue; Fire Chief Richard Marinucci, Farmington Hills, Michigan, Fire Department; Steward McMillan and Rod Carringer, Task Force Tips Inc., Valparaiso, Indiana; David Wunderlin; Robert Alderman; Russell Strickland, Maryland Fire and Rescue Institute, University of Maryland, College Park, Maryland; Insurance Services Office, Jersey City, New Jersey; National Fire Protection Association, Quincy, Massachusetts; Clow Corporation, Oskaloosa, Iowa; Akron Brass Company, Wooster, Ohio; Elkhart Brass Mfg. Co., Elkhart, Indiana; and Pierce Manufacturing, Appleton, Wisconsin. I also want to thank the staff at Pearson Education, in particular my editor, Katrin Beacom, and her assistant, Kierra Kashickey. Special thanks is extended to project manager Karen Ettinger.

About the Author

Gene Mahoney was released to inactive duty as a pilot from the U.S. Navy in 1946. He served an additional 18 years in the Reserve, retiring as a lieutenant commander. During his time with the navy, he flew both reciprocating engine and jet aircraft.

Gene joined the Los Angeles Fire Department as a firefighter in 1947. He retired as a battalion chief in 1969. During his time with the department, he was assigned to various areas of the city, including five years in the downtown area, five years in the harbor area, and five years in the south-central area of the city. As a battalion chief, he served five years in the most active fire-fighting battalion in the city, additional time in the high-rise area of the city, and as commander in charge of the fire-fighting forces at the Los Angeles International Airport. His special-duty assignments included several years in the training section. At the time of his retirement, he was responsible for the public relations section of the department.

Gene retired from the Los Angeles Fire Department to accept the position of fire chief for the city of Garden Grove, California. He was later advanced to the position of public safety director and then accepted the assignment as assistant city manager for public safety. In these positions, he was responsible for the operation of both the fire and police departments. He left the city of Garden Grove to accept the position of fire chief for the Arcadia, California, Fire Department. He retired from this position in 1975.

Gene, together with another Los Angeles Fire Department captain, was responsible for the development of the fire science curriculum at Los Angeles Harbor College, Wilmington, California, and served there as a part-time instructor for twelve years. He also taught fire administration courses at Long Beach State College, Long Beach, California, for two years. Upon retiring as fire chief from the city of Arcadia, he accepted the position of fire science coordinator at Rio Hondo College, Whittier, California. While there, he developed the fire science curriculum into one of the most complete programs in the United States. The program includes a Fire Academy, which provides all the training required for certification as a Fire Fighter I in California. He retired from Rio Hondo College as a professor of fire science in 1988.

While with the Los Angeles Fire Department, Gene attended the University of Southern California, where he received his B.S. degree in Public Administration with a minor in Fire Administration in 1956 and three years later his M.S. degree in Education.

In addition to authoring several articles in professional magazines, Gene has authored several textbooks and study guides in the field of fire science. The textbooks include *Fire Department Hydraulics, Introduction to Fire Apparatus and Equipment, Fire Department Oral Interviews: Practices and Procedures,* and *Fire Suppression Practices and Procedures.* The study guides include one for his text, *Introduction to Fire Apparatus and Equipment*; one entitled *Firefighters Promotion Examinations*; and one on *Effective Supervisory Practices.* He also had a novel published, entitled *Anatomy of an Arsonist.*

During his career, Gene has been very active in professional and service organizations. He served as:

District Chairman, Boy Scouts of America
President, United Way
District Chairman, Salvation Army
President, International Association of Toastmasters
President, Rio Hondo College Faculty Association

Introduction

The fire service is and probably always will be embedded in tradition. While the strength of tradition is stability, it also has its weakness. The weakness is that it is very difficult to make changes in any organization that thrives on tradition. When changes do occur, they usually occur over a long period of time in which the organization is engulfed in controversy. Despite this, however, changes do get implemented.

In the early years of the twentieth century, the objective of the fire service was to extinguish fires. Many fire officers believed that the best method of making the attack was to advance through the front door and wash the fire out the back door. Attacks were made primarily using smooth-bore tips in the belief that directing lots of water in the correct location was the best means of extinguishing a fire. While smooth-bore tips do a good job of extinguishing a fire, the runoff at times is as much as 90 percent. It was not unusual for the monetary loss to property caused by water damage to far exceed that caused by the fire itself. Insurance companies were so concerned with this that they started dispatching their own salvage companies to fires in buildings they insured to limit the water damage caused by the fire department.

Later, fire departments changed their objective of fire fighting from extinguishing the fire to extinguishing the fire with the least amount of loss to lives and property. To comply with this concept, tactics were changed and many departments started dispatching their own salvage companies to fires to limit the loss from water damage.

In the middle part of the twentieth century, fire attacks by hand lines were principally divided into three areas, as follows.

Small fires such as rubbish fires, small grass fires, and automobile fires were generally attacked using preconnected booster lines from a reel of ¾-inch or 1-inch hard rubber hose. The lines were equipped with a combination nozzle that could be manually adjusted to provide either a straight stream or a spray/fog stream. The source of water was from a tank carried on the apparatus. Then, as now, the majority of all fires fell into the category of small fires.

The normal attack for fires in dwellings and small commercial properties was to lay a single 2½-inch supply line from a hydrant to the fire and reduce it to two preconnected wyed lines of 1½-inch hose, each equipped with a combination straight stream and spray/fog nozzle. If the company was a triple combination company, the pumper would return to the hydrant and provide the necessary pressure to the nozzles. The general practice was to provide a pressure of 50 pounds per square inch (psi) to each nozzle, which discharged approximately 75 gallons per minute (gpm). The disadvantage of this layout was that the pumper was back at the hydrant, which made it difficult for firefighters to secure additional equipment needed from the apparatus. Of course, if the company was a two-piece company, the hose wagon would remain at the fire to provide additional equipment as needed.

Larger exterior fires and those in larger commercial and industrial buildings demanding handheld lines were normally attacked using 2½-inch lines equipped with smooth-bore tips. Most departments equipped the apparatus with three different size tips, a 1-inch tip, a 1⅛-inch tip, and a 1¼-inch tip. The 1-inch tip discharged approximately 210 gpm, the 1⅛-inch tip approximately 265 gpm, and the 1¼-inch tip approximately 325 gpm, all at a nozzle pressure of 50 psi.

At first glance it appeared that the company officer had a choice of which tip to use, depending upon his or her estimate of the amount of water needed to effectively extinguish the fire. Unfortunately, that was not the way things usually worked.

The general principle that seemed to apply was whichever tip was originally selected was the one used during the entire extinguishing operation. To change to a different tip would generally result in the necessity of shutting down the line. Few fire officers chose to do this, especially with the fire still in progress. The tip selected generally depended upon the layout of the original supply line.

Usually one of the nozzles was in its original configuration and ready to be connected when a forward layout was made. Another of the nozzles was equipped with a double male in preparation for use when a reverse layout was made. Consequently, the tip selected depended upon whether the supply line was from the hydrant to the fire or from the fire to the hydrant.

The three tips generally available for a master stream were a 1½-inch tip, which furnished approximately 600 gpm, a 1¾-inch tip for 800 gpm, and a 2-inch tip for approximately 1100 gpm, all at a nozzle pressure of 80 psi. At times these were arranged in a stacked formation with the 2-inch tip connected to the appliance, the 1¾-inch tip to the 2-inch tip, and the 1½-inch tip on the end. Unless a definite decision was made to use the 1¾-inch or the 2-inch tip prior to loading the line, the initial attack was made using the 1½-inch tip. To change this to one of the other tips required that the appliance be shut down and the 1½-inch tip be removed. This was seldom done.

A few years prior to the dawn of the twenty-first century, several improvements were introduced to the fire service which offered a tremendous improvement in operations. Combined, they completely changed the way many layouts and attacks were conducted. Three significant changes were the introduction and use of 1¾-inch hose to replace the previously used 1½-inch hose for interior fire fighting, the introduction and use of large-diameter hose for supply lines and to feed heavy stream appliances, and the development and use of the automatic nozzle on booster lines, inside lines, and heavy-stream appliances. The standard acceptable nozzle pressure for these nozzles was 100 psi.

No combination of new equipment has ever had as great an impact on the fire service. The three changes not only improved operations, but helped simplify the pump operator's task. The automatic nozzle was also referred to variously as the thinking nozzle, the intelligent nozzle, and the pump operator's best friend. The nozzle did not, however, in any way change the need for all firefighters, and especially fire officers and pump operators, to be familiar with and understand the basic principles and application of fire service hydraulics. In fact, the National Fire Protection Association (NFPA) in its performance requirements for a pump operator as outlined in standard number 1002, Standard for Fire Apparatus Driver/Operator Professional Qualifications (1998 edition), states in several places that requisite knowledge includes "hydraulic calculations for friction loss and flow using both written formulas and estimation methods."

The objective of this book is to provide the reader with the knowledge necessary to establish a foundation for understanding and applying time-honored principles of fire hydraulics. The three changes affecting fire department operations, together with changes in the use of hydraulic formulas, have been taken into consideration and adopted in this edition.

Principles of Fire Department Hydraulics

Objectives

Upon completing this chapter, the reader should:

- Understand the various characteristics of water—the basic element of fire department hydraulics.
- Be able to discuss the effect that steam expansion has on the ability to purge an area of smoky and noxious gases.
- Be able to discuss the advantages that water has as an extinguishing agent.
- Be able to recognize some of the disadvantages that water has when used as an extinguishing agent.
- Understand and be able to discuss the four characteristics of water that affect its use in fire protection.
- Know the weight of 1 cubic foot of water, the number of gallons of water in 1 cubic foot, the weight of 1 gallon of water, the number of cubic inches in 1 cubic foot, the number of cubic inches in 1 gallon of water, the weight of a column of water measuring 1 inch by 1 inch in base by 1 foot high, and the weight of a column of water measuring 1 inch by 1 inch in base by 2.304 feet high.
- Be able to define both force and pressure.
- Know the differences among static pressure, flow pressure, and residual pressure.
- Be able to explain the six basic rules governing the primary characteristics of pressure in liquids.
- Be able to define "head."
- Be able to determine the pressure when the head is known.
- Be able to determine the head when the pressure is known.
- Understand the effect of elevation on pressure.
- Be able to define back pressure and forward pressure.
- Be able to determine the back pressure or forward pressure when lines are laid either uphill or downhill.
- Be able to determine the force on the base of a container.
- Understand and be able to work problems involving the force on clapper valves.

◆ WATER—THE PRIMARY EXTINGUISHING AGENT

Hydraulics is the branch of physics dealing with the mechanical properties of water and other liquids and the application of these properties in engineering. For the purpose of this text, **fire department hydraulics** is defined as the portion of general hydraulics that pertains to water and its use in fire fighting and fire protection.

Fire department hydraulics is not a precise science. Empirical formulas are in general use for solving problems. Many numbers used in problem solving are rounded off for convenience and simplification. Despite this fact, it is important that firefighters and others engaged in fire protection activities be familiar with fire department hydraulic principles and their application to fire fighting and fire protection.

Water is the primary extinguishing agent used in fire protection. In its pure state it is a colorless, odorless, and tasteless substance. It is a relatively stable chemical compound composed of two atoms of hydrogen and one atom of oxygen. Although it has the ability to extinguish fire under certain circumstances by smothering, diluting, and emulsifying, its primary use as an extinguishing agent is because of its heat absorption capability and its availability.

Several properties of water affect its heat-absorbing qualities. Water freezes at 32°F (Fahrenheit) at a normal sea level pressure of 14.7 psi. The ice that forms when water freezes will melt commencing at 32°F. When 1 pound of water freezes, 143.4 Btu of heat is released. When 1 pound of ice melts into water, 143.4 Btu of heat is absorbed. This is referred to as the **latent heat of fusion**. The latent heat of fusion is defined as the amount of heat absorbed or released by a substance as it passes between the solid and liquid phases. Latent heat is measured in Btu's or in calories per unit weight. One Btu is defined as the amount of heat required to raise the temperature of 1 pound of water 1°F. One **calorie** is defined as the amount of heat required to raise the temperature of 1 gram of water 1°C (Celsius). One Btu is equal to 252 calories.

The **specific heat** of a substance is the amount of heat required to raise its temperature 1°F, or the number of calories required to raise the temperature of 1 gram of the substance 1°C. From a fire protection standpoint, the specific heat of a substance should be thought of as its thermal capacity, or its ability to absorb heat. The specific heat of water is 1.0. The specific heat of water is higher than that of other substances.

To raise 1 pound of water from a temperature of 60°F to 212°F requires 152 Btu. When water reaches 212°F at a sea level pressure of 14.7 psi, it is ready to start the change from a liquid to a vapor. Additional heat is absorbed as water changes from a liquid to steam; however, the temperature of the liquid does not increase. The absorption of the additional heat reduces the volume of the liquid, with volume reduction continuing until the last drop of water has been converted to steam. The amount of heat required to convert 1 pound of water into steam is 970.3 Btu. This is referred to as the **latent heat of vaporization**. The latent heat of vaporization is defined as the amount of heat absorbed or given off as a substance passes between the liquid and gaseous phases.

Water weighs approximately 8.33 pounds per gallon (8.35 is generally used for most hydraulic calculations). When the information concerning the heat absorption qualities of water is combined with information on the weight of water, it is easy to see that water is much more effective as an extinguishing agent during the process of changing from a liquid to steam than in raising its temperature from an ambient temperature of 60°F to the boiling point. As an example, the amount of heat required to raise the temperature of 1 gallon of water from 60°F to 212°F is approximately 1266 Btu (8.33 × 152). The amount of heat required to change 1 gallon of water from a liquid to steam is approximately 8082 Btu (8.33 × 970.3). Therefore, it should be noted that nearly six and one-half times as much heat is absorbed during the process of vaporization as is absorbed during the process of raising the temperature of water from 60°F to 212°F. In addition, the maximum effectiveness of water as a cooling agent can only be obtained when the entire amount of water discharged at a fire has been totally converted to steam. To put these figures in perspective, wood burning generally produces between 8000 and 9000 Btu per pound.

Another factor of importance that occurs when water is changed to steam is its tremendous expansion. The expansion ratio is approximately 1700 to 1 at a normal atmospheric pressure of 14.7 psi and a temperature of 212°F. The expansion ratio itself is a function of the temperature of the fire area. At 500°F, the ratio is approximately 2400 to 1. When the temperature in the fire area reaches 1200°F, the ratio increases to approximately 4200 to 1. This expansion forces smoky and noxious gases from the involved structure and reduces the amount of oxygen available to support combustion.

The effect that steam expansion has on the ability to purge an area of smoky and noxious gases can best be illustrated by referring to a fire condition. Taken at 90 percent efficiency, 50 gallons of water converted to steam at various temperatures can be expected to occupy the following amounts of space:

Temperature (°F)	Cubic Feet of Steam	Sample Room Size
212	10,000	8′ by 25′ by 50′
400	12,000	8′ by 25′ by 62½′
800	17,500	8′ by 25′ by 87½′
1000	20,000	8′ by 25′ by 100′

It should be noted that the primary value of the steam expansion is the purging of the air and the resultant reduction of the available oxygen needed to support combustion; however, fires in ordinary combustible materials are normally extinguished by the absorption of heat, not by the smothering effect created by the steam. The smothering effect has the tendency to suppress flaming, but it is the cooling effect that extinguishes the fire.

Whereas water is effective as an extinguishing agent primarily because of its cooling ability, it can also be used occasionally to extinguish fires by emulsification or dilution. When used for emulsification, it is generally applied to the surface of a viscous flammable liquid in a relatively strong, coarse water spray. Care is necessary to avoid violent frothing.

The use of water to extinguish a fire by dilution is limited. Generally, large amounts of water are required, which restrict the use of the dilution method to those situations where the danger of overflow is not a problem. The percentage of water required and the time necessary for extinguishment vary with the liquid to be diluted.

Although water has a number of advantages for use as an extinguishing agent, it also has the following disadvantages:

1. Surface tension, which limits its ability to penetrate some materials such as upholstery fabric and baled cotton.
2. Conductivity, which creates the possibility of electrical shock for firefighters working around high-voltage lines.
3. Violent reaction with certain chemicals.
4. Low viscosity, which allows rapid runoff and therefore limits the ability to blanket a fire.

5. Ability to freeze.
6. Potential damaging effects on certain products, particularly electrical equipment.

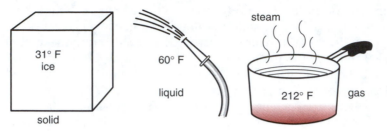

FIGURE 1.1 ◆

In addition to its extinguishing capabilities, there are several characteristics of water that affect its use in fire protection. Some of them are as follows:

1. Water may be found as a solid, a liquid, or a gas, depending upon the temperature to which it is exposed. It is normally a liquid or gas when used in firefighting operations; however, its ability to freeze cannot be ignored. Constant thought must be given to the possibility of water supplies becoming unavailable during freezing temperatures (Figure 1.1).

2. Water seeks its own level. As an example, as long as the top of each portion of the container is open to the atmosphere, water poured into a container at point *A* in Figure 1.2 will eventually be of equal depth with that in other portions of the container. As water seeks its own level, it creates a pressure at any point where it is stopped from accomplishing this objective. In the example shown in Figure 1.3, pressure has been created at

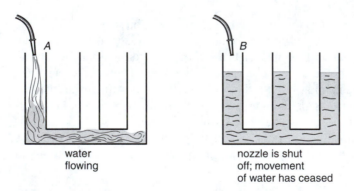

FIGURE 1.2 ◆

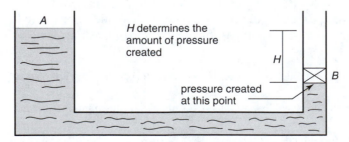

FIGURE 1.3 ◆

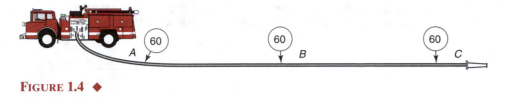

FIGURE 1.4 ◆

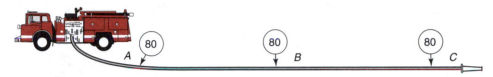

FIGURE 1.5 ◆

point *B*. The amount of pressure created depends upon the difference in elevation between point *A* and point *B*.

3. Water is practically incompressible. A pressure of approximately 30,000 psi is required to reduce the volume of water 1 percent. Due to the incompressibility, an increase in pressure applied to water at rest in a confined container will be transmitted equally in all directions. This concept can be illustrated best by a pumping situation. If the pumper shown in Figure 1.4 is pumping at a discharge pressure of 60 psi with the line laid at ground level and the hose is full of water with the nozzle closed, the pressure of 60 psi will be transmitted equally throughout the hose line; therefore, **pressure gages** tapped in at points *A*, *B*, and *C* will read 60 psi. If the pump discharge pressure is increased to 80 psi, then the increase in pressure will be transmitted throughout the entire length of hose and the pressure reading at points *A*, *B*, and *C* will be 80 psi (see Figure 1.5). Because water is incompressible, whenever it is contained in such form that movement is possible, a difference of pressure between two points will cause a movement of the water. This concept is illustrated in Figure 1.6.

4. The **density** of a substance is defined as its weight per unit volume. The density of water varies with the temperature. Water is at its maximum density (weight) at a temperature of 39.2°F. At this temperature it weighs approximately 62.425 pounds per cubic foot. A weight of 62.5 pounds per cubic foot is generally used in the fire service. The density (weight) of water is less than 62.425 pounds per cubic foot at temperatures both above and below 39.2°F. Some sample densities of water at various temperatures are shown in Table 1.1.

FIGURE 1.6 ◆ Pressure at point *A* causes the water to move up to point *B* in the tube.

TABLE 1.1 ◆ Density of Water at Various Temperatures			
Temperature (°F)	*Weight per Cubic Foot*	*Temperature (°F)*	*Weight per Cubic Foot*
32.0	62.416	80	62.217
39.2	62.425	90	62.118
50.0	62.408	100	61.998
60.0	62.366	150	61.203
70.0	62.300	200	60.135

This table is for freshwater. Saltwater weighs approximately 64 pounds per cubic foot at a temperature of 39.2°F.

BASIC FIRE DEPARTMENT HYDRAULIC FIGURES

A container measuring 1 foot by 1 foot by 1 foot has a volume of 1 cubic foot (Figure 1.7).

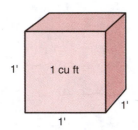

FIGURE 1.7 ◆

If a 1 cubic foot container is filled with water, the water will weigh approximately 62.5 pounds (Figure 1.8).

If the water from the 1 cubic foot container is poured into 1-gallon bottles, it will fill approximately 7.48 bottles (Figure 1.9).

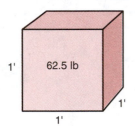

FIGURE 1.8 ◆

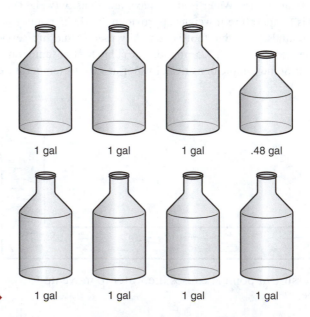

FIGURE 1.9 ◆

The water in each of the 1-gallon bottles will weigh approximately 8.35 pounds (Figure 1.10). This can be determined by dividing 62.5, the approximate number of pounds in 1 cubic foot, by 7.48, the number of gallons in 1 cubic foot.

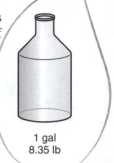

1 gal
8.35 lb

FIGURE 1.10   ◆

One cubic foot, measured in inches, is 12 inches by 12 inches by 12 inches. The total number of cubic inches in 1 cubic foot is 1728 ($12 \times 12 \times 12$) (Figure 1.11).

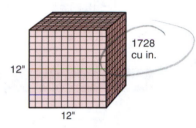

1728 cu in.

12"

12"

FIGURE 1.11   ◆

One gallon of water contains approximately 231 cubic inches (Figure 1.12). This can be found by dividing 1728, the number of cubic inches in 1 cubic foot, by 7.48, the number of gallons in 1 cubic foot.

If 1 cubic foot is divided into units measuring 1 inch by 1 inch at the base by l foot high, it will contain 144 of the units (Figure 1.13). This can be determined by finding the area of the top of the cube (12 inches $\times$ 12 inches).

1 gal
231 cu in.

FIGURE 1.12   ◆

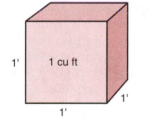

1' | 1 cu ft

1'

1'

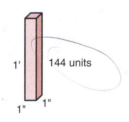

1' | 144 units

1" | 1"

FIGURE 1.13   ◆   One cubic foot contains 144 of 1 inch $\times$ 1 inch $\times$ 1 foot units.

The weight of water in each of the 144 units in 1 cubic foot is .434 pounds (Figure 1.14). This can be determined by dividing 62.5 pounds, the approximate weight of 1 cubic foot of water, by 144, the number of units in 1 cubic foot.

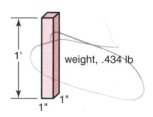

1'

weight, .434 lb

1"

1"

FIGURE 1.14   ◆

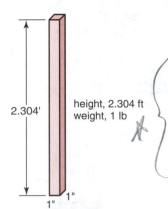

2.304'

height, 2.304 ft
weight, 1 lb

1" 1"

FIGURE 1.15 ◆

A column of water that measures 1 inch by 1 inch at the base will have to be 2.304 feet in height in order to weigh 1 pound (Figure 1.15). This can be found by dividing 1 by .434. A summary of useful numbers follows.

1. Water weighs approximately 62.5 pounds per cubic foot.
2. There are approximately 7.48 gallons in 1 cubic foot.
3. One gallon of water weighs approximately 8.35 pounds.
4. There are 1728 cubic inches in 1 cubic foot.
5. One gallon of water contains approximately 231 cubic inches.
6. A column of water measuring 1 inch by 1 inch in base by 1 foot high weighs approximately .434 pounds.
7. A column of water measuring 1 inch by 1 inch in base by 2.304 feet high weighs approximately 1 pound.

◆ PRINCIPLES OF PRESSURE IN WATER

Force (F) refers to the amount of energy applied at a given point or points. Force can be given in different units, but it is usually expressed in pounds.

Pressure (P) may be defined as force per unit area. If a force *F* is applied to the surface of a fluid and acts over an area *A* perpendicular to it, then the pressure can be expressed as

$$P = \frac{F}{A}$$

The force can be expressed as

$$F = PA$$

Pressure as calculated in the fire service is usually expressed in pounds per square inch (psi) and may be considered as the measurement of energy in water.

It is important that the relationship between force and pressure be clearly understood. A column of water 2.304 feet high that measures 1 inch by 1 inch at the base creates a pressure of 1 psi on the base. A container having a 1-foot-square base that is 2.304 feet in height houses 144 of the 1-inch by 1-inch columns of water (12 × 12). Since the pressure on the base of the container is 1 psi, the force is 144 pounds on the total area (*F = PA*, so *F* = 1 psi × 12 in. × 12 in. = 144 lb).

Static pressure is the pressure of water when it is not in motion. A gage attached to a hydrant when the water is not moving, or to a hose line when the nozzle is closed, registers static pressure.

Flow pressure is the pressure of water after it has been placed in motion. Flow pressure is less than static pressure due to friction loss in the water carrier, twists and bends in the carrier, restrictions to the flow of water, and other restrictions.

Residual pressure is a term used to express the pressure remaining at the hydrant outlet after the water is flowing.

PRINCIPLES OF PRESSURE IN LIQUIDS

Six basic rules govern the primary characteristics of pressure in liquids.

1. Fluid pressure is perpendicular to any surface on which it acts.
2. The pressure at any point in a fluid at rest is of the same intensity in all directions.

3. External pressure applied to a confined liquid is transmitted undiminished in all directions.

4. The downward pressure of a liquid in an open container is directly proportional to the depth of the liquid.

5. The downward pressure of a liquid in an open container is directly proportional to its density.

6. The downward pressure of a liquid on the bottom of an open container is independent of the shape or size of the container.

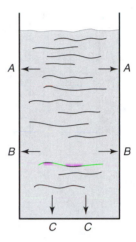

FIGURE 1.16 ◆

Figure 1.16 illustrates the principle that fluid pressure is perpendicular to any surface on which it acts. While the pressure at points *A*, *B*, and *C* is different as a result of the differences in depths of the water, the pressure at each of the points acts perpendicular to the surfaces, as shown by the direction of the arrows. If the pressures were not acting perpendicular to the surfaces, there would be a tendency for the water to move down the sides of the container; this would result in constant motion.

Figure 1.17 illustrates a pumping situation in which the nozzle is closed while the **pumper** is pumping at a pressure of 100 psi. As the nozzle is closed, the water in the hose line is at rest (not moving). Gages placed at various places in the hose line will indicate the same pressure as that at the pump as long as the line is laid at level ground. This illustrates the principle that a fluid at rest is of the same intensity in all directions.

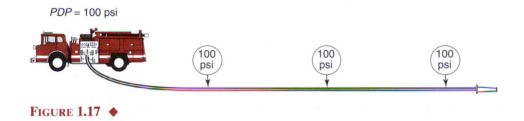

FIGURE 1.17 ◆

Figure 1.4 shows a pumping situation in which the line is laid at ground level with the nozzle closed while the pumper is pumping at 60 psi. The 60 psi is transmitted throughout the hose line and the pressure of the water at rest is of the same intensity in all directions.

Figure 1.5 illustrates the same condition, except the pressure at the pump has been increased to 80 psi. The increase in pressure does not result in any movement of water because the nozzle is closed. The 20-psi increase in pressure is transmitted undiminished throughout the length of the hose line, which results in a reading of 80 psi on each of the gages.

Figure 1.18 demonstrates the principle that the downward pressure of a liquid in an open container is directly proportional to the depth of the liquid. The pressure at point *A*, which is located 20 feet below the surface of the water, is 8.68 psi (.434 × 20). The pressure at point *B*, which is located 40 feet below the surface of the water, is 17.36 psi (.434 × 40), or twice the pressure at the 20-foot level. The pressure at these locations acts perpendicular to the surface (side of the container).

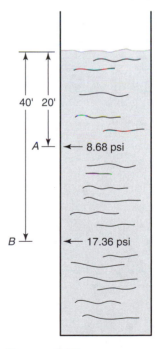

FIGURE 1.18 ◆

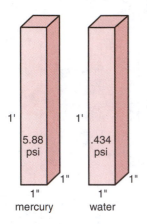

5.88
psi

.434
psi

1'   1'

1"   1"

1"   1"
mercury   water

FIGURE 1.19 ◆

Figure 1.19 shows two identical containers that measure 1 inch by 1 inch by 1 foot. Both containers are full, one with mercury and one with water. Because mercury is 13.546 times as heavy as water, the pressure at the base of the container full of mercury is 13.546 times as great as the pressure created at the base of the container filled with water. This illustrates the principle that the downward pressure of a liquid in an open container is directly proportional to its density.

In Figure 1.20, three differently shaped containers are shown, each having a base of 1 square inch. The level of the water in each of the containers is the same. Because the depth of the water in each of the containers is the same, the downward pressure on the bottom of the containers is the same. This is because the downward pressure is independent of the shape of the container. This principle could also be demonstrated by different shapes of water tanks in general use. The pressure created at the point of use of water from these tanks is a function of the distance of the water level above the point of use and is completely unaffected by the shape of the tank.

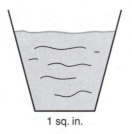

1 sq. in.   1 sq. in.   1 sq. in.

FIGURE 1.20 ◆

◆ HEAD

The amount of pressure created mechanically by a pump can be controlled by the pump operator. The amount of pressure created by gravity depends upon the height of the level of the water above the point of reference. For example, a column of water 34 feet high generates a pressure of about 15 psi at its base; therefore, the pressure of 15 psi can be stated as a pressure head of 34 feet.

Head is the vertical distance from the surface of the water being considered to the point being considered. Neither the size of the body of water nor the horizontal distance from the body of the water to the point being considered affects the head. As an example, the head is 120 feet in the two illustrations in Figure 1.21.

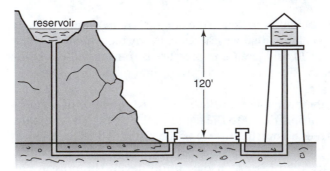

reservoir

120'

FIGURE 1.21 ◆

In the illustration of the reservoir in Figure 1.21, the body of water is relatively large and there is a considerable distance from the reservoir to the hydrant. In the illustration of the **gravity tank**, the body of water is relatively small and there is a relatively short distance from the water to the hydrant. As the vertical distance from the surface of the water to the hydrant is the same in both illustrations, the head is the same.

Head is important because the amount of head determines the amount of pressure created. Head pressure occurs because (a) for each foot of head, water exerts a pressure of .434 psi and (b) each 2.304 feet of head develops a pressure of 1 psi.

DETERMINING THE PRESSURE WHEN THE HEAD IS KNOWN

The principle of converting head to pressure is used in water systems by collecting water in reservoirs and delivering it to water mains by gravity. The principle is also used to provide pressure to sprinkler systems and yard hydrants of industrial buildings. Tanks providing water for these systems are elevated to varying heights depending on the pressure requirements.

Pressure is directly proportional to the head. A column of water 1 inch square and 1 foot high weighs .434 pounds, thus a pressure of .434 psi is exerted at the base of the column. The pressure increases by .434 psi for each additional 1-foot increase in head. In Figure 1.22, the head is 120 feet. The pressure created can be determined by using the following formula:

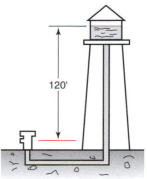

120'

FIGURE 1.22 ◆

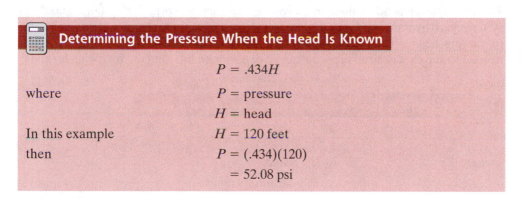

Determining the Pressure When the Head Is Known

$$P = .434H$$

where

$$P = \text{pressure}$$
$$H = \text{head}$$

In this example

$$H = 120 \text{ feet}$$

then

$$P = (.434)(120)$$
$$= 52.08 \text{ psi}$$

QUESTION The surface of the water in a reservoir is 176 feet above a hydrant. What will be the static pressure on the hydrant created by the head?

ANSWER

$$P = .434H$$

where

$$H = 176 \text{ feet}$$

Then

$$P = (.434)(176)$$
$$P = 76.38 \text{ psi}$$ ■

QUESTION A 40-foot-tall rectangular tank is three-fourths full of water. The bottom of the tank is 50 feet above the roof of a ten-story building (Figure 1.23). Three **sprinkler heads** have opened due to a fire on the fourth floor. The heads are located 8 feet above the floor. What was the static pressure on the heads prior to opening? Consider that the floors are 12 feet apart.

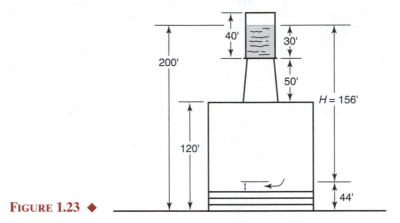

FIGURE 1.23 ◆

ANSWER It is always best to draw a diagram for this type of problem (Figure 1.23).

$$P = .434H$$

where $\qquad\qquad H = 156$ feet

Then $\qquad\qquad P = (.434)(156)$

$$P = 67.7 \text{ psi}$$ ■

DETERMINING THE HEAD WHEN THE PRESSURE IS KNOWN

The principle of head is used in providing pressure to sprinkler systems, **standpipe** systems, yard hydrants, and such. The tank or storage of water for these systems is elevated to varying heights depending on the pressure requirement of the system. The head needed to supply a desired pressure can be determined by using the following formula:

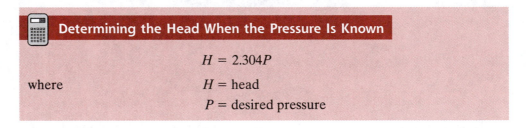

Determining the Head When the Pressure Is Known

$$H = 2.304P$$

where $\qquad\qquad H = \text{head}$

$\qquad\qquad P = \text{desired pressure}$

QUESTION What head is necessary in order to provide a pressure of 75 psi (Figure 1.24)?

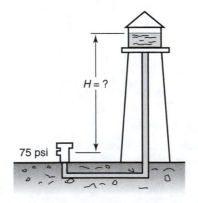

FIGURE 1.24 ◆

ANSWER

$$H = 2.304P$$

where

$$P = 75 \text{ psi}$$

Then

$$H = (2.304)(75)$$

$$H = 172.8 \text{ feet}$$ ■

QUESTION The sprinkler heads on the third floor of a three-story building are located 30 feet above ground elevation (Figure 1.25). How high above ground elevation must the surface of the water in a gravity tank be to provide a pressure of 50 psi at the sprinkler heads on the third floor?

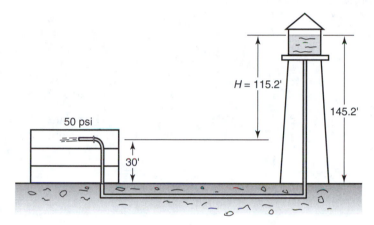

50 psi

$H = 115.2'$

145.2'

30'

FIGURE 1.25 ◆

ANSWER First, determine the head required above the sprinklers:

$$H = 2.304P$$

where

$$P = 50 \text{ psi}$$

Then

$$H = (2.304)(50)$$

$$H = 115.2 \text{ feet}$$

The sprinkler heads are located 30 feet above elevation, so the surface of the water must be 30 feet + 115.2 feet, or 145.2 feet, above ground level.   ■

◆ **ELEVATION PRESSURE**

Elevation pressure refers to a pressure loss or gain whenever a nozzle or other discharge opening is located above or below the pressure source. For the purpose of solving hydraulic problems in this book, elevation pressure is referred to as back pressure when the discharge of water is located above the pump and forward pressure when the discharge of water is located below the pump.

In this chapter, both the **back pressure** and **forward pressure** will be determined by the actual pressure created by the head ($P = .434H$). This has been done to provide a firm foundation for understanding what actually happens in elevated situations. Later in the book the .434 will be rounded to .5 for the purpose of simplifying elevation problems.

BACK PRESSURE

Back pressure, as the term is used in the fire service, is the pressure necessary for a pumper to overcome resistance that is due to the head when lines are laid uphill, into standpipes, into deck guns, or into elevated platforms or whenever the discharge of water is above the level of the pump. It also refers to similar pressure working against hydrants or other sources of water. Back pressure is encountered in many situations and can be determined by using the following formula:

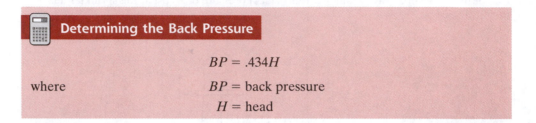

Determining the Back Pressure

$$BP = .434H$$

where

$$BP = \text{back pressure}$$
$$H = \text{head}$$

QUESTION What pressure must a pumper provide to overcome the back pressure when water is discharged from a nozzle tip located 85 feet above the level of the pump (Figure 1.26)?

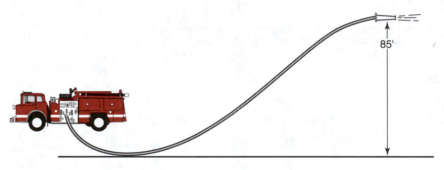

85'

FIGURE 1.26　◆

ANSWER

$$BP = .434H$$

where

$$H = 85 \text{ feet}$$

Then

$$BP = (.434)(85)$$
$$= 36.89 \text{ psi}$$

MULTISTORY BUILDINGS

Multistory buildings are mentioned in problems in a number of locations throughout this book. Unfortunately, there is no standard for the distance between stories in actual multistory buildings in the United States. However, for solving hydraulic problems, the International Fire Service Training Association (IFSTA) allows 5 psi per story for elevation losses. This most nearly converts to the distance between stories of 12 feet ($12 \times .434 = 5.2$). However, for the purpose of achieving some type of standardization for multistory buildings, this book will assume a distance of 10 feet per story, divided as shown in Figure 1.27.

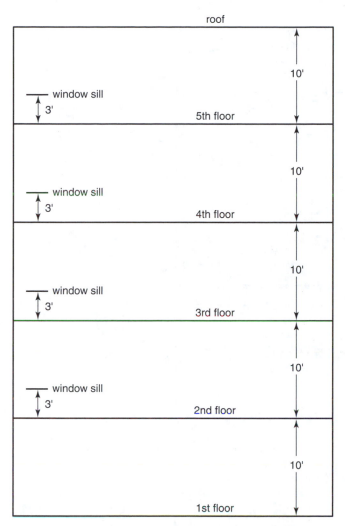

FIGURE 1.27 ◆ Dimensions for a multistory building. Use 5 psi per story for calculating back pressure.

QUESTION A pumper is pumping into the standpipe of an eight-story building. The outlet on the roof of the building is 3 feet above the roof. If this outlet is open, disregarding friction loss in the hose and in the standpipe, what is the minimum pump pressure required to cause water to flow from the outlet? For this problem, consider each story to be 10 feet high and the roof to be 10 feet above the top floor. Pumping to the roof is the same as pumping to the ninth floor. The actual height above ground level is one floor less or 8×10 plus 3 feet to the outlet (Figure 1.28).

ANSWER

$$BP = .434H$$

where

$$H = 83 \text{ feet}$$

Then

$$BP = (.434)(83)$$
$$= 36.02 \text{ psi}$$

■

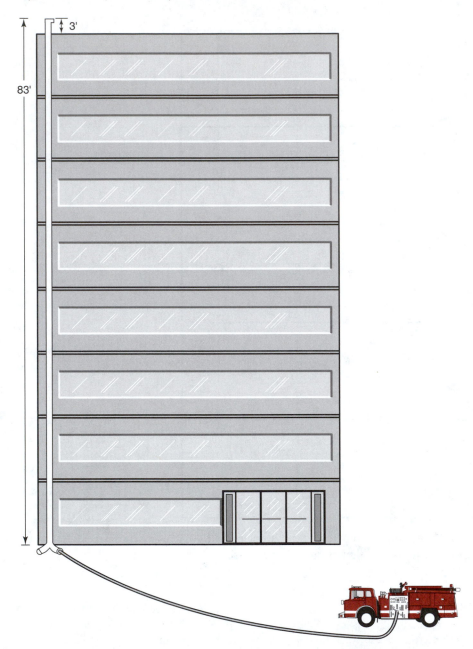

FIGURE 1.28 ◆ What pressure is required for back pressure to cause water to flow from the roof outlet of an eight-story building?

This is the technical answer arrived at by using the back pressure formula. Later in the book, 5 psi per story will be used along with .5 psi per foot for any elevation above the roof. The answer under the change would be:

$$BP = (5)(8) + 1.5$$
$$= 41.5\,\text{psi}$$

FORWARD PRESSURE

The term forward pressure is used in the fire service to refer to the pressure aid given to a pump, a hydrant, or another source of water pressure when the discharge of the water is below the level of the pressure source. Forward pressure is the opposite of back pressure, but it is determined by using a similar formula:

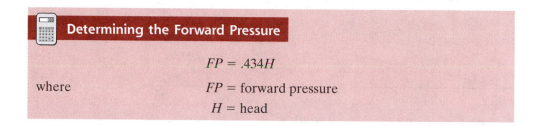

Determining the Forward Pressure

$$FP = .434H$$

where

FP = forward pressure

H = head

QUESTION   A pumper is pumping through a single 2½-inch line laid downhill to a point 143 feet below the level of the pump (Figure 1.29). What is the forward pressure aiding the pump?

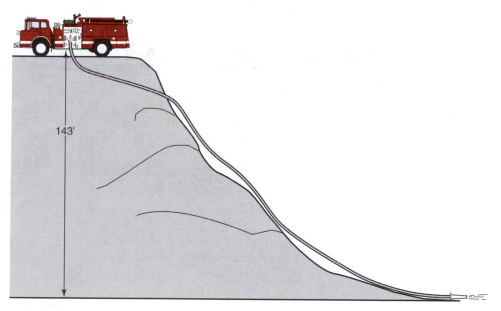

143'

FIGURE 1.29 ◆

ANSWER

$$FP = .434H$$

where

H = 143 feet

Then

$$FP = (.434)(143)$$
$$= 62.06 \text{ psi}$$

LINES LAID UPHILL OR DOWNHILL

Lines laid uphill or downhill are generally said to be laid up or down a given **grade**, respectively. For example, a line may be laid up a 10 percent grade or down a 12 percent grade. Grade is the vertical rise or decline of elevation for every 100 feet of hose laid out. The phrase "up a 10 percent grade" refers to an elevation rise of 10 feet in 100 feet. The phrase "down a 15 percent grade" means a decline of 15 feet in 100 feet. This concept is illustrated in Figure 1.30.

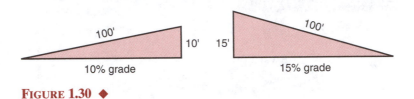

FIGURE 1.30 ◆

Lines laid up or down grades result in back pressure against the pump or forward pressure aiding the pump, respectively. The amount of back pressure or forward pressure is determined by the resultant head. Head is the vertical distance from the pump to the place where the water is being discharged, and can be determined by using the following formula:

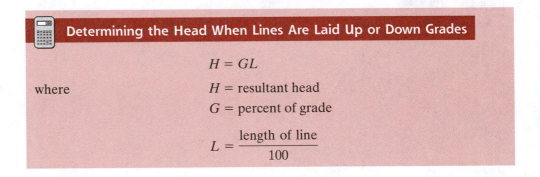

Determining the Head When Lines Are Laid Up or Down Grades

$$H = GL$$

where
H = resultant head
G = percent of grade

$$L = \frac{\text{length of line}}{100}$$

QUESTION A pumper is pumping through 700 feet of a single 2½-inch hose. The line is laid up a 12 percent grade (Figure 1.31). What is the back pressure against the pump?

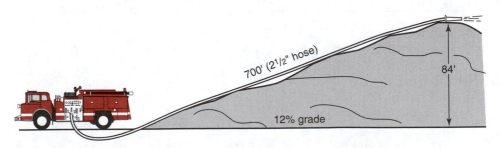

FIGURE 1.31 ◆

ANSWER

$$H = GL$$

where

$$G = 12 \text{ percent}$$
$$L = 700/100 = 7$$

Then

$$H = (12)(7)$$
$$= 84 \text{ feet}$$

Therefore

$$BP = .434H$$
$$= (.434)(84)$$
$$= 36.46 \text{ psi} \qquad ■$$

QUESTION A pumper is pumping through two **siamesed** 2½-inch lines each 400 feet in length. At their ends these lines are jointly attached to a single 2½-inch line that is 400 feet in length and equipped with a 1¼-inch tip. The layout is down a 15 percent grade (Figure 1.32). What is the forward pressure aiding the pump?

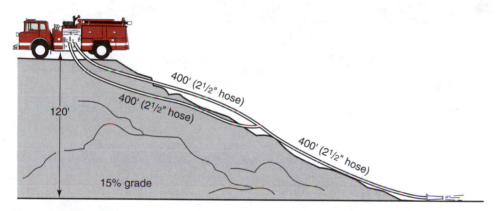

FIGURE 1.32 ◆

Note: Siamesed lines are two or more lines that are attached to different outlets of a pumper and eventually join at a common point.

ANSWER

$$H = GL$$

where

$$G = 15$$
$$L = 800/100 = 8 \text{ (from pump outlets to nozzle tip)}$$

Then

$$H = (15)(8)$$
$$= 120 \text{ feet}$$

Therefore

$$FP = .434H$$

where

$$FP = \text{forward pressure}$$
$$H = \text{head}$$

Then

$$FP = (.434)(120)$$
$$= 52.08 \text{ psi.} \qquad ■$$

COMBINATION ELEVATIONS

Occasionally lines may be laid up and down grades in the same **layout**. They also may be laid up or down grades into **ladder pipes**, **elevated platforms**, standpipes, and such. Back pressure or forward pressure is determined by the resultant head. The resultant head is the vertical distance from the pump to the point where the water is being discharged.

QUESTION A pumper is pumping through 1200 feet of a single 2½-inch hose. The line is laid up a 15 percent grade for 800 feet, over the crest of the hill, and then down 400 feet of a 12 percent grade on the other side of the hill (Figure 1.33). What is the back pressure or forward pressure on the pump?

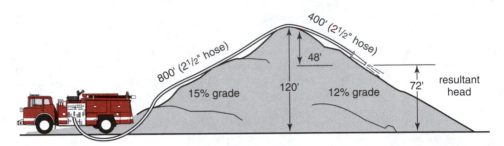

800' (2½" hose) 400' (2½" hose)

48'

15% grade 120' 12% grade 72' resultant head

FIGURE 1.33 ◆

ANSWER

Uphill	**Downhill**
$H = GL$	$H = GL$
$G = 15$	$G = 12$
$L = 8$	$L = 4$
$H = (15)(8)$	$H = (12)(4)$
$= 120$ feet	$= 48$ feet

Now,

$$\text{Resultant head} = 120 - 48 = 72 \text{ feet}$$

Then

$$BP = (.434)(72)$$
$$= 31.25 \text{ psi}$$

◆ **FORCE ON THE BASE OF A CONTAINER**

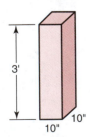

3'

10"
10"

FIGURE 1.34 ◆

The pressure created by gravity is directly proportional to the head, and the pressure (in psi) at the bottom of the container is unrelated to the shape of the container; thus, the force on the base of a container can be found by multiplying the area of the base (in square inches) by the pressure created by the head. For example, the water in a full container that measures 10 inches by 10 inches at the base by 3 feet in height will exert a pressure at the base of (Figure 1.34)

$$P = .434H$$

where

$$H = 3 \text{ feet}$$

Then

$$P = (.434)(3)$$
$$= 1.302 \text{ psi}$$

Since the area of the base of the container is 100 square inches (10 in. × 10 in.), the force on the base of the container will be given as follows:

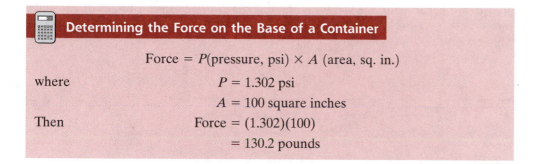

Determining the Force on the Base of a Container

$$\text{Force} = P(\text{pressure, psi}) \times A \text{ (area, sq. in.)}$$

where
$$P = 1.302 \text{ psi}$$
$$A = 100 \text{ square inches}$$

Then
$$\text{Force} = (1.302)(100)$$
$$= 130.2 \text{ pounds}$$

QUESTION The odd-shaped container in Figure 1.35 is 5 feet high. The base of the container measures 2 feet by 4 feet, and thus has a total area of 8 square feet. Since there are 144 square inches in a square foot, there are 1152 square inches in the base (8 × 144). What is the force exerted on the base of the container?

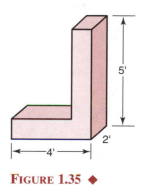

FIGURE 1.35 ◆

ANSWER
$$\text{Force} = P \text{ (in psi)} \times A \text{ (in sq. in.)}$$

where
$$P = .434H$$
$$H = 5 \text{ feet}$$

Thus
$$P = (.434)(5)$$
$$= 2.17 \text{ psi}$$
$$A = 1152 \text{ square inches } (8 \times 144)$$

Then
$$\text{Force} = (2.17)(1152)$$
$$= 2499.84 \text{ pounds}$$

◆ **FORCE ON CLAPPER VALVES**

A **clapper valve** is an automatic valve installed in hydraulic systems which permits the flow of water in one direction only. Clapper valves are used in sprinkler systems, standpipes, and siamese assemblies and in piping in pumping systems. Clapper valves are open or closed depending upon the difference in force on the two sides of the valve. The valve on the top in Figure 1.36 is closed, so the flow of water has stopped. The valve on the bottom has been opened by the force of the water.

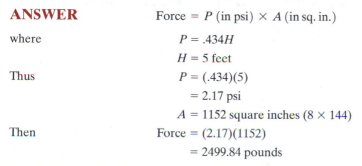

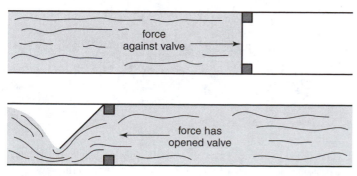

FIGURE 1.36 ◆

Some hydraulic systems are designed to permit water pressure to be exerted on both sides of the clapper valve at all times. The valve will open or close, depending on the force exerted against each of its sides. The valves are designed so that there is a greater area exposed to the water on one side than on the other. Thus, when the valve is closed, a force greater than that on the closed side is required in order to open it. This principle is shown in Figure 1.37.

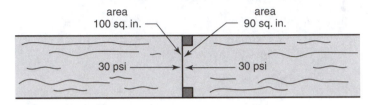

FIGURE 1.37 ◆ The area on the left side of the valve is greater; the valve will remain closed.

QUESTION The clapper valve shown in Figure 1.37 has an area of 100 square inches on one side and 90 square inches on the other side. If the pressure holding the valve closed is 30 psi, what pressure will be required on the other side to open it?

ANSWER To open the valve, a force greater than that holding it closed must be provided. The force holding the valve closed is

$$\text{Force} = P \times A$$

where
$$P = 30 \text{ psi}$$
$$A = 100 \text{ square inches}$$

Then
$$\text{Force} = (30)(100)$$
$$= 3000 \text{ pounds}$$

The pressure *P* required to open the valve is found as follows:

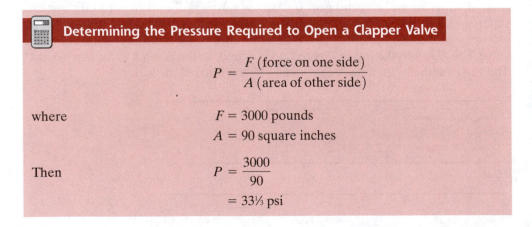

Determining the Pressure Required to Open a Clapper Valve

$$P = \frac{F \text{ (force on one side)}}{A \text{ (area of other side)}}$$

where
$$F = 3000 \text{ pounds}$$
$$A = 90 \text{ square inches}$$

Then
$$P = \frac{3000}{90}$$
$$= 33\tfrac{1}{3} \text{ psi}$$

Note: The pressure of 33⅓ psi will result in a balance of forces on both sides of the clapper valve. It will be necessary to increase the pressure slightly above the 33⅓ psi in order to open the valve. ■

Summary of Chapter Formulas

$$P = .434H$$

where P = pressure

          H = head

$$H = 2.304P$$

where H = head

          P = pressure

$$BP = .434H$$

where BP = back pressure

          H = head

$$FP = .434H$$

where FP = forward pressure

          H = head

$$H = GL$$

where H = resultant head

          G = percent of grade

          $L = \dfrac{\text{length of line}}{100}$

Force (F) = $P \times A$ (pressure × area)

$$\text{Pressure required to open a clapper valve} = \frac{\text{force on one side}}{\text{area of the other side}}$$

Review Questions

1. What is the weight of 1 cubic foot of water, as used in calculations by the fire service?
2. How many gallons are there in 1 cubic foot?
3. How many cubic inches are there in 1 cubic foot?
4. How is the number of cubic inches in a cubic foot determined?
5. How many cubic inches are there in 1 gallon?
6. How is the number of cubic inches in 1 gallon determined?
7. What is the weight of water in a column that measures 1 inch by 1 inch in base by 1 foot high?
8. How was the weight of water in the column in Question 7 determined?
9. What is the height of a column of water that measures 1 inch by 1 inch in base and weighs 1 pound?
10. How was the height of the column of water in Question 9 determined?
11. Define flow pressure.

12. Define static pressure.
13. Define residual pressure.
14. What is the definition of head as used in considerations of water pressure?
15. Give the formula for finding the pressure when the head is known.
16. Give the formula for finding the head when the pressure is known.
17. Define back pressure.
18. Give the formula for finding back pressure.
19. Define forward pressure.
20. Give the formula for finding forward pressure.
21. What is meant by a 10 percent grade? A 15 percent downgrade?
22. What is the formula for finding the head when the grade and the length of line are known?
23. What is the formula for finding the force on the base of a container?
24. How is the pressure required to open a clapper valve found when the valve is being held closed by a greater force on the opposite side?

■ ■

Test One

1. A hydrant is located 128 feet below the surface of the water supplying it. What is the pressure on the hydrant?

2. A 30-foot-high tank is located 20 feet above the roof of an eight-story building. The tank is two-thirds full of water. What will be the pressure at a sprinkler head on the third floor if the sprinkler head is located 8 feet off the floor? Consider that the tank feeds the sprinkler system and that there is 12 feet to each floor.

3. What head is necessary in order to provide a pressure of 85 psi?

4. Outlets on a wet standpipe system are 4 feet above the floor. There is a need for a static pressure of 60 psi at the outlet on the second floor. Consider that the building is ten stories high and there is 12 feet to a floor. How high above the roof must the water level in a tank be in order to supply the required pressure?

5. A pumper delivering water through a standpipe is working against a head of 112 feet. What is the back pressure exerted against the pump?

6. A pumper is pumping through a single 2½-inch line that is laid downhill to a point 145 feet below the level of the pump. What is the forward pressure aiding the pump?

7. A pumper is pumping through 650 feet of single 2½-inch hose. The line is laid up a 15 percent grade. What is the back pressure against the pump?

8. A pumper is pumping through 1000 feet of a single 2½-inch hose. The line is laid up a 12 percent grade for a total of 700 feet, over the crest of the hill, and then 300 feet down the 15 percent grade on the other side of the hill. What is the back pressure or forward pressure on the pump?

9. A container is 6 feet high and has a base of 650 square inches. What is the force on the bottom of the container if it is filled with water?

10. A clapper valve has an area of 22 square inches on one side and 18 square inches on the other side. A pressure of 50 psi is working against the side having an area of 22 square inches. How much pressure is required on the other side of the valve to balance the two forces?

Water Tanks and Hose Capacity

2 CHAPTER

Objectives

Upon completing this chapter, the reader should:

- Understand the various types of tanks used in the fire service.
- Be able to determine the area of a square, the area of a rectangle, and the area of a circle when the dimensions are given.
- Be able to determine the volume of a rectangular container.
- Be able to determine the volume of a cylindrical container.
- Be able to determine the gallon capacity of either a rectangular or cylindrical container.
- Be able to determine the weight capacity of either a rectangular or cylindrical container.
- Be able to determine either the weight of the water or the amount of water in hose lines of various sizes.

Water tanks are used in the field of fire protection on fire apparatus, in industrial establishments, and within the water supply systems of many cities. Recommended construction standards for water tanks used on fire apparatus are contained in National Fire Protection Association (NFPA) standard 1901. The standard recommends that water tanks on fire apparatus have a minimum capacity of 500 gallons, except on apparatus equipped with aerial ladders or elevated platforms; here the minimum capacity may be 150 gallons. Initial **attack pumpers** carry a minimum of 200 gallons and mobile water-supply vehicles a minimum of 1000 gallons.

The tanks used on pumpers are generally rectangular in shape. Most in use today are made of plastic material. Rectangular tanks are seldom used for water storage in industrial establishments or as a portion of the water supply systems of cities. The gravity or suction-type water tanks are most often used for these systems.

Gravity tanks may be cylindrical with pitched roofs, elliptical, or hemispherical. Steel gravity tanks are available in standard sizes ranging from 30,000 to 500,000 gallons in capacity. Standard capacities for wooden gravity tanks range from 30,000 to

100,000 gallons. Older gravity tanks that are situated so that hose lines can be used directly from the hydrants supplied by the tank should have a minimum capacity of 30,000 gallons, with the installation being designed so that the bottom of the tank is at least 75 feet above the ground.

Steel suction tanks used for water supply are generally cylindrical. Common sizes of these storage tanks range from 50,000 to 1,000,000 gallons in capacity.

While there is little practical value in fire department personnel being able to determine the capacity of water tanks when the dimensions are known, the capability of being able to determine this information helps provide a basic foundation for the understanding of fire service hydraulics. This chapter is devoted to laying such a foundation. Basic knowledge is not presented as something new, but instead as a starting point in proceeding from the simple to the complex and from the known to the unknown. Basic information is presented for determining the volume and the capacity of rectangular and cylindrical containers and the volume and the capacity of a fire hose, which is a cylindrical container whose dimensions are measured in inches rather than feet.

◆ SQUARE FEET AND CUBIC FEET

SQUARE FOOT

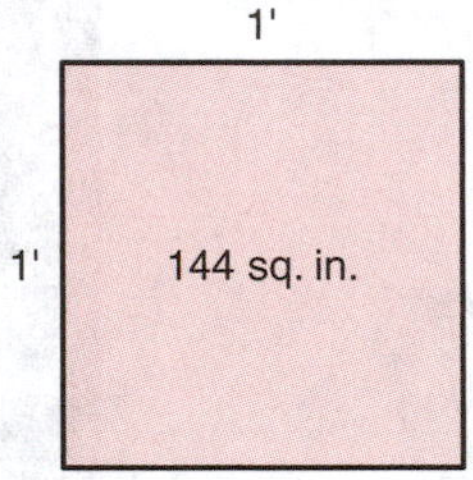

A square foot is a plane surface with four sides measuring 1 foot each and having four right (90°) angles. A square inch is a plane surface with four sides measuring 1 inch each and having four right (90°) angles. There are 144 square inches in 1 square foot (see Figure 2.1).

Any plane surface with an area equal to the area of a square foot as just defined is said to have an area of 1 square foot. Any plane surface having an area equal to the area of a square inch as just defined is said to have an area of 1 square inch.

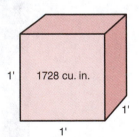

FIGURE 2.1 ◆ A square foot.

CUBIC FOOT

A cube is a special form of rectangular prism. Its length, width, and height are equal. A cubic foot is 1 foot long, 1 foot wide, and 1 foot high. A cubic inch is 1 inch long, 1 inch wide, and 1 inch high. There are 1728 cubic inches in 1 cubic foot (see Figure 2.2).

Any container having a volume the same as the volume in a cubic foot as just defined is said to have a volume of 1 cubic foot. Any container having a volume the same as the volume of a cubic inch as just defined is said to have a volume of 1 cubic inch.

FIGURE 2.2 ◆ A cubic foot.

◆ SQUARES, RECTANGLES, AND CIRCLES

SQUARES

A **square** is a plane surface with four equal sides and four right (90°) angles (Figure 2.3). Area is the surface extent of a flat surface. The area of a square can be determined by using the following formula:

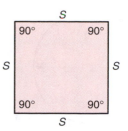

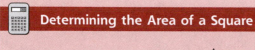
Determining the Area of a Square

$$\text{Area} = S^2$$

where $\quad\quad\quad\quad\quad S = \text{side}$

FIGURE 2.3 ◆ A square.

In the square shown in Figure 2.4,

$$\text{Area} = S^2$$

where $\quad\quad\quad\quad S = 3 \text{ feet}$

Then $\quad\quad\quad\quad \text{Area} = (3)(3)$

$\quad\quad\quad\quad\quad\quad\quad\quad = 9 \text{ square feet}$

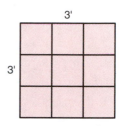

FIGURE 2.4 ◆ A 3-foot square.

RECTANGLES

A **rectangle** is an oblong plane surface with parallel sides and four right (90°) angles (Figure 2.5). The area of a rectangle can be determined by using the following formula:

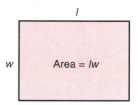

FIGURE 2.5 ◆ A rectangle.

Determining the Area of a Rectangle

$$\text{Area} = lw$$

where $\quad\quad\quad\quad l = \text{length}$

$\quad\quad\quad\quad\quad\quad w = \text{width}$

In the rectangle shown in Figure 2.6,

$$\text{Area} = lw$$

where $\quad\quad\quad\quad l = 6 \text{ feet}$

$\quad\quad\quad\quad\quad\quad w = 4 \text{ feet}$

Then $\quad\quad\quad\quad \text{Area} = (6)(4)$

$\quad\quad\quad\quad\quad\quad\quad\quad = 24 \text{ square feet}$

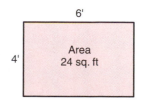

FIGURE 2.6 ◆ A 4-foot by 6-foot rectangle.

CIRCLES

A **circle** is a plane figure bounded by a curved line, all points on the curved line being equally distant from the center (Figure 2.7). The area of a circle, as used in calculations in the fire service, is usually expressed in square feet or square inches and can be determined by using one of the following two formulas:

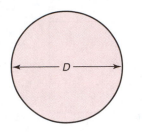

FIGURE 2.7 ◆ A circle.

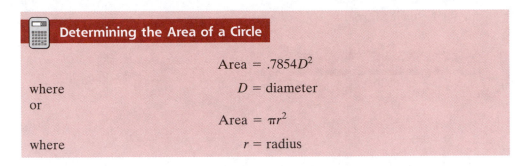

Determining the Area of a Circle

$$\text{Area} = .7854D^2$$

where
$$D = \text{diameter}$$

or

$$\text{Area} = \pi r^2$$

where
$$r = \text{radius}$$

The formula commonly used in the fire service is $\text{Area} = .7854D^2$. This probably came about because of the use of the formula for a circle in the development of the formula for the discharge of water from a nozzle tip. It is apparent that the discharge formula should be expressed in terms of the nozzle diameter rather than the radius of the nozzle tip. Sizes of **smooth-bore tips** are expressed in terms of the diameter in inches. Examples are ¾-inch, ⅝-inch, 1-inch, and 1¼-inch tips.

In the circle shown in Figure 2.8,

$$\text{Area} = .7854D^2$$

where
$$D = 3 \text{ inches}$$

Then
$$\text{Area} = (.7854)(3)(3)$$
$$= 7.07 \text{ square inches}$$

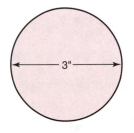

FIGURE 2.8 ◆

QUESTION How many square inches in a ½-inch tip (see Figure 2.9)?

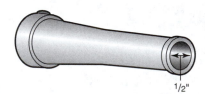

FIGURE 2.9 ◆ 1/2"

ANSWER

$$\text{Area} = .7854D^2$$

where
$$D = .5 \text{ inches}$$

Then
$$\text{Area} = (.7854)(.5)(.5)$$
$$= .1964 \text{ or } .2 \text{ inch}$$

QUESTION How many square inches in a 1-inch tip (see Figure 2.10)?

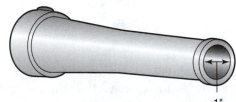

FIGURE 2.10 ◆ 1"

ANSWER   Area $= .7854D^2$

where $D = 1$ inch
Then   Area $= (.7854)(1)(1)$
                                        $= .79$ or $.8$ inch   ■

Note: When the size of a tip is doubled, its area becomes four times greater. For example,

Area of 1-inch circle $= .8$ inch
Area of ½-inch circle $= .2$ inch

Then

$$\frac{.8}{.2} = 4$$

◆ **VOLUME CAPACITY OF CONTAINERS**

As used in this book, **volume** refers to the amount of space included within the bounding surfaces of rectangular or cylindrical containers. The volume of water tanks is normally expressed in cubic feet, while the volume of a fire hose is expressed in cubic inches.

DETERMINING THE VOLUME OF CONTAINERS

The volume of both rectangular and cylindrical containers is equal to the area of the base of the container times the height of the container. As the base of a rectangular container is a rectangle and the base of a cylindrical container is a circle, the volume of the two containers can be expressed in formula form as follows:

 Determining the Volume of a Rectangular Container (Figure 2.11)

Volume $= lwh$

where $l =$ length
                   $w =$ width
                   $h =$ height

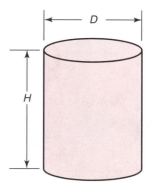

FIGURE 2.11 ◆ A rectangular container.

 Determining the Volume of a Cylindrical Container (Figure 2.12)

Volume $= .7854D^2H$

where $D =$ diameter
                   $H =$ height

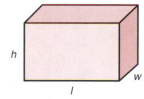

FIGURE 2.12 ◆ A cylindrical container.

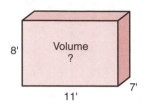

8'   Volume ?

11'   7'

FIGURE 2.13 ♦

QUESTION What is the volume of a rectangular container that measures 11 feet long by 7 feet wide by 8 feet high (Figure 2.13)?

ANSWER   Volume = lwh

where l = 11 feet
                                   w = 7 feet
                                   h = 8 feet
Then   Volume = (11)(7)(8)
                           = 616 cubic feet   ■

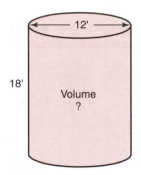

|← 12' →|

18'   Volume ?

FIGURE 2.14 ♦

QUESTION What is the volume of a cylindrical container that has a 12-foot diameter and is 18 feet high (Figure 2.14)?

ANSWER   Volume = $.7854D^2H$

where D = 12 feet
                                   H = 18 feet
Then   Volume = (.7854)(12)(12)(18)
                           = 2035.76 cubic feet   ■

GALLON CAPACITY OF CONTAINERS

The gallon capacity of containers whose measurements are in feet can be determined by finding the volume of the container and multiplying it by 7.48, the number of gallons in a cubic foot. This is expressed in the following formula:

 Determining the Gallon Capacity of a Container

$$\text{Gallon capacity} = 7.48V$$

where V = volume in cubic feet

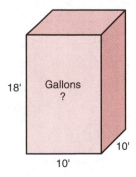

18'   Gallons ?

10'   10'

FIGURE 2.15 ♦

QUESTION How many gallons of water can a rectangular container hold if its base measures 10 feet by 10 feet and it is 18 feet in height (Figure 2.15)?

ANSWER   Gallon capacity = 7.48V

where V = lwh
and l = 10 feet
                                   w = 10 feet
                                   h = 18 feet

First, determine the volume:

                Volume = (10)(10)(18)
                       = 1800 cubic feet

Then

$$\text{Gallon capacity} = (7.48)(1800)$$
$$= 13{,}464 \text{ gallons} \quad ■$$

QUESTION   A neighborhood swimming pool is being considered for use as an emergency source of water (Figure 2.16). The pool is rectangular in shape and is 40 feet long and 15 feet wide. The average depth is 5 feet. Approximately how much water does the pool hold?

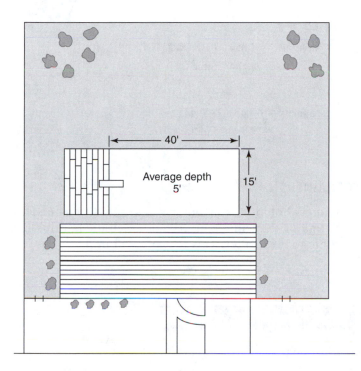

Average depth 5'

40'

15'

FIGURE 2.16 ◆

ANSWER

Gallon capacity $= 7.48V$

where

$$V = lwh$$
$$l = 40 \text{ feet}$$
$$w = 15 \text{ feet}$$
$$h = 5 \text{ feet}$$

First, determine the volume:

$$\text{Volume} = (40)(15)(5)$$
$$= 3000 \text{ cubic feet}$$

Then

$$\text{Gallon capacity} = (7.48)(3000)$$
$$= 22{,}440 \text{ gallons} \quad ■$$

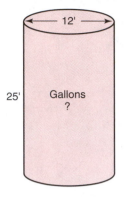

25' Gallons ?

FIGURE 2.17 ◆

QUESTION How many gallons of water can a cylindrical container hold that has a diameter of 12 feet and a height of 25 feet (Figure 2.17)?

ANSWER

Gallon capacity $= 7.48V$

where

$$V = .7854D^2H$$

and

$$D = 12 \text{ feet}$$

$$H = 25 \text{ feet}$$

First, determine the volume:

$$\text{Volume} = (.7854)(12)(12)(25)$$

$$= 2827.44 \text{ cubic feet}$$

Then

$$\text{Gallon capacity} = (7.48)(2827.44)$$

$$= 21,149.25 \text{ gallons}$$ ■

WEIGHT CAPACITY OF CONTAINERS

The weight capacity of containers holding fresh water whose measurements are in feet can be determined by finding the volume of the container and multiplying it by 62.5, the weight of 1 cubic foot of water. This is expressed in the following formula:

 Determining the Weight Capacity of a Container

Weight capacity $= 62.5V$

where

$V = $ volume in cubic feet

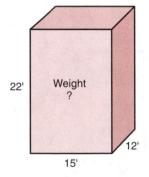

22' Weight ?

15'   12'

FIGURE 2.18 ◆

QUESTION What is the weight of water in a rectangular container that has a base measuring 12 feet by 15 feet and is 22 feet in height if the container is filled with water (Figure 2.18)?

ANSWER

Weight capacity $= 62.5V$

where

$$V = lwh$$

and

$$l = 15 \text{ feet}$$

$$w = 12 \text{ feet}$$

$$h = 22 \text{ feet}$$

First, determine the volume:

$$\text{Volume} = (15)(12)(22)$$

$$= 3960 \text{ cubic feet}$$

Then

$$\text{Weight capacity} = (62.5)(3960)$$

$$= 247,500 \text{ pounds}$$ ■

QUESTION What is the weight of water in a full cylindrical container that has a diameter of 15 feet and a height of 30 feet (Figure 2.19)?

ANSWER   Weight capacity $= 62.5V$

where $V = .7854D^2H$

and $D = 15$ feet

                         $H = 30$ feet

First, determine the volume:

$$\text{Volume} = (.7854)(15)(15)(30)$$
$$= 5301.45 \text{ cubic feet}$$

Then

$$\text{Weight capacity} = (62.5)(5301.45)$$
$$= 331{,}340.63 \text{ pounds}$$

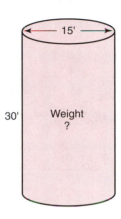

FIGURE 2.19 ◆

◆ **FIRE HOSE**

Hose used in the fire service ranges in size from ¾ inch to 6 inches; the most common sizes for general use for handheld lines, other than booster lines, are 1½, 1¾, 2, and 2½ inches. The actual inside diameter of the hose when used in firefighting operations is greater than specified because of the expansion caused by pressure. The amount of expansion depends primarily on the construction of the jacket and the amount of pressure applied. The construction of the jacket also affects the weight of the hose itself; hoses with synthetic fiber jackets are much stronger and lighter than the older cotton-jacketed hoses. One objective of this chapter is to offer some insight into the capacity of various sizes of hose lines. Understanding the capacity provides a much firmer foundation for evaluating problems involving movement of loaded lines, particularly the movement of lines to the upper floors of multistory buildings.

VOLUME OF HOSE LINES

A section of hose is simply a long cylindrical container. The primary difference between this type of cylindrical container and one used for the storage of water is the unit of measurement. The volume of a cylindrical water tank is expressed in cubic feet and the volume of a cylindrical length of fire hose is normally expressed in cubic inches. Otherwise, the same formula is used for determining the volume.

QUESTION What is the volume of a 50-foot section of 2½-inch hose (Figure 2.20)?

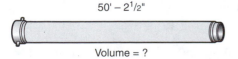

50' – 2¹/₂"

Volume = ?   **FIGURE 2.20** ◆

ANSWER   Volume $= .7854D^2H$

where $D = 2.5$ inches

                                $H = 600$ inches ($50\,\text{ft} \times 12\,\text{in.}$)

Then   Volume $= (.7854)(2.5)(2.5)(600)$

                             $= 2945.25$ cubic inches   ■

GALLON CAPACITY OF HOSE LINES

The gallon capacity of hose lines is found by finding the volume in cubic inches and dividing the volume by 231, the number of cubic inches in a gallon. This is expressed by the following formula:

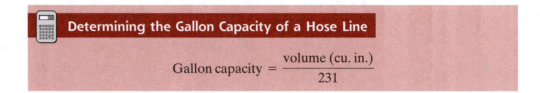

Determining the Gallon Capacity of a Hose Line

$$\text{Gallon capacity} = \frac{\text{volume (cu. in.)}}{231}$$

QUESTION   How many gallons of water does 100 feet of $1\frac{1}{2}$-inch hose hold (Figure 2.21)?

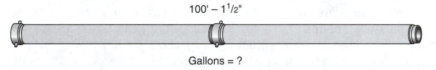

100' – 1¹/₂"

Gallons = ?

FIGURE 2.21 ◆

ANSWER $\text{Gallon capacity} = \dfrac{\text{volume (cu. in.)}}{231}$

where   Volume $= .7854D^2H$

                                $D = 1.5$ inches

                                $H = 1200$ inches ($100\,\text{ft} \times 12\,\text{in.}$)

First, determine the volume:

                     Volume $= (.7854)(1.5)(1.5)(1200)$

                            $= 2120.58$ cubic inches

Then

                     $\text{Gallon capacity} = \dfrac{2120.58}{231}$

                            $= 9.18$ gallons   ■

QUESTION   How many gallons of water does 100 feet of 3-inch hose hold (Figure 2.22)?

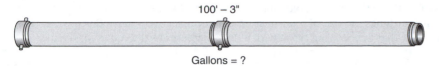

100' – 3"

Gallons = ?

FIGURE 2.22 ◆

ANSWER

$$\text{Gallon capacity} = \frac{\text{volume (cu. in.)}}{231}$$

where

$$\text{Volume} = .7854 D^2 H$$
$$D = 3 \text{ inches}$$
$$H = 1200 \text{ inches } (100 \text{ ft} \times 12 \text{ in.})$$

First, determine the volume:

$$\text{Volume} = (.7854)(3)(3)(1200)$$
$$= 8482.32 \text{ cubic inches}$$

Then

$$\text{Gallon capacity} = \frac{8482.32}{231}$$
$$= 36.72 \text{ gallons}$$

■

Note: A 3-inch hose holds four times as much water as a 1½–inch hose (36.72/9.18 = 4). This illustrates a principle that should be remembered. If the size of the hose is doubled, it has four times as great a carrying capacity.

QUESTION How many gallons of water does a 100-foot section of 6-inch hose hold (Figure 2.23)?

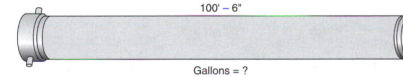

100' – 6"

Gallons = ?

FIGURE 2.23 ◆

ANSWER

$$\text{Gallon capacity} = \frac{\text{volume (cu. in.)}}{231}$$

where

$$\text{Volume} = .7854 D^2 H$$
$$D = 6 \text{ inches}$$
$$H = 1200 \text{ inches } (100 \text{ ft} \times 12 \text{ in.})$$

First, determine the volume:

$$\text{Volume} = (.7854)(6)(6)(1200)$$
$$= 33929.28 \text{ cubic inches}$$

Then

$$\text{Gallon capacity} = \frac{33929.28}{231}$$
$$= 146.88 \text{ gallons}$$

■

Note: Observe first that a 6-inch hose holds about 1½ gallons of water per running foot. Second, a 6-inch hose is four times the size of a 1½-inch hose, but carries 16 times as much water. This illustrates the point that the carrying capacity of a hose increases as the square of the increase in size:

$$\text{Increase} = \frac{6}{1.5} = 4$$

Squaring the increase given $4^2 = 16$.

WEIGHT CAPACITY OF FIRE HOSE

The weight of water in a fire hose is determined by finding the number of gallons of water in the hose and multiplying by 8.35 pounds, the weight of 1 gallon of water.

Determining the Weight Capacity of a Hose Line

$$\text{Weight capacity} = \text{Gallons} \times 8.35$$

QUESTION   How much does the water filling 100 feet of 2½-inch hose weigh (Figure 2.24)?

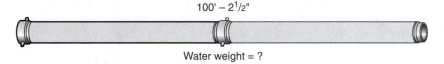

100' – 2¹/₂"

Water weight = ?

FIGURE 2.24 ◆

ANSWER

$$\text{Weight capacity} = \text{Gallons} \times 8.35$$

$$\text{Gallons} = \frac{V \,(\text{cu. in.})}{231}$$

where

$$V = .7854D^2H$$
$$D = 2.5 \text{ inches}$$
$$H = 1200 \text{ inches} \ (100 \text{ ft} \times 12 \text{ in.})$$

First, determine the volume of the hose:

$$\text{Volume} = (.7854)(2.5)(2.5)(1200)$$
$$= 5890.5 \text{ cubic inches}$$

Next, determine the number of gallons of water in the hose:

$$\text{Gallons} = \frac{5890.5}{231}$$
$$= 25.5$$

Then

$$\text{Weight capacity} = (25.5)(8.35)$$
$$= 212.93 \text{ pounds}$$

QUESTION How much does the water filling 100 feet of 5-inch hose weigh (Figure 2.25)?

100' – 5"

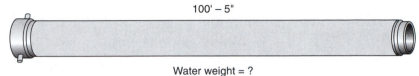

Water weight = ?

FIGURE 2.25 ◆

ANSWER

$$\text{Weight capacity} = \text{Gallons} \times 8.35$$

$$\text{Gallons} = \frac{V \text{ (cu. in.)}}{231}$$

where

$$V = .7854D^2H$$

$$D = 5 \text{ inches}$$

$$H = 1200 \text{ inches } (100 \text{ ft} \times 12 \text{ in.})$$

First, determine the volume of the hose:

$$\text{Volume} = (.7854)(5)(5)(1200)$$

$$= 23562 \text{ cubic inches}$$

Next, determine the number of gallons of water in the hose:

$$\text{Gallons} = \frac{23562}{231}$$

$$= 102$$

Then

$$\text{Weight capacity} = (102)(8.35)$$

$$= 851.7 \text{ pounds} \qquad ■$$

Note: A 5-inch hose holds about 1 gallon of water per foot. The weight of water in a hose increases at the same ratio as the carrying capacity. If the size of the hose is doubled, the weight of the water increases four times. The weight of water in 100 feet of 5-inch hose (851.7) is four times greater than that in 100 feet of 2½-inch hose (851.7/212.93 = 4).

■■■

Summary of Chapter Formulas

Area of a Square

$$\text{Area} = S^2$$

where S = side

Area of a Rectangle

$$\text{Area} = lw$$

where l = length
          w = width

Area of a Circle

$$\text{Area} = .7854D^2$$

where D = diameter

Volume of a Rectangular Container

$$\text{Volume} = lwh$$

where l = length

w = width

h = height

Volume of a Cylindrical Container

$$\text{Volume} = .7854D^2H$$

where D = diameter

H = height

Gallon Capacity of a Container

$$\text{Gallon capacity} = 7.48V$$

where V = volume in cubic feet

Weight Capacity of a Container

$$\text{Weight capacity} = 62.5V$$

where V = volume in cubic feet

Gallon Capacity of Hose

$$\text{Gallon capacity} = \frac{\text{volume (cu. in.)}}{231}$$

Weight Capacity of Hose

$$\text{Weight capacity} = \text{gallons} \times 8.35$$

Review Questions

1. What formula is used for finding the volume of a cylindrical container?
2. What is the formula for finding the volume of a rectangular container?
3. How can the gallon capacity of a rectangular container be determined?
4. How can the gallon capacity of a cylindrical container be determined?
5. How can the volume capacity of a hose line be determined?
6. How can the weight capacity of a rectangular container be determined?
7. How can the weight capacity of a cylindrical container be determined?
8. How can gallon capacity be changed to weight capacity?
9. How can weight capacity be changed to gallon capacity?

Test Two

1. What is the area of a rectangle that has a length of 14 feet and a width of 9 feet?
2. What is the volume of a rectangular container that measures 18 feet by 11 feet by 13 feet?
3. What is the area of a circle with a 13-foot diameter?
4. What is the volume of a cylindrical container that has a diameter of 15 feet and a height of 22 feet?
5. How many gallons of water can a rectangular container hold that measures 12 feet by 13 feet by 17 feet?
6. What is the gallon capacity of a cylindrical container with a diameter of 14 feet and a height of 20 feet?
7. What is the weight of water in a full rectangular container measuring 22 feet by 18 feet by 30 feet?
8. What is the weight of water in a full cylindrical container with a diameter of 30 feet and a height of 60 feet?
9. If a container has a capacity of 1246 gallons of water, what is the weight of water in the container when it is full?
10. If a cylindrical container, when full, holds 28,346 pounds of water, how many gallons of water can it hold?
11. What is the weight of the water in a 50-foot section of 2-inch hose?
12. How many gallons can 100 feet of 4-inch hose hold?
13. What is the weight of the water in a 50-foot section of 3-inch hose?
14. How many gallons of water are there in a 50-foot section of 3½-inch hose?
15. How many gallons of water are there in a 50-foot section of 1¾-inch hose?

Water Supply and Testing Procedures

3 CHAPTER

```
■ ■ ■ ■ ■ ■ ■ ■ ■ ■ ■ ■ ■ ■ ■ ■ ■ ■ ■ ■ ■ ■ ■ ■ ■ ■ ■
```

Objectives

Upon completing this chapter, the reader should:

- Be able to explain the characteristics of a gravity water system, a pumping water system, a combination water system, and a high-pressure water system.
- Be able to define average daily consumption, maximum daily consumption, and peak hourly consumption.
- Be able to discuss needed fire flow.
- Understand the difference between the system adequacy and the system reliability of a water system.
- Be able to discuss the distribution of a water system including the primary feeders, secondary feeders, and distribution mains.
- Understand the various types of pipes used in a water distribution system.
- Know the differences among the various types of hydrants used in a water system.
- Understand the recommended markings for fire hydrants.
- Understand the maintenance and testing procedures for fire hydrants.
- Be able to identify various types of emergency water supplies.

```
■ ■ ■ ■ ■ ■ ■ ■ ■ ■ ■ ■ ■ ■ ■ ■ ■ ■ ■ ■ ■ ■ ■ ■ ■ ■ ■
```

Water is the lifeblood of a fire department. Without an adequate water supply, a fire department cannot effectively carry out its primary function—fire extinguishment. It is important that each fire department member assigned to the firefighting force be thoroughly familiar with the water supply and distribution system in his or her community.* This is especially true for those who are responsible for supplying effective streams on the fire ground.

Regardless of his or her experience as a firefighter, an officer can be at a total loss at times if he or she lacks knowledge about the water supply system. Due to an

---

*In this chapter the word *community* refers to a city, a town, a fire protection district, or any other government jurisdiction offering fire protection to its citizens.

unanticipated loss of water, buildings have been destroyed by fire while another usable source, unknown to the **officer-in-charge** at the fire, remained untouched. Portions of communities, even entire communities, have been reduced to ashes when water mains suddenly ruptured by earthquakes left the fire department without water. In some of these instances, adequate preparation might have reduced the damages significantly. Predisaster installations of emergency water supplies, together with the establishment of relay operations, may have been necessary; by taking these steps, the firefighter can avoid the frustration of standing by helplessly while a fire burns.

This chapter outlines the basic elements of a water supply system and presents a practical method for testing the capacity of various sections of distribution systems. Although reference is made to ideals and recommended standards, few communities exist that do not have some areas where the limitations of the water system can lead to serious challenges when major fires occur. Only through planning can disasters be averted in these instances.

One of the first things to be done in making such plans is to identify areas of the city where problems may be encountered. Identification of deficient areas should include both those areas where problems may be encountered on initial response and those where larger fires are more likely to occur and, consequently, large volumes of water will be needed.

In those areas where residual pressures are low and it may be difficult to obtain sufficient flows from hydrants, consideration should be given to responding with tankers/tenders on the first-alarm assignment. The tankers/tenders can be used to ensure a reasonable amount of water for initial attacks while supplementary measures are taken to boost the flow available from hydrants. Supplementary supplies such as swimming pools, lakes, rivers, and water from adjacent service levels should be identified and plans made for their use. Such plans might include **relay operations** and the addition of equipment capable of moving water from the source of supply to pumpers, deficient distribution areas, or the fire itself.

◆ TYPES OF WATER SYSTEMS

The two basic types of water systems are the **gravity system** and the direct pumping system. The distribution system of many communities depends upon a combination of the two.

GRAVITY SYSTEMS

In a gravity system, water is stored at some point that is elevated above the distribution system and flows into the distribution system by gravity. The most reliable type of gravity system is one in which the water is collected in impounding reservoirs and fed to the distribution system without having to depend at any point upon the use of pumps. Unfortunately, few communities are so situated that this type of system is possible. Gravity systems depend upon the difference of elevation between the storage source and the distribution point to provide sufficient working pressures for both fire protection and domestic use.

PUMPING SYSTEMS

Pumping systems depend upon pumps to provide the necessary pressures to overcome friction loss in the system and provide the desired pressures for fire protection. The pumping plant for such a system is normally located near the source of supply. While a gravity system is more reliable than a pumping system, a pumping system is an effective provider.

COMBINATION SYSTEMS

Most communities take their water supply from rivers, other bodies of water, or wells and use a combination of gravity and pumping systems for distribution. In some systems, water is taken from various sources and pumped into storage facilities; from there it flows into the distribution system by gravity. Other systems collect water in impounding reservoirs that may be located many miles from the distribution system and depend upon pumps to provide the pressures in the distribution system. Another effective type of combination system depends upon gravity to provide pressures for routine operations, but uses pumps to increase the pressure when the demands on the system are extraordinary. Many combination systems are designed so that the pumps cut in whenever the pressure reaches a predetermined minimum. Some provide for manual boosting of the pressure when requested by fire officials.

HIGH-PRESSURE SYSTEMS

In some communities, water distribution systems for fire protection are completely divorced from the domestic distribution system. Many of these separate fire protection systems are designed as **high-pressure systems**. High-pressure systems provide high pressures at hydrant outlets at all times. This permits the effective use of hose streams directly from hydrants without depending upon the use of fire department pumpers to boost the pressure. These high-pressure systems may be gravity type, pumping type, or a combination of the two. Because of the high pressures in these systems, specially designed mains, valves, and hydrants are used.

◆ **CAPACITY OF SYSTEMS**

With the exception of those communities that provide a separate distribution system for fire protection, the water system serves the dual purpose of supplying water for both domestic consumption and fire protection. While the double-duty system runs counter to good fire protection principles, it is the type most prevalent throughout the United States.

Water for domestic use includes that necessary for dwellings, commercial occupancies, industrial plants, and other types of occupancies. Water in these occupancies is used for sanitation, industrial processes, agriculture, and so forth. These dual-purpose water systems should be able to supply sufficient water for fire protection and at the same time meet the maximum consumption demands for domestic purposes. Consumption for domestic purposes generally falls into three categories: **average daily consumption**, **maximum daily consumption**, and **peak hourly consumption**.

AVERAGE DAILY CONSUMPTION

The **average daily consumption** of water for a community is determined by dividing the total amount of water used in a one-year period by the number of days in the year.

MAXIMUM DAILY CONSUMPTION

The **maximum daily consumption** is the maximum amount of water used in a community during any twenty-four hour period in a three-year period. A day is not considered in determining the maximum daily consumption if an excessive amount of water is used due to an unusual situation, such as refilling a reservoir after cleaning. The maximum daily consumption is normally between 150 and 200 percent of the average daily consumption.

In some communities, the water system may be supplied from two or more sources located at different elevations above the community. For example, one area of the community may be supplied from a reservoir located 3000 feet above the community, whereas another area may be supplied by a reservoir located 2300 feet above the community. These separate reservoirs are referred to as service levels and provide different flow pressures in the community. Whenever a system has two or more service levels (as in the foregoing example), consideration should be given during an evaluation period to the maximum daily consumption of the service level being evaluated.

PEAK HOURLY CONSUMPTION

The **peak hourly consumption** is the maximum amount of water that can be expected to be used in any given hour of a day. The peak hourly consumption can vary between 150 and 400 percent of the average hourly consumption during a peak day.

REQUIRED FIRE FLOW

For years, the term *required fire flow* was used by various professional organizations such as the National Fire Protection Association (NFPA), the American Water Works Association (AWWA), and the Insurance Service Office (ISO) to define the water flow standards for fire protection needed for an individual piece of property or a detailed section of a municipality. However, in the new ISO **Fire Suppression Rating Schedule**, the term *needed fire flow* is used for this purpose. The author agrees with this change. Whereas *required* denotes a mandatory arrangement for which a penalty is associated for noncompliance, *needed* suggests a recommendation for the development of an adequate system for fire protection. Therefore, this book will use the term *needed fire flow* in lieu of *required fire flow* when outlining the fire protection standards for a community. However, the reader should keep in mind that for all practical purposes, the terms *required fire flow* and *needed fire flow* are synonymous.

Water systems should be designed to supply the needed fire flow to every section of the community while the domestic use is at the maximum daily consumption rate. The needed fire flow is the amount of water deemed required for firefighting purposes in order to confine a major fire to the buildings within a block or other group complex. The needed fire flow varies from one section of a community to another depending upon the size, construction, occupancy, and exposures of buildings within or surrounding a block or group complex. The minimum needed fire flow for any area within a community for a single fire is 500 gpm, while the maximum is 12,000 gpm. The 500 gpm minimum is basically for uncongested areas of small dwellings, while the 12,000 gpm is for large industrial areas and downtown areas of large cities. Additional flows from 2000 to 8000 gpm are normally recommended for simultaneous fires.

Although the needed fire flow may seem excessive at times, records show that actual usage of water at fires occasionally exceeds the generally accepted needed fire flow. For example, approximately 43,500 gpm was used at a lumber mill fire in Seattle on May 20, 1958, and 50,000 gpm was used at the stockyards fire in Chicago on

May 19, 1934. At the stockyards fire in Chicago, a total of 106 pumpers were in use at one time. This all-time record for pumpers at a peacetime fire was almost matched during the Los Angeles riots in 1965.

SYSTEM ADEQUACY

A water system is considered adequate when it can deliver the needed fire flow for the recommended duration of hours while the domestic consumption is at the maximum daily rate. The recommended duration of hours varies with the needed fire flow. For example, the recommended duration is two hours when the needed fire flow is 2500 gpm or less, while the recommended duration is ten hours when the needed fire flow is 10,000 gpm or greater.

The ability to provide the desired flow depends largely upon the pipe size used in the distribution system. While 6-inch pipe is considered to be the smallest size recommended for general use, the increased capacity of larger pipe should be evaluated. The following is an example of the effect of increasing pipe size:

If the diameter is	*carrying capacity is increased*
doubled	6 times
tripled	18 times
quadrupled	40 times
quintupled	73 times

SYSTEM RELIABILITY

A water system is considered reliable when it can supply the needed fire flow for the number of recommended duration hours with the domestic daily consumption at the maximum daily rate under certain emergency or unusual conditions. By necessity, there must be a certain amount of duplication of a system in order for it to be considered reliable. For example, the system must be designed so that the needed fire flow can still be supplied if a reservoir is shut down for cleaning, a pump is out of service for repair, or the primary piping has to be shut down due to breaks in the lines. A good reliable system has two or more primary feed lines extended by separate routes from the source of supply to the main districts of the community.

PRESSURES

From a fire protection standpoint, the objective of a water system is to supply water in sufficient volumes and at such pressures to adequately provide the fire streams necessary for use at a fire in any location within the community. In most cases, fire department pumpers are used to take water from the hydrants at whatever pressures are available and then boost the pressures to those necessary for working hose streams. In order to overcome friction loss in the hydrant branch, the hydrant itself, and the **suction hose** from the hydrant to the pumper, the generally accepted minimum flow pressure from a hydrant is 20 psi.

While minimum acceptable pressures can be as low as 20 psi, the normal pressure range in most communities is 65 to 75 psi. Pressures in this range are adequate for ordinary consumption in most buildings up to ten stories and are also sufficient to provide for sprinkler systems in buildings up to five stories.

When pressures increase above 75 psi, there is a tendency to cause leaks in domestic plumbing. Additionally, ordinary water supply systems are designed for maximum working pressures of 150 psi, so pressures greater than 150 psi are considered excessive. When such pressures are encountered, it is generally necessary to use pressure-reducing valves in order to keep pressures within safe limits.

◆ DISTRIBUTION SYSTEMS

In any water system used for fire protection, a distribution system must be developed for transferring water from a source or storage supply to fire hydrants that are strategically placed throughout the community. Fire department pumpers then connect to the hydrants and increase the pressure to that needed for **effective fire streams**. In smaller systems it may be possible to transfer water directly from its source to the water mains to which the hydrants are connected. However, in larger systems the water will move through three different classifications of mains as it travels from the source to the hydrant. The three classes of mains are generally referred to as **primary feeders**, **secondary feeders**, and **distribution mains**.

PRIMARY FEEDERS

Primary feeders are large pipes that are used for moving water from the source of supply or storage area to the secondary feeders. The size of the primary feeders may vary from 48 inches in larger cities to 12 inches in smaller communities. Regardless of size, primary feeders should have a capacity sufficient to deliver the needed fire flows to all built-up areas of a community when the domestic consumption is at the maximum daily consumption rate. A water supply should have enough primary feeders that a temporary shutdown of a single feeder line will not result in a critical loss of water to the distribution system. Where possible, primary feeders should be taken from more than one source so that maximum reliability of the system is assured. If only one source of water is available, then a minimum of two primary feeders should be used to transfer water from the source of supply to the distribution system. These primary feeders should follow separate paths from the source to the primary built-up area of the community.

SECONDARY FEEDERS

Secondary feeders are smaller than primary feeders, but larger than the distribution mains used in the grid system. The function of the secondary feeders is to reinforce the distribution system. Secondary feeders tie the grid system to the primary feeders so as to aid in the concentration of the needed fire flow at any point within the grid network.

Whenever possible, secondary feeders should be looped so that water is supplied to the distribution mains from two directions. The terms *looped* and *cross-connected* are synonymous. Looping the secondary feeders increases both the capacity to the distribution system and the reliability of the entire system. Secondary feeders in well-designed systems should not be spaced more than 3000 feet apart.

DISTRIBUTION MAINS

Distribution mains are used to supply water directly to fire hydrants and to various occupancies within a community for domestic purposes. Mains should be cross-connected so as to supply a grid throughout the built-up areas of the community. In a well-designed system, water is supplied to each hydrant from at least two directions. In addition, the looping of mains substantially doubles capacity and enhances reliability.

Twelve-inch mains are recommended on all principal streets, with 8-inch mains cross-connected every 600 feet in business districts. The recommended minimum size for mains in residential areas is 6 inches. These should also be cross-connected at intervals of not more than 600 feet.

Although it is desirable that all mains be a minimum of 8 inches, mains 6 inches and smaller often exist in various sections of not only older communities, but also newer residential areas. **Dead-end mains**, which supply water from only one direction, exist in many communities, limiting the amount of water available for fire protection purposes. While fire department officials have to learn to live with their water systems, it is essential that the deficiencies of these systems be identified and plans be made for adjusting to fire flow demands that exceed normal day-to-day operations. Fire department officials may have to make extensive plans in order to cope with existing deficiencies, but this should in no way prevent them from establishing control standards that comply with recommended minimums for new construction.

TYPES OF PIPES

Pipes used for mains are generally made of asbestos–cement, cast-iron and ductile-iron, reinforced concrete, plastic, and steel. All pipes used for mains should conform with standards established by the AWWA. The piping used in a water distribution system depends upon the working pressures that will be encountered and the conditions under which the piping will be installed.

Asbestos–Cement Pipe

Asbestos–cement pipe was widely installed, generally in sizes of 4 to 12 inches in diameter, in American water systems in the 1950s, 1960s, and 1970s as a noncorrosive and economical alternative to cast-iron pipe. When it became evident that a fatal lung disease often resulted from breathing airborne asbestos fibers, severe restrictions were mandated for working with asbestos. Asbestos–cement pipe went out of favor and plastic or polyvinyl chloride (PVC) pipe became the pipe of choice for those diameters in most water systems. Although asbestos–cement pipe rarely is installed today, thousands of miles of it remain in water systems throughout the country.

Cast-Iron Pipe

Cast-iron pipe, when used in water mains, is selected on the basis of maximum working pressures and laying conditions. Pipe is available for working pressures as high as 350 psi. The pipe is subject to corrosion from water; however, the initial rate of corrosion is not as great as that of steel. After a few years of exposure, the corrosion rate is little different from that of steel.

Ductile-Iron Pipe

Ductile-iron pipe has the corrosion-resistance quality of cast iron, while its strength and ductility approach those of steel. Under ordinary conditions, its use is preferred over steel primarily because of the high installation costs of steel pipe.

Reinforced Concrete Pipe

Reinforced concrete pipe is not normally used in the distribution system; however, it is used on long conducts and aqueducts. Because it is used primarily for transferring large quantities of water, it is generally only available in sizes of 24 inches or larger.

Steel Pipe

Steel is used for water mains because of its high tensile strength. It is particularly advantageous where the distribution system may be subjected to impact pressures from railroad tracks, highways, industrial machinery, and such or in areas where the danger of earthquake shock is great. Steel pipe is also advisable where unstable soil conditions or steep slopes are found. The primary disadvantage of using steel pipe is high installation costs.

Plastic Pipe (PVC)

Plastic pipe used in water systems should be constructed and used in accordance with the AWWA specifications, which outline piping criteria for 100-, 150-, and 200-pound pressures. Approved plastic pipe is particularly useful in areas where severe corrosion problems are a possibility. Polyvinyl chloride pipe is not subject to corrosion or build-up from corrosion by-products. It is also advantageous in areas subject to earthquake shocks or where live loading, shifting, or movement of the earth is anticipated.

Some communities require that all new pipe between 4 and 18 inches in diameter be manufactured from PVC or ductile iron.

Valves

Control valves should be installed in water distribution systems at strategic locations and at such intervals that portions of the system can be shut down or the flow controlled as the needs dictate. Gate valves (or butterfly valves) that can be operated with a key wrench are normally used (Figure 3.1).

Spacing of valves should allow small segments of the system to be shut down for repair without reducing the fire protection over a wide area. It is recommended that enough valves be installed that a break or failure of the system will not require shutting down a length of pipe greater than 500 feet in commercial areas, 800 feet in residential areas, or ¼ mile for arterial mains.

Under normal conditions, valves in water distribution systems should be operated only by trained members of the water department. However, some fire departments carry tools on their ladder or utility companies that can be used to close the valves in an emergency. Before ordering these valves closed, however, an officer should be fully aware of what will result from this action.

HYDRANTS

Hydrants are the terminal points of the fire protection part of a water supply distribution system. Hydrants must be properly designed and strategically located in order to fulfill the potential of the distribution system. Despite this fact, many hydrants are poorly located and improperly set for efficient operations. Hydrants can be found with outlets turned in the wrong direction, located too far from a paved road, set too low for proper use of hydrant wrenches, and obstructed by trees and poles. It is important that fire department officials and water department officials coordinate their efforts to ensure that these conditions are corrected.

Standards for hydrants are prepared by the AWWA. A standard hydrant is designed for a working pressure of 150 psi and is hydrostatically bench-tested at 300 psi. An installed hydrant with a single 2½-inch outlet should be capable of a flow of a minimum of 250 gpm with a pressure loss of not more than 1 pound between the street main and the outlet. Those hydrants with two 2½-inch outlets should be capable of a flow of a minimum of 500 gpm with a pressure loss of not more than 2 pounds between

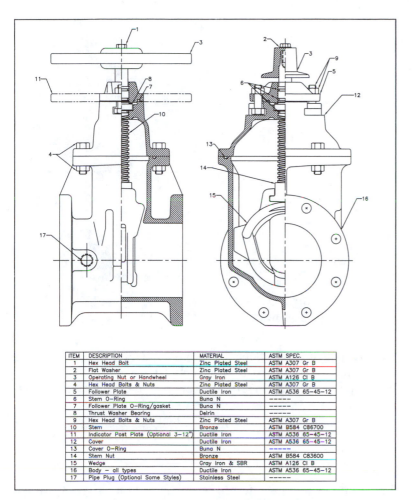

ITEM	DESCRIPTION	MATERIAL	ASTM SPEC.
1	Hex Head Bolt	Zinc Plated Steel	ASTM A307 Gr B
2	Flat Washer	Zinc Plated Steel	ASTM A307 Gr B
3	Operating Nut or Handwheel	Gray Iron	ASTM A126 Cl B
4	Hex Head Bolts & Nuts	Zinc Plated Steel	ASTM A307 Gr B
5	Follower Plate	Ductile Iron	ASTM A536 65-45-12
6	Stem O-Ring	Buna N	-----
7	Follower Plate O-Ring/gasket	Buna N	-----
8	Thrust Washer Bearing	Delrin	-----
9	Hex Head Bolts & Nuts	Zinc Plated Steel	ASTM A307 Gr B
10	Stem	Bronze	ASTM B584 C86700
11	Indicator Post Plate (Optional 3-12")	Ductile Iron	ASTM A536 65-45-12
12	Cover	Ductile Iron	ASTM A536 65-45-12
13	Cover O-Ring	Buna N	-----
14	Stem Nut	Bronze	ASTM B584 C83600
15	Wedge	Gray Iron & SBR	ASTM A126 Cl B
16	Body - all types	Ductile Iron	ASTM A536 65-45-12
17	Pipe Plug (Optional Some Styles)	Stainless Steel	-----

FIGURE 3.1 ◆ Gate valves are used to control the flow to various parts of the system. SBR, Styrene Butadiene Rubber. *Courtesy Clow Valve Co., Oskaloosa, Iowa.*

the street main and the outlet. Four-inch outlets should be capable of a minimum flow of 1000 gpm with a pressure loss of not more than 5 pounds between the street main and the outlet. Hydrants should be installed so that they are set plumb, with the lower outlet the proper distance above ground level to allow for use of hydrant wrenches. It is also good practice to maintain a clearance around hydrants of at least 3 feet.

While almost all new hydrants installed conform to the AWWA standards, many hydrants in service throughout the United States do not. Fire officials should check with water department officials to determine which hydrants do not comply, as the hydrants are not visually identifiable. Excessive friction losses should be expected on these hydrants. Such losses can be determined during the regular testing of the system.

Threads on hydrant outlets should conform to national standards. All hydrant outlets should be protected by caps when not in use. In addition to protecting the threads, caps on **dry-barrel hydrants** prevent foreign material from entering the barrel and prevent the discharge of water from outlets that are not in use when the main valve is opened.

It is recommended that all hydrants have a minimum of two outlets in the event that one is damaged. It is better that one of the outlets be a size large enough for use

Figure 3.2 ◆ A California wet-barrel hydrant.
Courtesy Clow Valve Co., Oskaloosa, Iowa.

of the pumper suction. Although two outlets are recommended, a great number of single-outlet hydrants are found in some parts of the country.

There are two basic types of hydrants in service throughout the United States: **wet-barrel** and dry-barrel. Some of the variations of these two with which fire department members should be familiar are the flush type, the high-pressure type, and dry hydrants.

Wet-Barrel Hydrants

The wet-barrel hydrant (sometimes referred to as the California type) is filled with water at all times, making water available at the hydrant outlet as soon as the hydrant valve is opened (Figure 3.2).

The use of wet-barrel hydrants is restricted to areas where there is no danger of freezing. These hydrants are generally equipped with a compression-type valve for each outlet; however, some models are designed with a single valve in the bonnet (or top), which controls the water flow to all outlets.

The simplest type of wet-barrel hydrant is one that is sometimes referred to as a built-up hydrant. A built-up hydrant consists of common piping (normally 4 in. in size) for a **riser**, topped with an angle valve (Figure 3.3). Some built-up hydrants use 6-inch piping for the riser and are equipped with both a 2½-inch outlet and a pumper outlet. If a built-up hydrant is only provided with one outlet, it is recommended that it be a pumper outlet.

Manufactured wet-barrel hydrants are available in various sizes and shapes; however, basically they all operate in the same manner (see Figure 3.4). Some hydrants have been manufactured to comply with the aesthetic demands of city officials, but do comply with the general requirements for hydrant capacity.

Wet-barrel hydrants are bolted to the footpiece as a unit. This permits easy replacement if the hydrant gets damaged. Damage to hydrants by careless drivers is not unusual. Unfortunately, when wet-barrel hydrants are knocked over by a vehicle,

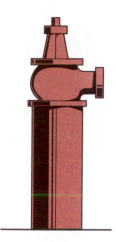

FIGURE 3.3 ◆ A built-up wet-barrel hydrant.

it usually results in a geyser-like discharge and the loss of a large quantity of water. In addition to the water loss, there is always the possibility of water damage to surrounding property.

At least one manufacturer has designed a wet-barrel hydrant that will give a *dry break* in the event the hydrant is knocked over (Figure 3.5).

Figure 3.6 shows a typical installation of a wet-barrel hydrant. The street connection should not be less than 6 inches in diameter for ordinary use and 8 inches in diameter for hydrants equipped with pumper outlets. A valve is needed between the street main and hydrant so that the hydrant can be shut down for repair without shutting down the entire main. It is usually better to bolt the valve directly to the main-line tee. This avoids the possibility of a break between the main and the valve, which would require a shut down of the main until the pipe can be repaired.

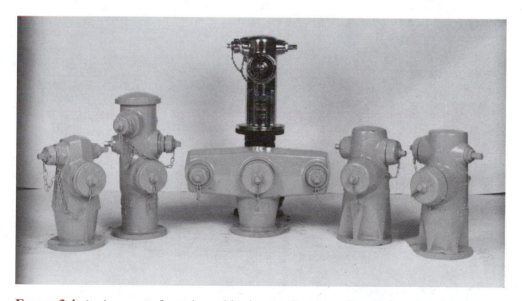

FIGURE 3.4 ◆ A group of wet-barrel hydrants. *Courtesy Clow Valve Co., Oskaloosa, Iowa.*

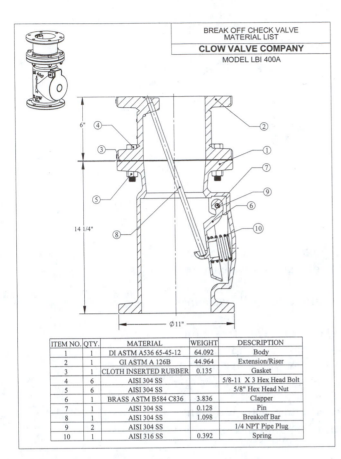

BREAK OFF CHECK VALVE
MATERIAL LIST
CLOW VALVE COMPANY
MODEL LBI 400A

ITEM NO.	QTY.	MATERIAL	WEIGHT	DESCRIPTION
1	1	DI ASTM A536 65-45-12	64.092	Body
2	1	GI ASTM A 126B	44.964	Extension/Riser
3	1	CLOTH INSERTED RUBBER	0.135	Gasket
4	6	AISI 304 SS		5/8-11 X 3 Hex Head Bolt
5	6	AISI 304 SS		5/8" Hex Head Nut
6	1	BRASS ASTM B584 C836	3.836	Clapper
7	1	AISI 304 SS	0.128	Pin
8	1	AISI 304 SS	1.098	Breakoff Bar
9	2	AISI 304 SS		1/4 NPT Pipe Plug
10	1	AISI 316 SS	0.392	Spring

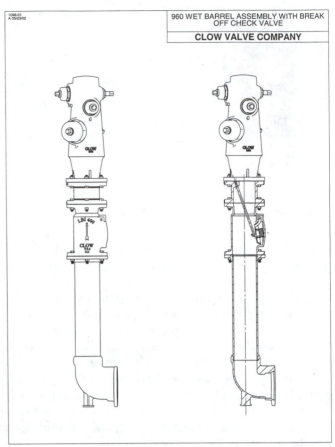

960 WET BARREL ASSEMBLY WITH BREAK
OFF CHECK VALVE
CLOW VALVE COMPANY

FIGURE 3.5 ◆ A positive break-off check valve for a wet-barrel hydrant. *Courtesy Clow Valve Co., Oskaloosa, Iowa.*

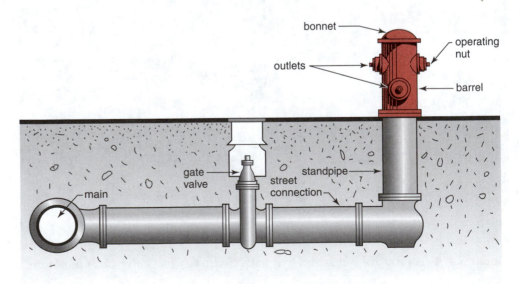

FIGURE 3.6 ◆ A typical wet-barrel hydrant installation.

Dry-Barrel Hydrants

The dry-barrel hydrant is the most commonly used. On this hydrant, the valve controlling the water is located below the frost line; thus, the barrel of the hydrant is dry except when in use. At the base of the hydrant there is a drain valve, which permits the hydrant barrel to be drained after use. The drain valve is designed so that it closes as the main valve is opened and opens as the main valve is closed.

It is important that this type of hydrant be installed in soil that allows reasonable drainage. If water does not drain properly, it might freeze in the barrel and render the hydrant inoperative. If drainage problems are anticipated when a new hydrant is installed, the hydrant should be placed in a pit of small stones. The pit, which should measure about 2 feet in diameter and 2 feet below the base of the hydrant, should be filled completely with coarse gravel or stones and placed around the barrel of the hydrant to a level several inches above the drain outlet. This type of installation assists in proper drainage and helps prevent clogging of the drain outlet by dirt or sand.

One type of drain valve consists of a ball that rolls against a seat when water enters the hydrant under pressure. The seated ball prevents water from leaking out of the drain line. When the hydrant is shut down, the ball rolls away from the seat, allowing the water to drain from the barrel. If the drain valve is located below the ground water table, water and slush can enter the hydrant and cause clogging; in addition, there is a possibility of the water freezing within the barrel (see Figure 3.7).

A dry-barrel hydrant has another advantage over the common wet-barrel hydrant in addition to the advantage of eliminating the possibility of water freezing within the barrel. If an accident results in a dry-barrel hydrant being torn from its base, the control valve remains closed and no water is lost. However, there is a slight drawback to this particular feature. In some cases, breaks have occurred between the upper and lower barrels of dry-barrel hydrants while the hydrants remained in place; the hydrants were rendered entirely ineffective without the knowledge of fire officials. It is possible that such a break could go unnoticed until a hydrant is needed for use at a fire. In most wet-barrel hydrants, a break of this type would not go unnoticed.

Flush-Type Hydrants

Flush-type hydrants are those in which the outlets and the control valves are located below ground level. A metal plate is generally used to cover the cast-iron box or manhole in which the hydrant is located. The words *fire hydrant* might be found on the metal cover. The cover must be strong enough to bear the weight of the anticipated traffic, but light enough to be easily removed for hydrant use.

Flush-type hydrants may be either wet-barrel or dry-barrel, depending upon where they are installed. Generally, the hydrant has only one outlet; however, two outlets are found on some.

Flush-type hydrants are used primarily in airports, loading ramps, and long corridors. While such hydrants are not recommended for general use, their availability in airports increases the capability of a fire department to cope with major crashes.

There are well-founded objections to flush-type hydrants for general use. One is the extra time it takes for a pumper to connect to the hydrant outlet; another is the difficulty of locating the hydrant when it is needed. Because it is below ground level, the

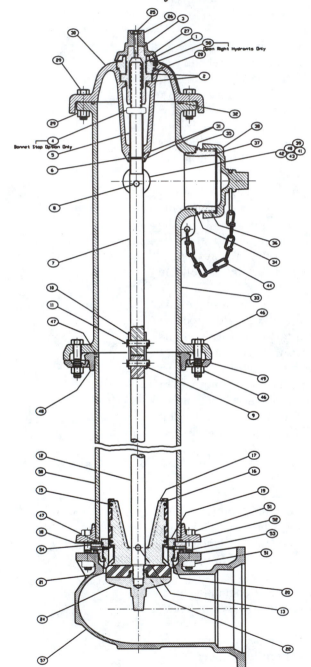

CLOW

Medallion Hydrant

DET.	QTY.	DESCRIPTION	MATERIAL
1	1	O-ring #131	Buna N
2	2	Thrust Bearing	Delrin
3	1	Operating Nut	Bronze ASTM B584 C86700
4	1	Jam Nut	Zinc Plated Steel
5	1	Stem Sleeve	Bronze ASTM B135
6	1	O-ring #119	Buna N
7	1	Upper Stem	1045 CRS
8	1	Upper Stem Pin	18-8 Stainless Steel
9	2	Cotter Pin	18-8 Stainless Steel
10	1	Safety Stem Coupling	Steel Tubing
11	2	Safety Coupling Pin	18-8 Stainless Steel
12	1	Lower Stem	1018 CRS
13	1	Lower Stem Pin	18-8 Stainless Steel
15	4	Spring Pin	316 Stainless Steel
16	2	Drain Valve Facing	Urethane Rubber
17	1	Upper Valve Plate	Bronze ASTM B584 C83600
18	1	O-ring #362	Buna N
19	1	Seat Ring	Bronze ASTM B584 C83600
20	1	O-ring #361	Buna N
21	1	Main Valve Seat	SBR/Urethane/EPDM
22	1	Lower Valve Plate Lock Washer	18-8 Stainless Steel
24	1	Lower Valve Plate	Gray Iron ASTM A126 Class B
25	1	Hold Down Screw	Zinc Plated Steel
26	1	Weather Cap	Gray Iron ASTM A126 Class B
27	1	Thrust Nut	Bronze ASTM B584 C83600
28	1	O-ring #151	Buna N
29A	4	Bolt	Zinc Plated Steel
29B	4	Nut	Zinc Plated Steel
30	1	Bonnet	Gray Iron ASTM A126 Class B
31	2	O-ring #217	Buna N
32	1	O-ring #444	Buna N
33	1	Nozzle Section	Gray Iron ASTM A126 Class B
34	1	Pumper Nozzle Lock	316 Stainless Steel
35	1	O-ring #250	Buna N
36	1	Pumper Nozzle	Bronze ASTM B584 C83600
37	1	Gasket	Rubber
38	1	Pumper Nozzle Cap	Gray Iron ASTM A126 Class B
39	2	Hose Nozzle Lock	316 Stainless Steel
40	2	O-ring #235	Buna N
41	2	Hose Nozzle	Bronze ASTM B584 C83600
42	2	Gasket	Rubber
43	2	Hose Nozzle Cap	Gray Iron ASTM A126 Class B
44	1	Chain	Zinc Plated Steel
46A	8	Safety Flange Bolts	Zinc Plated Steel
46B	8	Safety Flange Nuts	Zinc Plated Steel
47	2	O-ring #442	Buna N
48	1	Barrel Upper Flange	Ductile Iron ASTM A536 70-50-05
49	2	Safety Flange	Gray Iron ASTM A126 Class B
50	1	Barrel	Ductile Iron Pipe
51A	8	Shoe Bolt	Zinc Plated Steel
51B	8	Shoe Nut	Zinc Plated Steel
52	1	Barrel Lower Flange	Ductile Iron ASTM A536 70-50-05
53	1	O-ring #444	Buna N
54	1	Drain Ring	Bronze ASTM B584 C83600
57	1	Shoe	Ductile Iron ASTM A536 70-50-05
58	1	Socket Setscrew	18-8 Stainless Steel

FIGURE 3.7 ◆ A dry-barrel hydrant. *Courtesy Clow Valve Co., Oskaloosa, Iowa.*

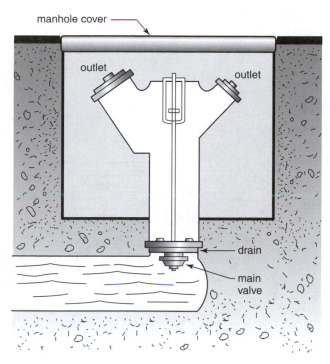

FIGURE 3.8 ◆ A flush-type hydrant.

hydrant can be blocked by something being placed on it, a vehicle parking over it, or snow or ice covering it (see Figure 3.8).

High-Pressure Hydrants

High-pressure hydrants are normally installed on high-pressure water mains that are independent of the domestic supply system. Pressures available on these hydrants vary among cities, but they normally will be in the range of 160 to 180 psi; in some cities, however, the pressures can range from 150 to 300 psi. The hydrants are designed to be extra heavy and may be provided with as many as four outlets. Hose lines used for fire-fighting operations can be laid directly from these hydrants without the use of fire department pumpers.

Occasionally, because of their particular location on the distribution system, extra high pressure may be found on ordinary distribution systems. Use of these hydrants by a pump operator who is not aware of the extra high pressures could result in damage to the pump, other equipment, or both. Some fire departments have identified these hydrants by special markings; one paints the bonnet (top) red.

Dry Hydrants

Dry hydrants are not connected to a positive-pressure source of water. These hydrants are referred to as suction hydrants in some areas of the country. Such hydrants are found in rural areas or near piers and are so located that water may be drawn from pools, lakes, rivers, and such. The appearance resembles that of a regular hydrant.

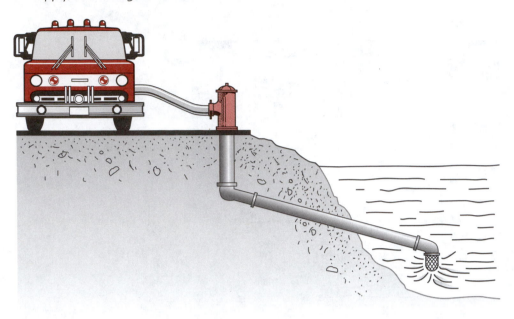

FIGURE 3.9 ◆ A dry-hydrant installation.

In essence, dry hydrants are permanently installed, hard-suction lines. A pipe extends from the hydrant to the body of water. A strainer is attached to the end of the piping. All fittings and connections are airtight. A pumper connects to the hydrant and proceeds to draft just as if a hard-suction line had been lowered into the water.

There are two main advantages of using a dry hydrant over taking water directly from draft. The first is that water is obtained more rapidly when drafting operations become necessary. Second, it is not necessary to couple together many sections of hard-suction hose in order to reach the water (see Figure 3.9).

There are several factors that should be considered when planning new dry hydrants. These hydrants can be a portion of the regular water supply system or be designed as a part of the emergency supply system. Thought should be given to whether the hydrant will be used to fill tankers or as direct protection for a particular occupancy or area. If used as a source of water for a particular occupancy or area, large-diameter hose may be sufficient to move the water from the hydrant to the fire or plans may be made to set up a relay operation.

When planning for the installation of a new hydrant, it is a good idea to restrict the lift from the water source to the hydrant outlet to 15 feet. There should be at least 2 feet from the water level to the strainer to help alleviate **whirlpools** when drafting. If the level of the water can change due to seasonal fluctuations, the lowest expected level should be considered as the standard for use. If possible, it is good practice to maintain at least 2 feet from the bottom of the strainer to the bottom of the water source.

When locating a new hydrant, such factors as plant growth and silt, ice and snow, droughts, inclines, and other items that might affect access to the hydrant and water source maintenance should be considered.

It is important that dry hydrants be identified in some manner so that mutual aid companies will not think they are pressured hydrants.

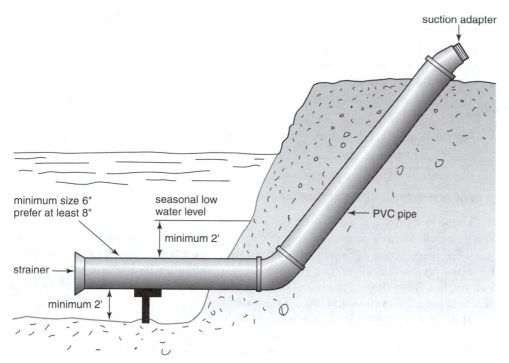

suction adapter

minimum size 6"
prefer at least 8"

seasonal low
water level

PVC pipe

minimum 2'

strainer

minimum 2'

FIGURE 3.10 ◆ A built-up hydrant.

Built-Up Dry Hydrants

Built-up dry hydrants have at least one advantage over standard dry hydrants. While all the items outlined for consideration of dry hydrants apply to built-up dry hydrants, built-up dry hydrants can be designed to help take the stress placed on a regular hydrant by the hard-suction hose. This is particularly important if the traditional rigid, hard-suction hose is used rather than the flexible type. Angling the hydrant as shown in Figure 3.10 and placing a protective item on the ground under the hard-suction hose results in considerable reduction in the stress placed on the hydrant. The hydrant should be designed so that adapters will not be needed to connect a hard-suction hose to it. Flexible hard-suction hose is preferred as standard equipment on apparatus rather than rigid hard-suction hose.

It is recommended that built-up dry hydrants be constructed of schedule 40 PVC using two elbows and airtight fittings. Piping should be a minimum of 6 inches in size, but preferably 8 inches or larger. Most built-up hydrants can supply a flow between 600 and 1000 gpm. In most cases pipe larger than 6 inches is needed if the lift is more than 10 feet. The seasonal low-water level should be used when considering the lift. Several manufacturers produce the necessary elbows and fittings required for construction of a built-up hydrant.

Hydrant Marking

Knowledge of the capacity of hydrants is extremely important in fire control. Hydrant marking may not be necessary in communities where adequate flows are available from all hydrants for the district protected; however, few cities enjoy this luxury. Most communities have water-deficient areas and most have a wide variety of flows from the same types of hydrants. NFPA standard 291 recommends a standard marking system for all hydrants based upon the capacity of the individual flow test of the hydrant. Flow tests of individual hydrants do not give as satisfactory results as group testing; however, knowledge of individual hydrant flow capabilities is valuable to those on the fire ground.

The NFPA recommendations are that the bonnet and nozzle caps be painted as follows:

For flows of 1500 or greater	Light blue
For flows of 1000 to 1499 gpm	Green
For flows of 500 to 999 gpm	Orange
For flows of less than 500 gpm	Red

These markings are based on flow tests of individual hydrants with the domestic consumption at an ordinary demand. The ratings are based on 20-psi residual pressures when initial pressures are over 40 psi and on half the initial pressure when the initial pressures are less than 40 psi. Flows are individual hydrant flows, and therefore smaller flows than indicated may be expected in some cases when a number of hydrants are used in the same area at the same time.

There is one other color that is recommended for hydrants. If a hydrant is permanently inoperative or not usable, all visible parts of the hydrant should be painted black. If a hydrant is temporarily out of service, it should be wrapped or otherwise prepared with a temporary indicator of its status.

The NFPA also recommends that nonmunicipal hydrants be painted a color that distinguishes them from municipal hydrants. Hydrants with nonpotable water should be painted violet (light purple) and private hydrants should be painted red.

Some communities also mark the location of hydrants that might be hidden from the sight of responding companies. Markings are placed on telephone poles or other objects located near the hydrant. The bands used for the markings are placed high enough to be seen over the tops of parked cars. There are at least two schools of thought on the colors that should be used for these indicator bands: one is that the bands should be the color of the hydrant, another is that the color of the band should indicate its flow capacity. At least one professional organization recommends that the hydrant bonnet of hydrants be colored as follows to indicate the main size:

Red	4 in.
Yellow	6 in.
Orange	8 in.
Green	10 in. or larger

These same colors could be used for indicator bands.

Some communities mark hydrant locations by placing traffic markers in or near the center of the street, immediately adjacent to each hydrant location. The reflectors are usually two or more colors on each side. If only one color is used on each side (for example, red on one side and yellow on the other), one color generally indicates that the hydrant is located to the right, whereas the other color generally indicates that the hydrant is located to the left. If two colors are used on each side, the second color can be used to indicate flow or main size.

These indicators are not as effective during daylight hours as they are at night, when the reflections are readily picked up by the headlights of responding apparatus.

The use of these reflectors permits hydrant locations to be easily identified even a block away. In addition to their value to initially responding companies, they are of particular value to companies arriving on greater alarms and to companies responding from adjacent cities on a mutual aid response. Some communities use reflector paint on hydrants to assist in locating the hydrant after dark.

In some cities, the size of the water main supplying the hydrant is stenciled on the hydrant barrel. Another simple type of marking is a number on the barrel indicating how many 250-gpm hose streams the hydrant can supply.

Hydrant Spacing

Recommendations for hydrant spacing are based on the fire flow demand of a given area. To reduce the length of hose lines, hydrants should be located not more than 300 or 400 feet from the building to be protected. As a general rule, hydrants should be placed at each street intersection; when blocks are extremely long, there should be additional hydrants in the center of the block. It is desirable that hydrants be not more than 800 feet apart in any area of a city and not more than 500 feet apart in built-up areas.

Consideration should be given to hydrant spacing for particular occupancies or target hazards. There should be a sufficient number of hydrants adequately located and spaced for the safety of the building to be protected. As a general rule, each of the hydrants should be within 500 feet of the building. However, where the fire department responsible for protecting the building is equipped with LDH on all its on-line apparatus, the Insurance Service Office will give credit for hydrants located within reach of the smallest LDH load that the fire department has on any of its on-line companies. The maximum length of LDH for which credit will be given is 1000 feet.

In areas where the fire flow demand is great, it may be necessary to increase the number of hydrants located at a street intersection. At some intersections in larger cities, hydrants may be found on each of the four corners of the intersection, with additional hydrants located in the center of the block.

It is desirable that some standardization be adopted for placing hydrants at intersections within a community. It is much easier for apparatus operators and company officers to become familiar with the distribution system when all hydrants are set on the same corner of the intersection (for example, always the southwest corner).

Maintenance and Testing

Hydrants should be inspected semiannually and after use. Preferably, these inspections should be conducted in the spring and fall of the year in those areas where freezing temperatures are encountered.

Where possible, inspections of hydrants should be made by members of the first-due engine companies. Maintenance tends to improve when inspections are made by those most likely to use the hydrant. Records of all tests should be kept by the fire department. A separate record card should be kept for each hydrant. The record card should include useful information such as the type of hydrant, main size, date of installation, test dates, repairs, and date of painting.

Inspection procedures should provide for an operating test, repair of any leaks, and the pumping out of dry-barrel hydrants when necessary. Dry-barrel hydrants should also be checked for proper drainage. The operating test should consist in removing the caps, opening the hydrant valve, and running the water. Care must be taken to ensure that the discharge does not damage any adjacent property. Some departments attach a cap with a short piece of garden hose to the outlet so that the flow can be controlled. The hydrants should be painted whenever necessary, but care must be taken not to restrict the removal of caps or the operation of valve stems. If any repairs or adjustments are needed or there is need to relocate the hydrant, a report should be filed with the water department upon completion of the test.

◆ EMERGENCY PROVISIONS

It is extremely important that fire department officials have a thorough knowledge of all sources of water that could be used should the main water supply be interrupted, destroyed, or need to be supplemented during unusual emergencies. Supplies that can be used during these conditions are referred to as auxiliary or emergency supplies. Emergency or auxiliary sources include private water supplies and swimming pools as well as rivers, canals, ponds, and other natural bodies of water. **Cisterns**, if available and not used on a routine basis, are also considered emergency supplies. Cisterns and other suction water sources should be capable of supplying a minimum of 250 gpm for at least two hours. Some of the sources classified as emergency supplies may be used as a source for the primary distribution system; however, in order to use these sources under emergency conditions, additional facilities and equipment must be provided. The development of these facilities and the responsibility for seeing that the needed equipment is available is properly a function of the fire department.

Provisions for the use of emergency supplies should be included in the operational procedures of the fire department. Periodic drills should be conducted using emergency water sources for fire streams. Where possible, drill conditions should simulate those that might be expected during emergencies where the normal water supply system is not available. Drills should include both drafting and relay operations.

PRIVATE WATER SUPPLIES

The largest source of private water will generally be found within the larger industrial plants. The mere fact that fire department officials are aware of the existence of such supplies is not enough. Knowledge regarding the amount of water available, procedures for the fire department to obtain water from these private sources, and the operational procedures necessary to move the water from the source of supply to the point of use are essential.

SWIMMING POOLS

In some communities, the total capacity of swimming pools is extensive. Swimming pools should never be overlooked as a source of emergency water. Unfortunately, most swimming pools are located so that pumpers will not be able to have a close approach. In most cases it is not possible to drop a suction hose directly from a pumper into a pool; consequently, many fire departments neglect these sources in their emergency operational plans.

It is the responsibility of the fire department to see that the necessary equipment is obtained to ensure the use of available water. Siphon ejectors can be used to remove water from a swimming pool even though the pool may be a considerable distance from the location where the pumper must be spotted. Siphon ejectors pick up water on a two-to-one basis. For every gallon of water pumped into the ejector, two will be returned. A large ejector is capable of supplying a single 2½-inch working line, while

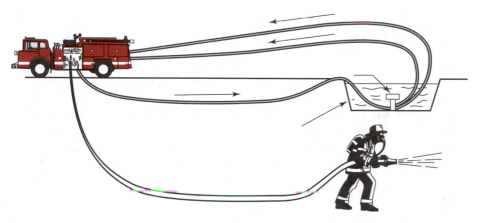

Figure 3.11 ◆ A siphon ejector removing water from a swimming pool.

the smaller ejectors can supply a single 1½-inch working line. Unfortunately, the departments equipped with ejectors normally do not carry the larger ones on the apparatus; however, in progressive departments, many of the smaller ejectors are routinely carried on engine companies (see Figure 3.11).

RIVER AND HARBOR SUPPLIES

The availability of water from these sources often depends on the fluctuation of the water level due to tides and seasonal changes. One sure method of ensuring a supply is to take water from these sources when it is available and store it for future use.

If piers are available, a pumper can be spotted at the water's edge and **drafting operations** commenced. Because piers are not always available, however, sometimes additional steps must be taken to get to the water. When the source of water is always deep enough for drafting, a hinged ramp leading down to a large float can provide the fire department with a means of using this supply source when the water level is low due to outgoing tides.

A more practical method of using these supplies is to provide pumps of the deep-well type with a discharge pipe leading to the location from which department pumpers can take water. It is better that a number of pumps be provided and a number of fire department outlets be provided; otherwise, an entire emergency supply could be eliminated due to damage to a single pump or piping system.

Fire boats also provide an excellent source of water for land companies. Consideration should be given to using boats other than those owned by the municipality in addition to those of the department. If extensive dependence for emergency water is to be placed upon fire boats, then it is better that the fire department have available one or more boat tenders. These tenders should carry a good supply of large hose that will permit the transfer of water from the river or harbor to locations a considerable distance into the city. Large-diameter hose is particularly useful for this purpose.

TANK TRUCKS/WATER TENDERS

Fire department officials should maintain a complete list of tank trucks/water tenders that could be placed into service in an emergency. The list should include trucks owned and operated by the municipality together with those available from private sources. The water department is the most likely source of tankers within a community. Building contractors and refineries are good outside sources.

Of course, the mere recording of the location of tank trucks is insufficient. Operational plans for the use of this equipment must be developed together with occasional drills with personnel from the private companies involved. It is essential that all pump operators know how to obtain water from trucks. In addition, couplings or adapters must be available on fire department apparatus for making the necessary connections to the outlets of the tank trucks.

Many rural fire districts dispatch their own tankers, referred to as mobile water supply apparatus, on the initial response to areas where hydrant distribution is not adequate or available. Such apparatus is equipped with large tanks having quick fill-and-drop connections. Many are also equipped with a pump.

CISTERNS

Cisterns are large storage tanks of water, normally located beneath the streets of municipalities. A well-planned cistern system provides cisterns where hydrants are located. Water can be taken from these sources either from a suction pipe provided for this purpose or by dropping a suction hose into the tank.

While cistern systems are not generally considered as a source of primary water, they are an excellent means of providing for emergency storage. Some method should be used by fire department officials to identify their location, and periodic drills should be conducted to ensure that all members are proficient in their use.

◆ FIRE-FLOW TESTS

Fire-flow tests indicate the amount of water that is available in various sections of a community. If sufficient water is not available in a particular section, the fire department should be aware of it in order to develop operational plans prior to the fire. Operations in weak or deficient areas of a community may involve responding with tankers, relaying, pumping from a stronger service level into a weaker area, or making use of emergency supplies. Even a well-equipped fire department with well-trained firefighters can do little toward bringing a fire under control without an adequate amount of water.

WHEN TO TEST

Fire-flow tests should be conducted on portions of a water supply system whenever any doubt exists as to the capability of the system to supply the amount of water necessary to combat a large fire in the area. Fire department officials should question the capability of the system whenever extensive construction is taking place in areas where the mains are known to be small or dead-end mains. The development of new housing tracts or shopping malls often pushes the community boundaries into the suburbs and makes it necessary for new water mains to be extended off those that may already be considered questionable. Another telltale sign of a possibly weakened system is the expansion of industrial plants or commercial complexes without a corresponding strengthening of the water system.

WHERE TO TEST

Tests should be conducted in every area of the community where doubt exists as to whether the water system can supply the volume of water necessary to cope with large-scale fires. Generally, it is not necessary to conduct tests in downtown sections of metropolitan areas; here, water mains are usually large, cross-connected, and part of a well-designed circulating system. The following areas should receive top priority.

1. Those where dead-end or small mains are prevalent.
2. Fringe areas of the community.
3. Spot developments, particularly those in outlying sections of the community.
4. Expansion of industrial plants.

HOURS FOR TESTING

From a fire protection standpoint, the ideal time for testing is when the domestic demand on the water system is the highest. Usually, this period is between 9:00 A.M. and 5:00 P.M. Monday through Friday, although it can vary. While these hours are generally good when considering an entire community, they may not be the best for testing a particular area of a community. For example, there are some industries where demand is extremely high while certain processes are being conducted. Fish canneries, as an example, use a tremendous amount of water during canning operations. Fire-flow tests in an area where a fish cannery is located, if conducted when canning operations are not in progress, may well indicate flows far in excess of those needed for the buildings to be protected. Another test during canning operations may show that the water system is extremely deficient.

PLANNING FOR TESTING

Once a fire chief has decided that a given area of a community should be tested, the tests should be done as soon as possible. Cooperation of water officials and insurance representatives should be sought. These officials are always interested in test results, in addition to having the expertise necessary to assist in the testing.

It should be remembered that a high degree of accuracy is not required. The objective of the tests is to determine the number of fire streams obtainable from the system. A standard 2½-inch, handheld line discharges 250 gpm. Fire-flow tests need only be calculated to the nearest 50 gpm for total quantities of less than 1000 gpm and to the nearest 100 gpm for quantities in excess of 1000 gpm.

The fire chief should plan in detail how the tests will be conducted before the actual field tests commence. Consideration should be given to the following areas:

1. Hydrants to be used in the tests.
2. Which hydrants will be designated as flow hydrants and which one will be designated as the pressure hydrant.
3. What readings will be taken.
4. Duties and responsibilities of each person participating in the tests.
5. Review of previous test results of the area to be tested.

EQUIPMENT REQUIRED

The amount and type of equipment needed for testing depends to a large degree upon the refinements desired. Since a high degree of accuracy is not required, fire officials can get satisfactory results using the following equipment:

1. One or more **Pitot blades** (see Chapter 4, p. 71) with 0- to 100-psi pressure gages. Pitot blades are available in several different shapes and designs, depending upon the manufacturer. Regardless of the type used, it is important that it be inserted in the flowing hydrant stream without

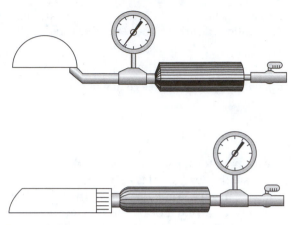

FIGURE 3.12 ◆ Pitot blades.

causing excessive wobble or spray. Readings should be as accurate as possible. Either a pressure-gage snubber or an air chamber is usually attached to the blade to remove any gage needle fluctuations. The pressure snubber is the slower of the two, but gives better results (see Figure 3.12).

2. A hydrant cap with a 0- to 200-psi pressure gage. This piece of equipment can be made by drilling a hole in a hydrant cap and inserting the pressure gage equipped with a petcock and bleeder connection. The petcock and bleeder connection is used to relieve the trapped air when first opening the hydrant and for relieving the hydrant pressure within the instrument after the hydrant has been shut down (see Figure 3.13).

3. Several hydrant wrenches.

4. A book for recording results.

5. A measuring scale marked in at least $\frac{1}{16}$ inch.

SELECTING THE HYDRANTS

Fire department officials should rely on the expertise of water department and insurance officials who participate in the tests to determine the total fire flow. Detailed engineering methods for determining the flow are explained in various other publications. The purpose of this section is to provide fire officials with a basic understanding of the testing procedures.

Two terms are used to identify the hydrants used in the testing. One hydrant is referred to as the **pressure hydrant**. The hydrant cap with the pressure gage is attached to this hydrant. The purpose of the pressure hydrant is to measure the static and residual pressures on the system. The other hydrants used in the tests are referred to as **flow hydrants**. Flow hydrants are used to determine the amount of water available from the system.

The hydrants to be used in the test should be identified after the area for testing has been selected. One of them should be designated as the pressure hydrant. The total number of hydrants to be used depends upon the strength of the system. Two hydrants may be sufficient in weak systems. Seven or eight hydrants may be required when the mains are large and the system is strong.

FIGURE 3.13 ◆ A hydrant cap with pressure gage.

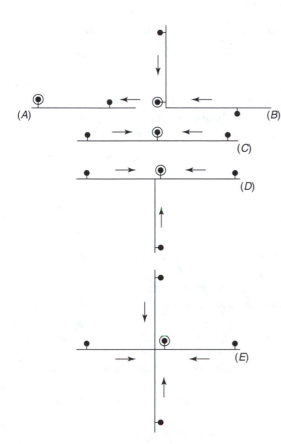

FIGURE 3.14 ◆ (*A*) A dead-end situation; (*B*) a situation in which the pressure hydrant is at the corner of a grid system; (*C*) the flow to the pressure hydrant is from two directions along a single main; (*D*) the flow to the pressure hydrant is from three directions; (*E*) the pressure hydrant is located in a well-designed loop system.

The hydrants chosen as flow hydrants should be located between the pressure hydrant and the larger mains that supply water to the testing area. Several configurations are possible. In Figures 3.14 and 3.15, the pressure hydrant is indicated by

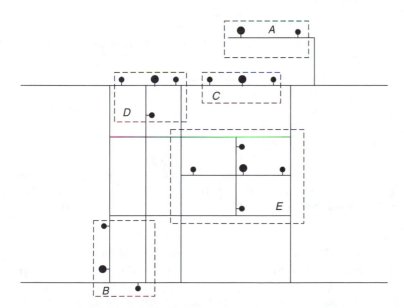

FIGURE 3.15 ◆ The situations from Figure 3.14 might appear this way in a distribution system.

a circle around the hydrant symbol. The direction of water flow is indicated by an arrow.

TEST PROCEDURES

The most basic test requires only two hydrants, a pressure hydrant and a flow hydrant, but the testing procedure is similar regardless of the number of hydrants involved. The following steps outline the testing procedure in its simplest form:

1. The cap on the pressure hydrant should be removed and replaced with the cap with the pressure gage and petcock. The petcock bleeder should be in the open position.
2. The pressure hydrant should be opened slowly for several turns. The flow of water will force any trapped air out through the petcock bleeder valve. The petcock should be closed when water is discharged in a steady stream from the bleeder valve. The hydrant should then be opened fully, using a slow, steady motion.
3. The reading on the gage attached to the pressure hydrant is the static pressure. This pressure should be recorded.
4. Remove a cap from a flow hydrant and measure the inside diameter of the discharge outlet. This procedure should be repeated for every hydrant outlet that will be used in the test. Each of the sizes should be recorded.
5. Feel the inside of the discharge opening and record the hydrant coefficient (see Chapter 5, p. 110). This should be repeated for each hydrant outlet that will be used in the test. Of course, this is not possible if the hydrants are of the wet-barrel type (see Figure 3.16).
6. Make sure the discharge from the hydrant will not do any damage to surrounding property, then open the first flow hydrant slowly. While opening the hydrant, watch for a signal from the person monitoring the pressure hydrant. As a safety measure in determining the flow that can be depended upon in the event of a fire, the pressure on the pressure hydrant outlet valve should not be reduced below a residual pressure of 20 psi. The discharge outlet valve should continue to be opened until the residual pressure reaches 20 psi or the outlet valve is fully opened, whichever occurs first.

   If more than one flow hydrant is being used in the test, each of the hydrants should be opened in the manner just described. Discharge from an additional hydrant should not be commenced until water from the previously opened hydrant is flowing steadily with the valve fully opened.

   If one outlet on each of the hydrants used in the test has been fully opened and the residual pressure is still above 20 psi, then additional outlets should be opened until the maximum amount of water available from all hydrants is flowing or the residual pressure has been reduced to 20 psi, whichever occurs first.
7. When the flow from a hydrant has stabilized itself, the Pitot blade should be inserted at the center of the stream, with the blade held a distance of half the diameter of the opening from the end of the outlet. (For example, with a 2½-inch outlet, hold the Pitot blade 1¼ inches from the end of the outlet.) If the needle of the gage fluctuates, use a reading midway between the high and low fluctuation points. Record the pressure reading. Also record the residual pressure on the pressure hydrant.
8. Close the valves on the flow hydrants, one hydrant at a time. No hydrant should be shut down until satisfactory pressure readings have been obtained from all flow hydrants. Replace the caps, then open the petcock bleeder on the Pitot blade and drain the unit.
9. Shut down the pressure hydrant. Open the petcock bleeder valve to relieve the pressure within the gage. Remove the cap and gage. Replace the hydrant cap.

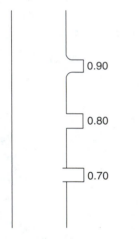

0.90

0.80

0.70

FIGURE 3.16 ◆ Hydrant coefficients.

The assisting engineers from the water department or insurance companies will be able to determine the total flow obtained during the tests with the information recorded. The total flow will indicate one of two conditions:

1. The total dependable flow for the area without reducing the residual pressure below 20 psi.
2. The total dependable flow for the area.

UNUSUAL RESULTS

Test results that seem exceptionally low for the size of the mains in the area, or flows that are considerably less than those obtained from previous testing, might be correctable. This fact should be relayed to water department officials, who should be able to diagnose the cause of the deficiencies and make the necessary corrections. Some of the possible causes are as follows:

1. Control valves in mains partly or totally closed.
2. Water mains clogged or partially blocked.
3. Sedimentation in mains, resulting in smaller effective diameters.
4. Map errors that list incorrect main sizes.
5. Serious leakage in the system.
6. Improperly operating hydrant valves.
7. Improperly operating check valves.
8. Frozen pipes or valves.
9. Existence of unmapped valves or meters.
10. High friction loss through meters.

◆ WATER SUPPLY MAPS

A map of the water system should be displayed in every fire station and in the offices of command officers. In small cities, the map should include the entire water system; in larger cities, the map should cover the area protected by the fire company on initial and second-alarm responses. Copies of the maps, together with additional information on the water supply system, should be available in the fire dispatching office.

Water supply maps on a larger scale and broken down into smaller areas should be carried on all fire apparatus and in all command cars. Pages of the map book should be properly indexed so that the area desired can be found rapidly. If the map of the city is broken down into grids, the grid number of the fire location can be relayed to responding companies at the time of dispatch.

Maps should include as much information as possible, including main sizes, types and locations of all hydrants, valve locations, and service boundaries. A service area is one that is supplied from a single source. Built-up areas of cities may encompass two or more service areas. Knowledge of boundaries of service areas is important, as it is possible to supplement a weak service area during an emergency by pumping water from a stronger adjacent area into the weaker area by using department pumpers and relay lines. Large-diameter hose is particularly useful for this purpose (see Figure 3.17).

FIGURE 3.17 ◆ Taking water from a strong service level and pumping it into a weak area.

Results of flow tests should be recorded on the maps in the fire stations and on those carried on apparatus and in command cars. Deficient areas should be indicated in red. If hydrants in the city are color-coded to indicate flows, the same color coding should be used to identify hydrants on station and apparatus maps.

It may not be possible to indicate the location of all available emergency supplies on maps; however, the primary sources of emergency supplies should be indicated. A serious attempt should be made to outline on the maps the location of as many sources of emergency supplies as possible in the deficient areas of the city. This is particularly important for maps carried on fire apparatus and in command cars.

Review Questions

1. What is one of the first things that needs to be done when making plans to cope with fires that will require the use of large volumes of water?
2. What are the two basic types of water systems?
3. What is the definition of a gravity water system?
4. What is the most reliable type of gravity water system?
5. What is the definition of a pumping water system?
6. What is a high-pressure water system?
7. How is the average daily consumption of a city determined?
8. Define (a) maximum daily consumption; (b) peak hourly consumption; (c) needed fire flow.
9. What is the minimum needed fire flow for any area of a community?
10. What are some of the record drafts that were used at fires?

11. When is a water system considered adequate?
12. How much greater is the carrying capacity of pipes when the diameter is doubled? Tripled? Quadrupled? Quintupled?
13. When is a water system considered reliable?
14. What is generally accepted as the minimum flow pressure from a hydrant in order to overcome friction loss in the hydrant branch, the hydrant itself, and the suction hose from the hydrant to the pumper?
15. What are the three classes of mains used in a water distribution system?
16. What are the primary feeders? Secondary feeders?
17. What are distributor mains?
18. What is the minimum-size main that should be used in residential areas?
19. What size mains are recommended on all principal streets in a community?

20. What are the different types of pipes used in distribution systems?
21. What is today's general policy regarding the use of asbestos–cement piping?
22. What is the basis for selecting cast-iron pipe for mains?
23. Where is PVC pipe particularly useful?
24. Where is steel pipe particularly advantageous in distribution systems?
25. Whose standards are used for PVC pipes?
26. Who prepares the standards for hydrants?
27. What are the pressure requirements on a standard hydrant?
28. What are the two basic types of hydrants in service in the United States?
29. Where are wet-barrel hydrants used?
30. What is the most common type of hydrant in service?
31. What are some of the precautions that should be taken when installing a new dry-barrel hydrant?
32. What are the primary advantages of a dry-barrel hydrant over a wet-barrel hydrant?
33. What is a flush-type hydrant?
34. Where are flush-type hydrants generally installed?
35. What are some of the objections to flush-type hydrants?
36. What are the pressure ranges on high-pressure hydrants?
37. What is a dry hydrant?
38. What are the primary advantages of dry hydrants?
39. What type of pipe should be used on a built-up dry hydrant?
40. What considerations should be given when planning a new dry hydrant?
41. What is the NFPA-recommended marking system for hydrants?
42. In addition to the NFPA marking systems for hydrants, what are some of the systems used by various cities?
43. As a general rule, where should hydrants be placed?
44. How far apart should hydrants be spaced in a city?
45. How often, and when, should hydrants be tested?
46. What are some of the sources for water for emergency purposes?
47. What is generally necessary in order to use swimming pools as a source of emergency water?
48. What are some of the methods that can be used to obtain emergency water from rivers and harbors?
49. What is a boat tender?
50. What is a cistern?
51. When should fire-flow tests be conducted by fire officials?
52. What areas of a city should receive top priority for fire-flow testing?
53. What are generally the best hours for conducting fire-flow tests?
54. What are some of the items that a fire chief should consider when planning fire-flow tests?
55. What equipment is needed for fire-flow tests?
56. Which gives the better results, a pressure-gage snubber or an air-chamber Pitot blade?
57. What is a pressure hydrant in a fire-flow test?
58. What is a flow hydrant in fire-flow tests?
59. Where should the pressure hydrant be located when flow testing a system?
60. Describe the general test procedure for fire-flow testing.
61. What are some of the correctable features that might account for test results that are exceptionally low for the size of mains being tested?
62. What type of information should water supply maps that are displayed in fire stations and carried on apparatus contain?
63. How should deficient areas be identified on water supply maps?

4 # Fire Streams

Objectives

Upon completing this chapter, the reader should:

- Be able to define a fire stream.
- Be able to recognize the common size of smooth-bore nozzles used in the fire service.
- Be able to explain some of the factors that affect the reach of a fire stream.
- Be able to explain the difference between the horizontal reach and the vertical reach of a solid stream.
- Be able to determine the approximate horizontal reach of various streams using the horizontal reach formula.
- Be able to determine the approximate vertical reach of various streams using the vertical reach formula.
- Understand the various factors affecting the penetration of a fire stream into a building.
- Be capable of working problems to determine the penetration of a stream into a building.
- Be able to work problems using the formula $C^2 = A^2 + B^2$.
- Know the difference between a direct attack and an indirect attack on a fire.
- Know the differences among automatic nozzles, fixed-gallonage nozzles, selectable-gallonage nozzles, and multipurpose nozzles.
- Be able to define the velocity flow of water moving in a hose line.
- Be able to determine the velocity flow when the head is known.
- Be able to determine the velocity flow when the pressure is known.
- Be able to compare the velocity flow in the hose with the velocity flow at the nozzle tip.
- Be able to determine the velocity flow from the nozzle tip when the velocity flow in the hose is known.
- Be able to determine the nozzle reaction from a solid-stream nozzle.
- Be able to determine the momentary nozzle reaction from a solid-stream nozzle.
- Be able to determine the nozzle reaction on fog nozzles.
- Be able to compare the nozzle reaction on various nozzle tips.
- Be able to describe some of the safety factors that should be followed when working streams from ladders and elevated master stream appliances.

There is little doubt that the rapid improvement in technology, the civil distur-
bances of the 1960s and 1970s, and the terrorist attacks in 2001 have had and will
have influences on the improvement of fire department equipment and operations.
However, the basic foundation of fire department operations has always been, and
probably always will be, the effective use of fire streams in extinguishment operations
(Figure 4.1). Perhaps the greatest impact in this arena in recent years has been the
adoption by many fire departments of 1¾-inch, 2-inch, and large diameter hoses.

The majority of all fires of any size, such as those in dwellings, have traditionally
been attacked using 1½-inch lines. Although firefighters have been able to do an effective job with these lines, because of the limited amount of water produced, the ability to quickly extinguish some of these fires has been restricted. With the adoption of
1¾-inch and 2-inch hoses, their ability to provide streams approximating or exceeding
that usually achieved with 2½-inch lines gives firefighters the tool to **knock down** fires
much more quickly. This together with the use of large diameter hose (LDH) has
greatly improved fire departments' ability to provide more effective fire streams. It is
important that fire department personnel have a basic knowledge foundation on
LDHs with their strengths and weaknesses.

It has been said that the use of LDH as a supply line provides fire officers with the
ability to "move" hydrants from a static position to a mobile position in front of a fire.
This in effect reduces the required length of attack lines. LDH lines can be used to lay
supply lines either from a hydrant to an attack pumper in front of the fire or from the
attack pumper to the hydrant. Because of low friction loss in the hose, supply lines can
be extended over long distances.

For example, moving 500 gpm through 5-inch hose only results in a friction loss of
2 psi per 100 feet. The friction loss in 1000 feet of hose moving 500 gpm is only 20 psi.
A 1500-gpm pumper connected to a strong hydrant could possibly move its entire

FIGURE 4.1 ◆ *Courtesy of Task Force Tips*

rated capacity to an attack pumper placed in front of a fire 1000 feet away. This is certainly a big improvement over the conventional use of 2½-inch hose.

While the use of LDH seems like an ideal situation, there are several considerations that should be given to its use. Laying a single supply line is putting all one's eggs in one basket. It is possible that the line could be destroyed or damaged, which would place those operating attack lines in a perilous position. For safety's sake, it is always good practice to not depend on a single supply line.

Small cars running over the hose present a particularly dangerous situation. A small car getting caught and spinning its wheels could tear a hole in the hose. Catalytic converters directly over the hose in a trapped car could burn a hole in the line. Consideration should also be given to the maneuvering of fire apparatus. One should avoid driving over couplings when moving apparatus and not drive over the hose unless it is absolutely necessary.

◆ FIRE STREAM TACTICS

Good **firefighting tactics** involve distributing personnel and equipment at the scene of a fire so effectively that the fire is extinguished with a minimum loss of life and property. The ultimate objective of extinguishment tactics is to place water on the seat of the fire in the form and quantity that will most easily extinguish the fire. This is accomplished through the use of fire streams.

A **fire stream** is defined as a stream of water from the time it leaves a nozzle until it reaches the point of intended use or until it reaches its projection limit, whichever occurs first. As it leaves the nozzle tip, a fire stream is affected by the discharge pressure, the nozzle design, and the nozzle adjustment. Once it leaves the nozzle tip, the stream is further affected by gravity, wind velocity, and air currents.

Fire streams are used to protect exposures, hold a fire in check, assist in ventilation, extinguish the fire, and save lives. Probably more lives have been saved at fires by an aggressive attack upon the fire with the use of hose streams than by any other method.

The selection of a fire stream for use at an emergency depends upon many factors: the size and intensity of the fire; the material or materials that are burning; whether the fire is free burning or confined; the accessibility of the fire; the amount of water available; the types of nozzles available; and the personnel and equipment available to accomplish the task at hand.

There is a wide variety of nozzles and nozzle types that can be used in firefighting operations, but all of them provide two basic types of streams: **solid streams** and **nonsolid streams**.

◆ SOLID STREAMS

Solid streams may be produced by smooth-bore nozzles or fog nozzles adjusted for solid-stream application. Solid streams are used primarily when a close approach cannot be made due to the size and intensity of the fire, on some types of unconfined fires, and wherever reach is a factor in protecting exposures or extinguishing the fire. Unconfined fires are those burning in a wide-open area, such as a lumberyard fire, or a fire in a structure that has burned through to the outside. Heat radiation from unconfined fires is usually intense. A large volume of water is usually required for these fires.

To be effective, solid streams must reach the fire in such form and amount that the water will absorb the heat from the fire faster than it can be generated. Good solid streams have the ability to reach long distances and can penetrate areas and materials that cannot be reached using nonsolid streams.

Smooth-bore nozzles commonly range in size from ½ to 3 inches. The ½-, ⅝-, and ¾-inch nozzles are used primarily for booster lines, whereas ⅞-, 1-, 1⅛-, and 1¼-inch nozzles are normally used for larger handheld lines. All of these tips are generally supplied at a **nozzle pressure** of 50 psi.

A tip discharging 350 gpm is about the maximum discharge that can be handheld by firefighting personnel. Discharges greater than 350 gpm produce a **nozzle reaction** too great to handle. Consequently, tips discharging water at a rate greater than 350 gpm are used on some types of master-stream appliances. These tips are normally supplied at a nozzle pressure of 80 psi.

REACH

There are a number of factors that affect the **reach** of fire streams. One of these is wind. Wind plays an important part in the reduction of the effectiveness of solid streams. Even a moderate breeze may reduce the effective range of a solid stream by 15 to 20 percent. A strong wind can render a stream completely ineffective, breaking it up into a spray and blowing the spray back into the faces of firefighters.

In addition to the wind, the reach of a stream is affected by other factors. As water leaves a nozzle tip, it is acted upon by gravity. The gravitational effect on the stream results in a curved conical form that produces a shorter reach than would be theoretically possible if the stream continued in a straight line. The air resistance to the movement of the water also affects the reach.

Reach is a measurement of the distance that a fire stream can be effectively thrown from a nozzle and still be classified as a good stream. The effective reach of a stream is generally expressed in feet (Figure 4.2).

Another factor affecting the reach of a solid stream is the nozzle pressure. The nozzle pressure of a fire stream can be determined only under test conditions. A Pitot gage is used for this purpose (Figure 4.3). Under field conditions, the nozzle pressure is estimated using field formulas to determine the pump pressure required to produce the desired nozzle pressure.

Whereas nozzle pressures of tips on small hoses may be satisfactory at 25 psi, the nozzle pressure of handheld lines of 1½-, 1¾-, 2-, and 2½-inch hoses should be maintained between 40 and 60 psi. Although pressures within this range will produce effective streams, most departments have standardized on the use of streams with a

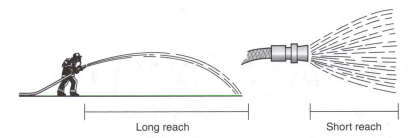

Long reach   Short reach

FIGURE 4.2 ◆ The effective reach of a straight stream (left) and a spray stream (right).

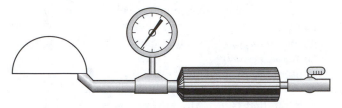

FIGURE 4.3 ◆ A Pitot gage.

nozzle pressure of 50 psi. Working nozzle pressures should be between 60 and 100 psi for effective solid streams on **master-stream** appliances. A nozzle pressure of 80 psi is used as standard by most departments.

Solid streams may be used at ground level or projected in upper floors of multi-story buildings. For purpose of identification, these streams are referred to as **horizontal streams** and **vertical streams**, respectively.

HORIZONTAL REACH

Knowledge of the reach of solid streams is important to effective firefighting operations. Reach is affected by nozzle size, hose size, nozzle pressure, and other factors that are within the selection control of the officer in charge at a fire. When it becomes necessary to throw a stream from the window of one building through the window of a building across the street, or when it becomes necessary to reach the seat of a fire that cannot be closely approached because of the extreme heat, it is important to know, prior to laying lines and selecting nozzles, that the stream will reach. The selection of the tip itself may mean the difference between reaching the fire or having the stream fall short of its intended mark.

A horizontal solid stream is one with an **angle of discharge** of 45 degrees or less. Theoretically, a solid stream has its greatest horizontal reach when the angle of discharge is 45 degrees. However, earlier studies found that under actual conditions, the greatest horizontal reach is obtained at an angle of approximately 32 degrees. As a result of these studies, some general rules have been developed for use on the fire ground for estimating the approximate reach of handheld horizontal streams. The approximate reach is one and three-fourths times the nozzle pressure at 20 psi; one and one-half times the nozzle pressure at 50 psi; one and one-fourth times the nozzle pressure at 75 psi; and equal to the nozzle pressure at 100 psi. The actual reaches determined by these studies for tips up to 1½ inches in size are shown in Table 4.1.

There are two formulas in general use for estimating the reach of a horizontal stream. Neither of these will result in the same answer as those obtained from the chart under all circumstances; however, either formula will give a result sufficiently accurate for most purposes.

The first formula is as follows:

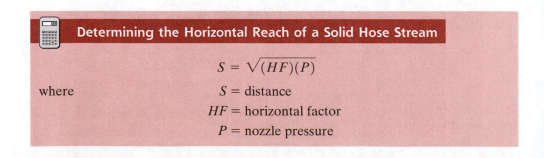

Determining the Horizontal Reach of a Solid Hose Stream

$$S = \sqrt{(HF)(P)}$$

where
S = distance
HF = horizontal factor
P = nozzle pressure

TABLE 4.1 ◆ Horizontal Reach of Straight Streams

Nozzle Pressure (psi)	Horizontal Reach (ft) at Given Nozzle Size (in.)				
	1	1¹⁄₈	1¹⁄₄	1³⁄₈	1¹⁄₂
20	37	38	39	40	42
25	42	44	46	47	49
30	47	50	52	54	56
35	51	54	58	59	62
40	55	59	62	64	66
45	58	63	66	68	71
50	61	66	69	72	75
55	64	69	72	75	78
60	67	72	75	77	80
65	70	75	78	79	82
70	72	77	80	82	84
75	74	79	82	84	86
80	76	81	84	86	88
85	78	83	87	88	90
90	80	85	89	90	91

Values of horizontal factors for different tip sizes are as follows:

½"	56	1¼"	92
⅝"	62	1⅜"	98
¾"	68	1½"	104
⅞"	74	1⅝"	110
1"	80	1¾"	116
1⅛"	86	2"	128

Note that the basis for these factors is a ½-inch tip with a factor of 56, and that there is a factor increase of 6 for each ⅛-inch increase in tip size.

QUESTION What is the approximate horizontal reach of a 1¼-inch tip when the discharge pressure is 40 psi?

ANSWER
$$S = \sqrt{(HF)(P)}$$
where
$$HF = 92$$
$$P = 40 \text{ psi}$$
Then
$$S = \sqrt{(92)(40)}$$
$$= \sqrt{3680}$$
$$= 60.7 \text{ feet}$$

The answer from Table 4.1 is 62 feet.

The second formula in general use is the following:

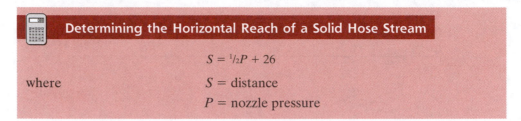

Determining the Horizontal Reach of a Solid Hose Stream

$$S = \tfrac{1}{2}P + 26$$

where $\qquad\qquad$ S = distance

P = nozzle pressure

The formula is applicable for tip sizes of ¾ inch and larger with nozzle pressures over 30 psi. For nozzle tips larger than ¾ inch, add 5 to the 26 for each ⅛-inch increase in tip size.

QUESTION What is the approximate horizontal reach of a 1⅛-inch tip when the discharge pressure is 50 psi?

ANSWER $\qquad\qquad$ $S = \tfrac{1}{2}P + 26$

where $\qquad\qquad$ $P = 50$ psi

Add 15 to the 26 due to the increase in nozzle size.

Then $\qquad\qquad$ $S = 50/2 + 26 + 15$

$$= 25 + 26 + 15$$

$$= 66 \text{ feet}$$

The answer from Table 4.1 is 66 feet. $\qquad\qquad$ ■

VERTICAL REACH

A vertical stream is one in which the angle of discharge is greater than 45 degrees. Vertical streams are used primarily for projecting water into a building from the outside. A general rule in the fire service is that full effectiveness of a handheld stream used for throwing water into a building from the outside cannot be achieved above the second floor; the third floor is the highest to which streams may be thrown with any degree of effectiveness from the street level. From these rules, it is apparent that vertical streams cannot be used to effectively direct water into a burning building using hand lines from the street level. Occasionally, however, water must be directed into buildings using vertical streams even though it is known that the effectiveness will be limited. Some officials agree that limited effectiveness is justified in a fire that is confined to a room on the outside wall of the building when it might take some time to move hand lines into position within the building to achieve extinguishment. Hitting these fires from the outside with the use of vertical streams may keep them from spreading and thus reduce the overall fire loss.

Master-stream appliances mounted on apparatus, ladder pipes, or elevated platforms are used more often than handheld lines for projecting streams into the upper floors of multistory buildings. For maximum effectiveness, the apparatus on which these appliances are mounted should be spotted across the street from the burning building. At one time, when the maximum reach of these apparatus was 100 feet, it was considered that effectiveness of street-directed streams could be achieved up to the tenth or the eleventh floor, depending upon the height to which the nozzle could be elevated above street level. With elevated platforms and aerial ladders in service that extend beyond 100 feet, these limitations have been extended.

The maximum vertical reach of a fire stream is obtained when the stream is discharged perpendicular to the ground; however, few occasions arise for this type of projection angle to be used at a fire. Vertical streams are seldom used at an angle greater than 75 degrees. The results of one study on vertical reach of straight streams is shown in Table 4.2.

A formula suggested for estimating the reach of vertical streams is given as follows. Although this formula seldom gives the exact result as shown in Table 4.2, it can produce results sufficiently accurate for ordinary purposes:

Determining the Vertical Reach of a Solid Hose Stream

$$S = \sqrt{(VF)(P)}$$

where
- S = distance
- VF = vertical factor
- P = nozzle pressure

Table 4.2   ◆   Vertical Reach of Straight Streams

Nozzle Pressure (psi)	Vertical Reach (ft) at Given Nozzle Size (in.)				
	1	$1\frac{1}{8}$	$1\frac{1}{4}$	$1\frac{3}{8}$	$1\frac{1}{2}$
20	35	36	36	36	37
25	43	44	44	45	46
30	51	52	52	53	54
35	58	59	59	60	62
40	64	65	65	66	69
45	69	70	70	72	74
50	73	75	75	77	79
55	76	79	80	81	83
60	79	83	84	85	87
65	82	86	87	88	90
70	85	88	90	91	92
75	87	90	92	93	94
80	89	92	94	95	96
85	91	94	96	97	98
90	92	96	98	99	100

Values of vertical factors for different tip sizes are as follows:

½″	85	1¼″	115
⅝″	90	1⅜″	120
¾″	95	1½″	125
⅞″	100	1⅝″	130
1″	105	1¾″	135
1⅛″	110	2″	145

Note that the basis for these factors is a ½-inch tip with a factor of 85, and there is a factor increase of 5 for each ⅛ inch in tip size.

QUESTION What is the approximate vertical reach from a 1-inch tip at a nozzle pressure of 40 psi?

ANSWER

$$S = \sqrt{(VF)(P)}$$

where

$$VF = 105$$
$$P = 40 \text{ psi}$$

Then

$$S = \sqrt{(105)(40)}$$
$$= \sqrt{4200}$$
$$= 64.8 \text{ feet}$$

The answer from Table 4.2 is 64 feet. ■

◆ STREAM PENETRATION

Stream penetration is the penetration of water from a hose stream into a building. The effectiveness of a solid stream discharging water into a burning building from the outside depends upon the amount of water that actually reaches the seat of the fire. In most instances, this requires that the water penetrate some distance into the building. Maximum penetration is obtained when the stream enters the building just above a window sill. Penetration is also greater at lower angles than at greater angles of discharge. Attempting any degree of penetration into a building is impractical at angles of discharge greater than 75 degrees (see Figure 4.4).

The actual penetration of a stream into a building is the horizontal distance from the outer wall of the building to the point where the water comes to rest within the building. How far the water travels after hitting the ceiling depends upon ceiling

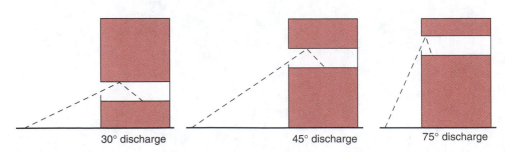

30° discharge   45° discharge   75° discharge

FIGURE 4.4 ◆

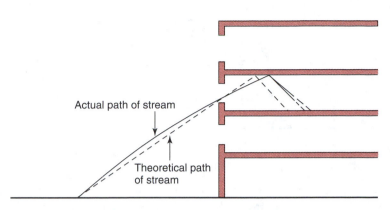

FIGURE 4.5 ◆

construction, ceiling height, and several other factors. After hitting the ceiling, the water will generally travel at least as far as the distance from the outer wall to the point of impact with the ceiling if it is not stopped by obstructions.

As previously mentioned, the actual path traveled by a fire stream is a slight curve rather than a straight line due to the action of gravity (see Figure 4.5). The actual path is advantageous when streams are thrown into upper floors of buildings, as a curved stream will hit further in on the ceiling than will a straight-line stream. Additionally, the deflection of the water will be at a lesser angle, which will result in greater penetration. Although it is not possible to determine the exact penetration of a stream within a building, a theoretical estimate can be made based on the knowledge that actual penetration is greater than theoretical penetration.

SIMILAR RIGHT TRIANGLES

Similar right triangles are two different-size right triangles that have identical angles. For example, triangle X and triangle Y in Figure 4.6 are similar right triangles.

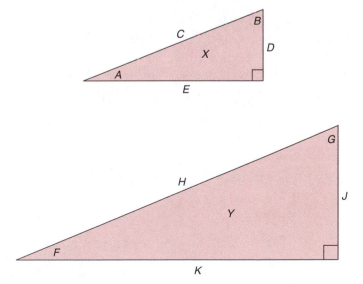

FIGURE 4.6 ◆ Similar right triangles.

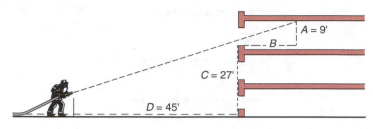

FIGURE 4.7 ◆

The number of degrees in angle A is identical to that in angle F, and the number of degrees in angle B is identical to that in angle G. With similar right triangles, proportions can be established to solve problems. For example,

$$\frac{D}{E} = \frac{J}{K} \quad \text{or} \quad \frac{C}{E} = \frac{H}{K} \quad \text{or} \quad \frac{D}{J} = \frac{E}{K}$$

When three of the values are known, it is easy to solve for the fourth. Examples are illustrated in the following problems.

Theoretically, the distance hit by the stream on the ceiling can be determined by comparing two similar right triangles. When comparing these triangles, assume that the height of a story is 12 feet and the height of a window sill above floor level is 3 feet. It then works out to be 9 feet from the window sill to the ceiling. The simple proportion solution to this type of problem can be seen by referring to Figure 4.7.

Determining the Unknown Side of One of Two Similar Right Triangles Used for Finding the Stream Penetration into a Building

$$\frac{A}{B} = \frac{C}{D}$$

where
A = height of the ceiling above the window sill

B = the distance from the outer wall to the point of contact on the ceiling

C = the height of the window sill above ground level

D = the distance from the nozzle to the building

From Figure 4.7,

$$A = 9 \text{ feet}$$
$$B = \text{unknown,}$$
$$C = 27 \text{ feet}$$
$$D = 45 \text{ feet}$$

Then

$$\frac{9}{B} = \frac{27}{45}$$

First, reduce the fraction 27/45 to its lowest terms by dividing both the numerator and denominator by 9:

$$\frac{9}{B} = \frac{3}{5}$$

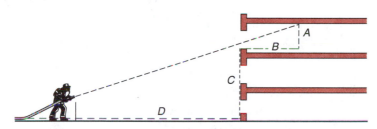

$$\underline{A} = \underline{4^{TH}}$$

By cross multiplication,

$$3B = (9)(5)$$
$$3B = 45$$
$$B = 15 \text{ feet}$$

In actual practice, the distance is greater than 15 feet. The assumption in this problem is that the stream travels in a straight line and the nozzle is on the ground. Actually, the stream travels in a slight downward curve due to the pull of gravity and the nozzle is normally held above ground level.

QUESTION Water is discharged from a nozzle located 60 feet from a building. The water just passes over the window sill on the third floor. How far in on the ceiling will the stream hit (see Figure 4.8)?

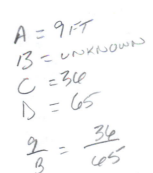

FIGURE 4.8 ◆

ANSWER

$$\frac{A}{B} = \frac{C}{D}$$

where

A = 9 feet (window sill to ceiling)
B = unknown
C = 27 feet (ground to third floor window sill)
D = 60 feet (nozzle to building)

Then

$$\frac{9}{B} = \frac{27}{60}$$

First, reduce the fraction 27/60 to its lowest terms by dividing both the 27 and the 60 by 3.

Then

$$\frac{9}{B} = \frac{9}{20}$$

By cross multiplication,

$$9B = (9)(20)$$

Canceling the 9 on both sides gives

$$B = 20 \text{ feet}$$ ■

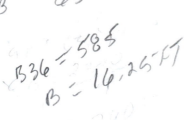

$A = 9 FT$
$B = UNKNOWN$
$C = 36$
$D = 65$

$$\frac{9}{B} = \frac{36}{65}$$

$B36 = 585$
$B = 16.25 FT$

QUESTION The nozzle on a ladder pipe is located 63 feet above ground level. Water from the ladder pipe just passes over the window sill on the eighth floor and hits the ceiling at a point

27 feet from the outside wall. What is the distance from the building to the nozzle tip on the ladder pipe (see Figure 4.9)?

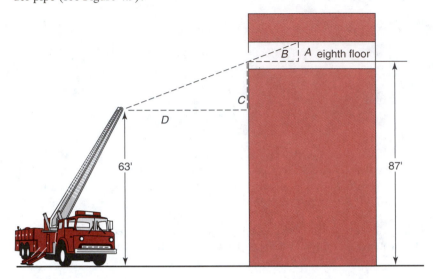

FIGURE 4.9 ◆

ANSWER

$$\frac{A}{B} = \frac{C}{D}$$

where

A = 9 feet (window sill to ceiling)

B = 27 feet (distance water is hitting in on ceiling)

C = 24 feet [floor of eighth story is 84 feet above ground (7 × 12), and 84 − 63 = 21 ft, plus 3 ft to window sill]

D = unknown

Then

$$\frac{9}{27} = \frac{24}{D}$$

First, reduce the fraction 9/27 to its lowest terms by dividing both the 9 and the 27 by 9.

Then

$$\frac{1}{3} = \frac{24}{D}$$

By cross multiplication,

$$D = (3)(24)$$
$$D = 72 \text{ feet}$$

■

THE FORMULA $C^2 = A^2 + B^2$

Occasionally two sides of a right triangle are known and the length of the third side must be found. These problems can be worked by the use of the following formula (see Figure 4.10):

$$C^2 = A^2 + B^2$$

where   C = the hypotenuse of a right triangle

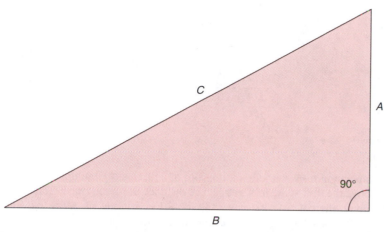

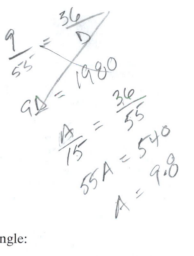

FIGURE 4.10 ◆ A right triangle. $C^2 = A^2 + B^2$.

A = one side of the right triangle

B = the other side of the right triangle

The formula can be rewritten to solve for individual sides of a right triangle:

Determining the Unknown Side of a Right Triangle

$$C = \sqrt{A^2 + B^2}$$
$$B = \sqrt{C^2 - A^2}$$
$$A = \sqrt{C^2 - B^2}$$

QUESTION A nozzle tip located 45 feet from a building discharges water through a window on the third floor. The water just passes over the window sill. What is the length of the stream from the nozzle tip to where it is passing over the window sill (see Figure 4.11)?

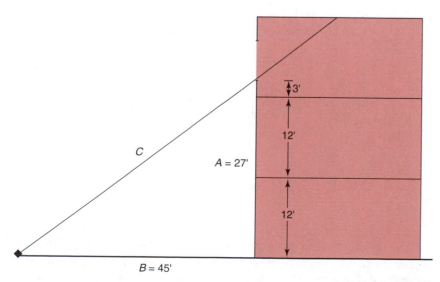

FIGURE 4.11 ◆

ANSWER $\qquad$ $C = \sqrt{A^2 + B^2}$

where $\qquad\qquad\qquad$ $A = 27$ feet

$\qquad\qquad\qquad\qquad\quad$ $B = 45$ feet

Then $\qquad\qquad\qquad$ $C = \sqrt{(27)^2 + (45)^2}$

$\qquad\qquad\qquad\qquad\qquad = \sqrt{729 + 2025}$

$\qquad\qquad\qquad\qquad\qquad = \sqrt{2754}$

$\qquad\qquad\qquad\qquad\qquad = 52.48$ feet ■

QUESTION A **portable monitor** is working on the roof of a three-story building. The nozzle tip is located 3 feet above the roof at the roof's edge. The distance of stream travel from leaving the nozzle tip to the point where it just passes over the sixth-floor window sill of the building across the street is 50 feet. What is the distance from the nozzle tip to the building across the street (see Figure 4.12)?

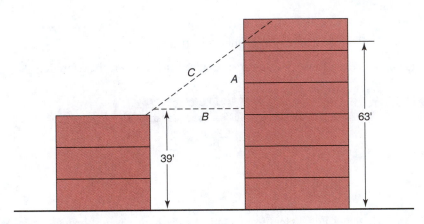

FIGURE 4.12 ◆

ANSWER $\qquad$ $B = \sqrt{C^2 - A^2}$

where $\qquad\qquad\qquad$ $A = 24$ feet $(63 - 39)$

$\qquad\qquad\qquad\qquad\quad$ $C = 50$ feet

Then $\qquad\qquad\qquad$ $B = \sqrt{(50)^2 - (24)^2}$

$\qquad\qquad\qquad\qquad\qquad = \sqrt{2500 - 576}$

$\qquad\qquad\qquad\qquad\qquad = \sqrt{1924}$

$\qquad\qquad\qquad\qquad\qquad = 43.9$ feet ■

◆ NONSOLID FIRE STREAMS

Nonsolid streams are used at fires far more often than are solid streams. Nonsolid streams are classified as **fog streams**. The form of the water droplets from fog nozzles and the fineness of the mist depend chiefly upon the design of the nozzle and the nozzle pressure.

The size of the water droplets from fog nozzles has a direct effect on the heat-absorption ability of the fire stream. The optimum diameter of a water droplet from an extinguishing viewpoint is in the range of 0.3 to 1.0 mm. Best results are obtained when the water droplets are uniform in size; however, no nozzle in general use is capable of discharging uniform droplets throughout the wide range of available nozzle pressures and discharge capacities.

The speed with which a fire can be extinguished depends upon the relationship of the rate of water application to the rate of heat generated, the degree of coverage, and the form and character of the water applied. The rate at which heat is absorbed by water in its liquid state is directly proportional to the surface exposed. This principle might best be illustrated using the example of a block of ice.

One cubic foot of ice exposes 6 square feet of surface area for the purpose of heat absorption. If the cubic foot of ice is divided into 1-inch cubes, there are 1728 of these cubes available for heat absorption. Each of the 1-inch cubes exposes a surface area of 6 square inches; therefore, the 1728 cubes expose a surface area of (6)(1728) or 10,368 square inches. This is equal to a surface area of 72 square feet. It can be seen, then, that if the cubic foot of ice is placed into a body of warm liquid and the 1728 cubic inches of ice is placed into an identical amount of warm liquid, the smaller cubes will absorb heat 12 times faster than the larger block of ice (72/6 = 12) (Figure 4.13). This same principle applies to the heat-absorption capacities of a solid stream as compared to a nonsolid stream.

Although the rate of heat absorption of water in its liquid state is directly proportional to the surface exposure, the perfect use of water as a fire extinguishing agent can be achieved only if water is applied in a form that converts the entire volume to steam. Water is approximately six and one-half times more effective as an extinguishing agent while it is changing from liquid at 212°F to steam than when the liquid is applied at 62°F and its temperature rises to 212°F.

Water can be applied to a fire in a direct attack or an indirect attack. In a **direct attack**, water is discharged directly onto the materials involved with the fire. Direct attacks can be made using either straight or fog streams; most fog nozzles are designed so that a selection can be made between a straight stream and a **spray stream** as the need dictates. Direct attacks are generally used when heat accumulation has not yet become a problem, where the entire amount of burning material can be readily hit with stream application, or on large fires, where the burning material is not confined.

In an **indirect attack**, water is applied to the heated atmosphere rather than to the burning material itself. Extinguishment depends upon the absorption of sufficient heat to bring the temperature of the entire area below that of the ignition temperature of the burning material. Skillful application of the water will not only absorb the heat in

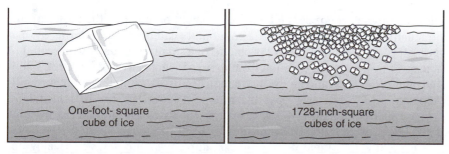

One-foot- square
cube of ice

1728-inch-square
cubes of ice

FIGURE 4.13 ◆

the immediate area of water application, but will produce large volumes of steam that will flow into remote areas and help control the fire.

The indirect method of attack was developed primarily as a result of investigations made by Chief Lloyd Layman while with the Parkersburg, West Virginia, Fire Department and during his tour of duty with the U. S. Coast Guard. Chief Layman's experiments were conducted primarily with high-pressure fog streams, and the results demonstrated the tremendous effectiveness of these streams when properly applied. The method is generally used in fighting structure fires whenever there is a great amount of heat accumulated and the fire is confined.

Whenever it is planned to use the indirect method of attacking a fire, considerable thought should be given to the safety factor for firefighters. Application of the streams upsets the thermal balance, decreases visibility, and can cause serious steam burns to firefighters.

REACH OF NONSOLID STREAMS

The reach of a fog nozzle depends upon the nozzle pressure, the type of nozzle, the stream pattern, and the size of water droplets. Reach may vary from a few feet to a considerable distance, depending upon the arrangement of several variables. In general, the wider the pattern, the less is the reach; however, if two streams have the same pattern and the water droplets of the two streams have the same initial velocity, the stream with the larger droplets will have the greatest reach. Although the reach of fog streams is not as great as that of solid streams, the nozzle operator does have the advantage of being able to change the reach of the stream; a mere turn of the nozzle results in a change of the stream pattern.

There are many variables in the design of nozzles producing fog patterns. Some nozzles provide one pattern only; others are designed so that water can be discharged as a **straight stream** or fog streams of various cone patterns. One type produces both a fog pattern and a straight stream at the same time. Those that produce both fog patterns and straight streams are referred to as combination nozzles. Those producing a fog pattern and a solid stream at the same time are referred to as multipurpose nozzles. Most combination nozzles have adjustable patterns ranging from 60 degrees to straight streams. Some nozzles are made with stops at fixed settings and also are provided with a measuring unit to discharge a constant flow of water. Others can operate at a constant pressure regardless of the setting.

◆ TYPES OF COMBINATION NOZZLES

Fire officials generally select nozzles and other equipment based upon the fire problems of their communities. The fire problems of a community can be stated in terms of *what there is to burn*. It is apparent that the fire problems of a small community with scattered buildings and no industry is quite different from those of a large city with high-rise buildings, a large commercial area, and an extensive industrial area. Nonetheless, all fire officials have the same selection among types of combination nozzles to use. Although there is a variety of so-called combination nozzles available, they are basically divided into four classifications: **automatic nozzles**; **constant-flow** or **fixed-gallonage nozzles**; **selectable-gallonage** or manually adjustable nozzles; and **multipurpose nozzles**. A specially designed nozzle that produces fog only is referred to as a **high-pressure fog nozzle**.

With the exception of some high-pressure fog nozzles that are of a different design, all of these nozzles are available in a variety of configurations for use on booster lines, handheld attack lines, and master-stream appliances. Many of them are designed to operate at 100 psi nozzle pressure; however some nozzles are designed to operate at a nozzle pressure of 50 psi, 75 psi, or some other pressure selected by the manufacturer or user. The lower pressure fog nozzles were developed for use in special situations such as fighting fires in high-rise buildings. The advantage of the lower pressure fog nozzles is that they reduce the effort required to either maintain or advance a hose line. The reduction in nozzle pressure results in a reduction of the nozzle reaction, which makes it easier to handle and maneuver a line.

AUTOMATIC NOZZLES

Automatic nozzles have been referred to as the thinking nozzle, the intelligent nozzle, the variable-gallonage nozzle, the self-adjusting nozzle, and the constant-pressure nozzle. They are designed to maintain a constant nozzle pressure regardless of the flow. This is normally a nozzle pressure of 100 psi; however, nozzles requiring less pressure are available. The nozzles are variable-flow types with the flow depending upon the nozzle operator's selection and the amount of water provided by the pump operator (Figure 4.14).

The father of this nozzle was Chief Clyde McMillan of the Gary, Indiana, Fire Department. He laid out the original design for the nozzle on a kitchen napkin the morning following a large fire in Gary, at which he was not happy with the results achieved with standard nozzles. His original concept was based on his principle of the **water triangle** (see Figure 4.15; this figure and the following information regarding the water triangle is provided through the courtesy of Task Force Tips, Valparaiso, IN).

Each side of the triangle represents one of three limits to any pumper setup. They are (1) the water supply, (2) the pumper power, and (3) the maximum allowable working pressure. Working the pumper to whichever of the limits is reached first produces the maximum possible delivery for that particular layout.

FIGURE 4.14 ◆ An array of automatic nozzles. *Courtesy of Task Force Tips*

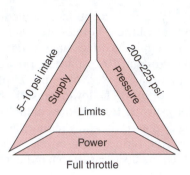

FIGURE 4.15 ◆ The "water triangle." *Courtesy of Task Force Tips*

When pumping into an automatic nozzle, the pump operator throttles out until he or she reaches a limiting side of the triangle. These limits appear as follows:

1. Water supply: Indicated by 5 to 10 psi on the inlet gage or by the suction hose going slightly soft. It can also show as the engine tending to run away (speed up erratically).
2. Pressure: Indicated by the limiting pressure, usually 200 to 225 psi showing on the pumper discharge pressure gage.
3. Power: Indicated by running out of throttle.

Although these limits will yield the maximum for a particular layout, this is not to say that the layout should not be improved! If one is working against the pressure limit, adding parallel or large-diameter lines will greatly increase the flow. If water supply is the problem, improvement is necessary on the suction side of the pump. This can be accomplished by using larger suction lines, using additional lines into the pump, or receiving water from an additional source (relay pumping). The power limit is reached only when a large volume of water is being discharged. This usually occurs when attempting to supply too much water to one or more streams. The load can be shared with a second pumper by shifting lines. The second pumper can be worked in tandem off the same hydrant with the first. Additional parallel or large-diameter lines can be used to reduce friction loss.

Although the same limits apply to a pumper when working with conventional tips, merely working to the system limit does not produce desired results unless the tip size is exactly correct. If the regular tip size is too large, a poor, underpressured stream is all that can be obtained. If the regular tip is too small, the stream will be overpressured, failing to deliver the volume available using the correct size tip. With automatic nozzles, the pump operator can achieve maximum efficiency as fast as he or she can adjust the throttle. The automatic nozzle will simultaneously adjust the tip size to best deliver the available water (Figure 4.16).

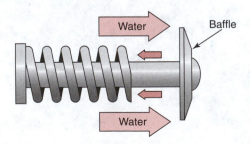

FIGURE 4.16 ◆

In addition to the principle of the water triangle, the automatic nozzle uses a principle very similar to that of a **relief valve**. As a relief valve senses an increase of flow, it generally opens a line to dump some of the water and therefore maintain the pressure for which it is set. An automatic nozzle senses the pressure at the base of the nozzle and makes adjustments to maintain the nozzle pressure for the flow that is being received. In effect, the nozzle is constantly changing the tip size to match the water being delivered. This permits the flow being supplied to be delivered at the proper nozzle pressure and correct velocity.

There is one thought that should be kept in mind with the use of an automatic nozzle. It is impossible for the nozzle operator to judge the exact amount of water being discharged from a nozzle. A stream may look good, but may not actually be supplying the flow intended. This is a concern to the pump operator, who should be constantly alert to making adjustments necessary by using his or her hydraulic knowledge to supply the proper pressure.

FIXED-GALLONAGE OR CONSTANT-FLOW NOZZLES

Whereas an automatic nozzle is designed to maintain a set nozzle pressure even though the flow may be changed, a **fixed-gallonage** or constant-flow nozzle is designed to provide the same flow at a specific nozzle pressure regardless of the setting of the flow pattern. For example, with a nozzle rated at a flow of 150 gpm, a nozzle operator has the capability to change the stream pattern from a straight stream to a narrow-angle stream or a wide-angle stream. Whichever selection he or she makes, the amount of water discharged will remain at 150 gpm.

Although the flow remains the same, most fixed-gallonage nozzles are designed to operate at a set nozzle pressure, generally 100 psi. However, some fixed-gallonage nozzles are available that operate at 75 psi nozzle pressure. As previously mentioned, these nozzles are designed for use at special types of operations such as fighting fires in high-rise buildings.

SELECTABLE-GALLONAGE OR MANUALLY ADJUSTABLE NOZZLES

Whereas a fix-gallonage nozzle is designed to provide a constant flow regardless of the stream pattern, a selectable-gallonage nozzle enables the nozzle operator to select a discharge rate that he or she deems best suitable for the task at hand (Figure 4.17). A series of flow settings is available that permits the selection of the desired flow at a predetermined nozzle pressure. In most cases the nozzle is designed to operate at 100 psi.

This type of nozzle demands that a close coordination be maintained between the nozzle operator and the pump operator. If the pump operator does not maintain the proper pressure, the flow will be different from that selected by the nozzle operator. It is necessary for the pump operator to know the setting selected at the nozzle in order to properly supply the pressure required.

MULTIPURPOSE NOZZLES

Multipurpose nozzles have the unique ability of providing combination fire streams that include a straight stream and a fog stream at the same time (Figure 4.20) or independently (Figure 4.18 and Figure 4.19). This capability provides both a penetration stream and a protective stream for firefighters advancing a line on an intense fire. One

FIGURE 4.17 ◆ A selectable-gallonage nozzle. This particular nozzle provides for a range of nozzle pressures and flow. *Courtesy of Akron Brass Company*

FIGURE 4.18 ◆ A saber jet nozzle in a fog position. *Courtesy of Akron Brass Company*

FIGURE 4.19 ◆ A saber jet nozzle in the solid bore position. *Courtesy of Akron Brass Company*

manufacturer produces a multipurpose nozzle that has the capability of operating efficiently at pressures as low as 50 psi and up to 100 psi. This feature is adaptable for interior, exterior, or high rise firefighting tactics.

HIGH-PRESSURE FOG NOZZLES

Although many high-pressure fog nozzles have the same appearance as other fog nozzles, some look and operate quite differently from standard nozzles. They are designed to provide a finely divided spray that has the appearance of fog. This requires an extremely high nozzle pressure, with some operating as high as 800 psi.

Using this type of nozzle for fighting structure fires is extremely limited. However, they can be very effective for certain types of fires. A good example is for fighting fires involving combustible metals, such as magnesium. A fire in magnesium is a fire in a molten mass. If water in solid form is applied, it penetrates the surface of the metal and expands at the rate of 1700 to 1, resulting in a violent reaction approaching that of an explosion. Using a high-pressure fog stream on this type of fire applies water to the surface in a fine mist without the danger of penetration and expansion.

There are other examples of the use of high-pressure fog nozzles. They are occasionally used for the extinguishment of fires involving combustible liquids contained in small tanks. The nozzle used for this purpose has been called an overshot nozzle. High-pressure fog nozzles have also been used successfully in the extinguishment of confined cellar fires. The nozzle used in this type of fire has been referred to as a cellar nozzle.

FIGURE 4.20 ◆ A saber jet nozzle in the combination streams position. *Courtesy of Akron Brass Company*

◆ SELECTING A NOZZLE

Fire officials give considerable thought to the fire problems in a community and the number of personnel assigned to a company whenever they think about the purchase of new nozzles. The task is not always easy because of the wide selection they have from which to choose. For example, smooth-bore tips are available in sizes of ½, ⅝, ¾, ⅞, 1, 1⅛, 1¼, 1⅜, 1½, 1¾, 2, 2¼, 2½, 2¾, and 3 inches. However, the task has been simplified to a degree by the changes that have been made in attack tactics. Nozzles in use for 1-, 1½-, 1¾-, and 2-inch lines are typically fog nozzles. The nozzles used on 2½-inch handlines and on heavy stream appliances may have smooth-bore tips or fog nozzles.

When considering the selection of a nozzle, thought should also be given to the flow desired and to who controls the flow. The amount of flow desired is closely related to the size of hose carried on a company. The flows for smooth-bore tips are fairly well fixed; however, the flows from fog nozzles vary considerably. Flows on nozzles for use on 1-inch hose vary from about 20 to 30 gpm. For 1½-inch hose, flows from 95 to 125 gpm can be expected. The amount increases for 1¾- and 2-inch hose to approximately 125 to 200 gpm.

The flow from the nozzle is controlled by the firefighter operating the nozzle when smooth-bore tips are used. It is also controlled at the nozzle on adjustable-gallonage fog nozzles. However, with the constant-flow and automatic nozzles, the flow is entirely controlled by the pump operator. In fact, to a large degree, the flow on all nozzles is controlled by the pump operator. A good example is with the use of an adjustable-gallonage nozzle. The amount of the flow is selected by the nozzle operator, but it is not available unless the pump operator supplies a sufficient amount of water and pressure to the nozzle. The same logic applies to an automatic nozzle. The nozzle pressure remains the same regardless of the flow, but the amount of water discharged is reduced unless the pump operator continues to supply the required amount to the nozzle.

◆ **VELOCITY FLOW**

The **velocity flow** of water is the speed with which it passes a given point while traveling in a given direction. The velocity flow of a stream issuing from an opening depends upon the discharge pressure or the head of water causing the flow. Velocity flow is usually measured in feet per second (fps), but may be measured in feet per minute (fpm), inches per second (ips), or inches per minute (ipm).

When velocity flow is known in one unit of measurement, it can readily be changed to another unit.

A. Feet per second (fps) can be changed to:
- Feet per minute (fpm) by multiplying by 60 (the number of seconds in 1 min).
- Inches per second (ips) by multiplying by 12 (the number of inches in 1 ft).
- Inches per minute (ipm) by multiplying by 720 (12 × 60).

B. Feet per minute (fpm) can be changed to:
- Feet per second (fps) by dividing by 60 (the number of seconds in 1 min).
- Inches per minute (ipm) by multiplying by 12 (the number of inches in 1 ft).
- Inches per second (ips) by dividing by 5 (multiplying by 12 and dividing by 60).

C. Inches per second (ips) can be changed to:
- Feet per second (fps) by dividing by 12 (the number of inches in 1 ft).
- Inches per minute (ipm) by multiplying by 60 (the number of seconds in 1 min).
- Feet per minute (fpm) by multiplying by 5 (dividing by 12 and multiplying by 60).

D. Inches per minute (ipm) can be changed to:
- Inches per second (ips) by dividing by 60 (the number of seconds in 1 min).
- Feet per minute (fpm) by dividing by 12 (the number of inches in 1 ft).
- Feet per second (fps) by dividing by 720 (12 × 60).

Two formulas are available for determining the velocity of flow of water discharging from an opening:

 Determining the Velocity Flow of the Water Discharging from an Opening When the Head Is Known

$$V = 8\sqrt{H}$$

where
V = velocity flow in feet per second (fps)
H = head

Determining the Velocity Flow of the Water Discharging from an Opening When the Pressure Is Known

$$V = 12.14\sqrt{P}$$

where
$$V = \text{velocity flow in fps}$$
$$P = \text{discharge pressure}$$

Note: The answer obtained from both formulas is in fps.

It is interesting to note how these two formulas were developed.

The velocity produced by a mass of water with pressure acting upon it is the same as if the mass of water were to start from rest and fall freely through a distance equal in feet to the pressure head. The potential energy in the mass is equal to the kinetic energy. The potential energy and the kinetic energy can be expressed by the following formulas:

$$PE = MGH$$

where
$$PE = \text{potential energy}$$
$$M = \text{mass}$$
$$G = \text{acceleration due to gravity}$$
$$H = \text{height or head}$$

and

$$KE = \tfrac{1}{2}MV^2$$

where
$$KE = \text{kinetic energy}$$
$$M = \text{mass}$$
$$V = \text{velocity}$$

Then

$$\tfrac{1}{2}MV^2 = MGH$$

Cancel

$$\tfrac{1}{2}\cancel{M}V^2 = \cancel{M}GH$$

and multiply both sides by 2 and cancel:

$$\cancel{(2)}\left(\frac{1}{2}\right)V^2 = 2GH$$

Then take the square root of each side:

$$\sqrt{V^2} = \sqrt{2GH}$$

Since $\sqrt{V^2} = V$, this gives

$$V = \sqrt{2GH}$$

The acceleration of gravity is 32 feet per second squared. Substitute this value for G:

$$V = \sqrt{2 \times 32H}$$
$$= \sqrt{64H}$$
$$= 8\sqrt{H}$$

Then substitute $2.304P$ into the formula for H (see p. 12):

(see p. 12)

$$V = 8\sqrt{2.304P}$$

Now, $\sqrt{2.304} = 1.518$, and so

$$V = (8)(1.518)\sqrt{P}$$
$$= 12.14\sqrt{P}$$

THE FORMULA $V = 8\sqrt{H}$

QUESTION What is the velocity flow of the water issuing from a hole in the side of a tank if the hole is located 85 feet below the surface of the water (Figure 4.21)?

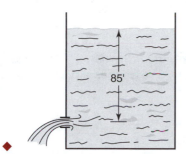

FIGURE 4.21 ◆

ANSWER

$$V = 8\sqrt{H}$$

where

$$H = 85 \text{ feet}$$

Then

$$V = 8\sqrt{85}$$
$$= (8)(9.22)$$
$$= 73.76 \text{ fps}$$

■

QUESTION A hydrant is located at the base of a gravity tank. The distance from the hydrant outlet to the level of the water in the tank is 125 feet (Figure 4.22). What is the velocity flow from the hydrant in fps?

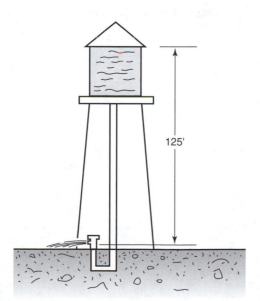

FIGURE 4.22 ◆

ANSWER

$$V = 8\sqrt{H}$$

where

Then

$$H = 125 \text{ feet}$$

$$V = 8\sqrt{125}$$

$$= (8)(11.18)$$

$$= 89.44 \text{ fps}$$

THE FORMULA $V = 12.14\sqrt{P}$

In the formula $V = 12.14\sqrt{P}$, P refers to the flow pressure. The flow pressure is the same as the nozzle pressure whenever a nozzle tip is in use. Whenever a nozzle tip is not being used, such as when water is flowing from a hydrant outlet, the flow pressure refers to the discharge pressure.

QUESTION A pumper is supplying water through 400 feet of a single 2½-inch hose equipped with a 1¼-inch nozzle tip. The nozzle pressure is 50 psi (Figure 4.23). What is the velocity flow from the tip?

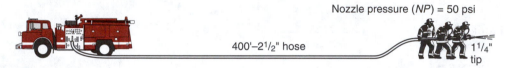

Nozzle pressure (*NP*) = 50 psi

400'–2½" hose

1¼" tip

FIGURE 4.23 ◆

ANSWER

$$V = 12.14\sqrt{P}$$

where

Then

$$P = 50 \text{ psi}$$

$$V = 12.14\sqrt{50}$$

$$= (12.14)(7.07)$$

$$= 85.83 \text{ fps}$$

QUESTION A hydrant is located approximately ½ mile from the reservoir supplying it. The level of water in the reservoir is 200 feet above the outlet of the hydrant. Due to friction loss in the mains, the pressure from the hydrant is 65 psi (Figure 4.24). What is the velocity flow from the hydrant in fps?

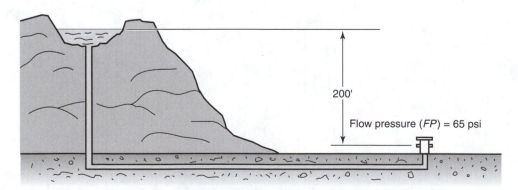

200'

Flow pressure (*FP*) = 65 psi

FIGURE 4.24 ◆

ANSWER $V = 12.14\sqrt{P}$

where $P = 65$ psi

Then $V = 12.14\sqrt{65}$

                                    $= (12.14)(8.06)$

                                    $= 97.85$ fps   ■

COMPARING THE VELOCITY FLOW IN THE HOSE WITH THE VELOCITY FLOW AT THE NOZZLE TIP

It is possible to determine the velocity flow of the water in the hose when the velocity flow from the nozzle tip is known and to determine the velocity flow of the water from the nozzle tip when the velocity flow of water in the hose is known. The relative speeds of the water in the hose and the water as it leaves the nozzle tip are in an inverse relationship with the relative areas of the hose and the nozzle tip. For example, if the area of the hose is four times as great as the area of the nozzle tip, then the water in the hose will travel four times more slowly than the flow from the nozzle tip (Figure 4.25).

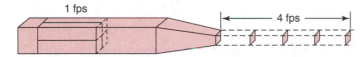

FIGURE 4.25 ◆

This relationship can be illustrated by the following formula:

Comparison of the Velocity Flow in a Hose with the Velocity Flow at the Nozzle Tip

$$VD^2 = vd^2$$

where V = velocity flow through the hose

            v = velocity flow from the nozzle tip

            D = diameter of the hose

            d = diameter of the nozzle tip

To find the velocity flow in the hose when the velocity flow from the nozzle tip is known, the formula can be rewritten as follows:

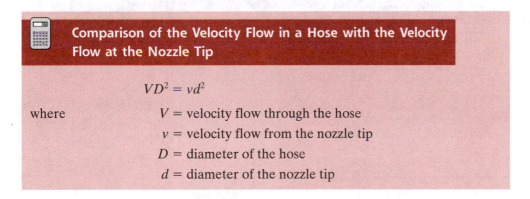
Determining the Velocity Flow in a Hose When the Velocity Flow from the Nozzle Tip Is Known

$$V = \frac{vd^2}{D^2}$$

QUESTION If the velocity flow from a 1¼-inch tip on a 2½-inch line is 200 fps, what is the velocity flow in the hose (Figure 4.26)?

2½" hose $V = ?$ 1¼" tip $V = 200$ fps

FIGURE 4.26 ◆

ANSWER

$$V = \frac{vd^2}{D^2}$$

where

$v = 200$ fps
$d = 1.25$ inches
$D = 2.5$ inches

Then

$$V = \frac{(200)(1.25)(1.25)}{(2.5)(2.5)}$$

$$= \frac{312.5}{6.25}$$

$$= 50 \text{ fps}$$ ■

To find the velocity flow from the nozzle tip when the velocity flow in the hose is known, the formula can be rewritten as follows:

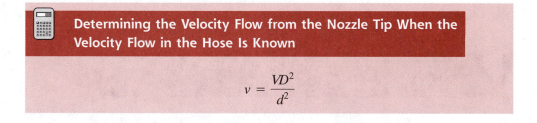

Determining the Velocity Flow from the Nozzle Tip When the Velocity Flow in the Hose Is Known

$$v = \frac{VD^2}{d^2}$$

QUESTION Water is flowing through a 3½-inch hose line at a rate of 30 fps. The line is equipped with a 1½-inch tip. What is the velocity flow from the tip (Figure 4.27)?

3½" hose $V = 30$ fps   1½" tip $V = ?$

FIGURE 4.27 ◆

ANSWER

$$v = \frac{VD^2}{d^2}$$

where

$V = 30$ fps
$D = 3.5$ inches
$d = 1.5$ inches

Then
$$v = \frac{(30)(3.5)(3.5)}{(1.5)(1.5)}$$

$$= \frac{367.5}{2.25}$$

$$= 163.33 \text{ fps} \qquad \blacksquare$$

These same formulas can be used to compare the velocity flow through different-size hose lines if the same amount of water is flowing in each line. When these formulas are used for this purpose,

$$V = \text{velocity flow in the larger hose}$$
$$v = \text{velocity flow in the smaller hose}$$
$$D = \text{diameter of the larger hose}$$
$$d = \text{diameter of the smaller hose}$$

QUESTION   If the velocity flow in a 2-inch hose is 300 fps, what is the velocity flow in a 3-inch hose if the same amount of water is flowing in each line (Figure 4.28)?

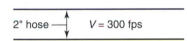

3" hose $V = ?$

2" hose — $V = 300$ fps

FIGURE 4.28 ◆

ANSWER

$$V = \frac{vd^2}{D^2}$$

where

$$v = 300 \text{ fps}$$
$$d = 2 \text{ inches}$$
$$D = 3 \text{ inches}$$

Then
$$V = \frac{(300)(2)(2)}{(3)(3)}$$

$$= \frac{1200}{9}$$

$$= 133.33 \text{ fps} \qquad \blacksquare$$

◆ NOZZLE REACTION

Nozzle reaction is a force moving in the opposite direction of water leaving a nozzle. The amount of this force is based on Newton's third law of motion that states for every action there is an equal and opposite reaction. This principle of physics applies to the discharge of water from a nozzle. The amount of force resulting from the jet action of discharge depends upon the amount of water discharged times the rate of change in velocity.

Some years ago, tests were conducted by E. M. Byington of the Boston Fire Department that led to the following formula for approximating the amount of reaction:

$$\text{Nozzle reaction (in lb)} = 1.5D^2P$$

where
$$D = \text{nozzle diameter}$$
$$P = \text{nozzle pressure}$$

The IFSTA training manual adjusts this formula slightly and uses the following formula:

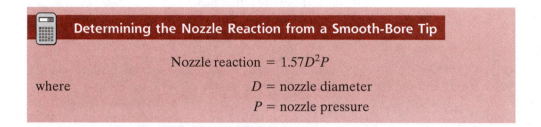

Determining the Nozzle Reaction from a Smooth-Bore Tip

$$\text{Nozzle reaction} = 1.57D^2P$$

where
$$D = \text{nozzle diameter}$$
$$P = \text{nozzle pressure}$$

This formula is used in this book.

QUESTION Water is being discharged from a 1¼-inch nozzle tip at a nozzle pressure of 50 psi. What is the nozzle reaction (*NR*) (Figure 4.29)?

1¼" tip

50 psi

NR = ?

FIGURE 4.29 ◆

ANSWER
$$NR = 1.57D^2P$$

where
$$D = 1.25 \text{ inches}$$
$$P = 50 \text{ psi}$$

Then
$$NR = (1.57)(1.25)(1.25)(50)$$
$$= 122.66 \text{ pounds}$$ ■

The force of this reaction is exerted in the direction opposite that of the flow from the nozzle tip. The reaction would be exerted against the pumper if the line were laid perfectly straight from the pumper discharge outlet to the nozzle; however, in actual firefighting situations the line lies on the ground and the nozzle is held by firefighters (Figure 4.30). The actual force that the firefighters must withstand is much less than that determined by formula. A good portion of the force is taken up by the hose line in contact with the ground.

If the direction of the stream is changed from a low angle of discharge toward a vertical angle of discharge, the resultant reaction changes: Much of the force that had

Some of force
being absorbed

FIGURE 4.30 ◆

been taken up by the ground is transmitted to the individual holding the line (Figure 4.31). This change also puts the individual in a very unfavorable position for resisting the force. Sudden changing of the discharge angle has resulted in lines **getting away** from firefighters. Firefighters must remain alert for this possibility.

Where the stream hits has no effect on the nozzle reaction. If the stream were moved from an open area to play water on the side of a building, the nozzle reaction would remain the same.

Result is a tendency for
the hose to bend

FIGURE 4.31 ◆

MOMENTARY NOZZLE REACTION

The pressure at the nozzle tip is equal to the pressure at the source if (1) the nozzle tip is closed and there is no flow of water in the hose line and (2) the line is laid at ground level so that no back pressure or forward pressure is involved. If a pumper is pumping at a pressure of 200 psi, then the pressure at the nozzle tip is 200 psi when the nozzle tip is closed. The entire 200 psi is applied to the discharge when the tip is first opened. This pressure diminishes rapidly until the nozzle is fully open and the flow pressure has been established.

It has been estimated that, during the opening of the nozzle, the **momentary nozzle reaction** is approximately 20 percent greater than the nozzle reaction of a stabilized flowing stream. This increase is included in the momentary nozzle reaction formula. The factor of 1.57 in the nozzle reaction formula given earlier is increased by 20%:

$$20\% \text{ of } 1.57 = (.20)(1.57)$$
$$= .31$$

Then $1.57 + .31 = 1.88$ for the factor in the momentary nozzle reaction formula:

 Determining the Momentary Nozzle Reaction from a Smooth-Bore Tip

$$\text{Momentary nozzle reaction} = 1.88D^2P$$

QUESTION What is the momentary nozzle reaction on a 1½-inch nozzle tip when the pressure is 70 psi?

ANSWER   Momentary nozzle reaction $= 1.88D^2P$

where $D = 1.5$
                                              $P = 70$ psi

Then   Momentary nozzle reaction $= (1.88)(1.5)(1.5)(70)$
                                                    $= 296.1$ pounds   ■

QUESTION If the nozzle reaction from a flowing tip is 130 pounds, what is the momentary nozzle reaction from the same tip?

ANSWER   Momentary nozzle reaction $= 120\%$ of NR

                                              $= (1.20)(130)$

                                              $= 156$ pounds   ■

NOZZLE REACTION ON FOG NOZZLES

The formula giving nozzle reaction as $1.57D^2P$ only applies to solid-stream nozzle tips. The nozzle reaction for fog nozzles is less because water is discharged in a wide pattern and all of the reaction is not directly received by the nozzle operator. With the increasing use of fog nozzles, it is important that a method be available for determining the nozzle reaction from these tips. A formula can be developed for use with fog nozzles by substituting into the formula for solid-stream tips. The commonly accepted formula for determining the **discharge** from a nozzle tip is

$$\text{Discharge} = 29.7D^2\sqrt{P}$$

where the answer is expressed in gpm. Therefore, the formula can be written

$$\text{gpm} = 29.7D^2\sqrt{P}$$

This formula can be manipulated to solve for D^2:

$$D^2 = \frac{\text{gpm}}{29.7\sqrt{P}}$$

The value of D^2 can then be substituted into the nozzle reaction formula:

$$NR = 1.57D^2P$$

where

$$D^2 = \frac{\text{gpm}}{29.7\sqrt{P}}$$

Then

$$NR = \frac{(1.57)(\text{gpm})(P)}{29.7\sqrt{P}}$$

Dividing 1.57 by 29.7 gives .0529, so

$$NR = \frac{(.0529)(\text{gpm})(P)}{\sqrt{P}}$$

Next multiply both the numerator and the denominator by $\sqrt{P}$:

$$NR = \frac{(.0529)(\text{gpm})(P)(\sqrt{P})}{(\sqrt{P})(\sqrt{P})}$$

Now, multiply out $(\sqrt{P})(\sqrt{P})$ in the denominator, recalling that $(\sqrt{P})(\sqrt{P}) = P$:

$$NR = \frac{(.0529)(\text{gpm})(P)(\sqrt{P})}{P}$$

Next, cancel out the P:

$$NR = \frac{(.0529)(\text{gpm})(\cancel{P})(\sqrt{P})}{\cancel{P}}$$

Then

$$NR = (.0529)(\text{gpm})(\sqrt{P})$$

Note: The commonly accepted formula is

$$NR = (.0505)(\text{gpm})(\sqrt{P})$$

This formula will be used for problems in this book.

This formula can be used for any fog nozzle, regardless of the nozzle pressure, as long as the flow is known and the nozzle pressure is known. However, as most fog nozzles are rated at 100 psi nozzle pressure, the formula can be simplified for field use as follows:

Determining the Nozzle Reaction from a Fog Nozzle

$$NR = (.05)(\text{gpm})(\sqrt{P})$$

where

$$P = 100 \text{ psi}$$

Then

$$NR = (.05)(\text{gpm})(\sqrt{100})$$
$$= (.05)(\text{gpm})(10)$$
$$= .5 \text{ gpm}$$

The nozzle reaction equals one-half of the flow. For example, the nozzle reaction on a 300-gpm constant-flow nozzle rated at 100 psi is maintained at 150 pounds, while the nozzle reaction of one rated at 500 gpm remains at 250 pounds.

It is interesting to note that if the formula is solved for a nozzle pressure of 75 psi, which is one of the other nozzle pressures commonly used for fog nozzles, the nozzle reaction is approximately one-third the flow. For field operations, then, a 300-gpm constant-flow nozzle rated at 75 psi has a constant nozzle reaction of approximately 100 pounds.

COMPARING THE NOZZLE REACTION OF VARIOUS NOZZLE TIPS

The increase in nozzle reaction from a smaller smooth-bore tip to a larger smooth-bore tip at the same nozzle pressure varies directly as the square of the diameters. This is shown in the following formula:

 Comparing the Nozzle Reaction of Different Nozzle Tips

$$\text{Factor of increase} = \frac{D^2}{d^2}$$

where

D = larger tip diameter

d = smaller tip diameter

Note: This formula was determined by comparing the *NR* for two different-size tips:

$$\text{Factor of increase} = \frac{1.57 D^2 \sqrt{P}}{1.57 d^2 \sqrt{P}}$$

where

D = larger tip diameter

d = smaller tip diameter

Then cancel out the 1.57 and the $\sqrt{P}$:

$$\text{Factor of increase} = \frac{\cancel{1.57}\, D^2 \cancel{\sqrt{P}}}{\cancel{1.57}\, d^2 \cancel{\sqrt{P}}}$$

QUESTION What is the nozzle reaction on a 2-inch tip if the nozzle reaction on a 1-inch tip is 120 pounds and the nozzle pressure is the same on both tips?

ANSWER

$$\text{Factor of increase} = \frac{D^2}{d^2}$$

where

D = 2 inch

d = 1 inch

Then

$$\text{Factor of increase} = \frac{(2)(2)}{(1)(1)}$$

$$= 4$$

Then

$$NR \text{ on 2-inch} = (4)(120)$$

$$= 480 \text{ pounds}$$

■

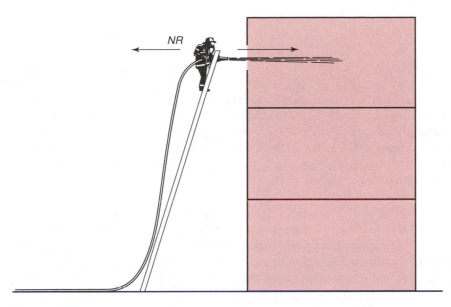

FIGURE 4.32 ◆

NOZZLE REACTION ON LADDERS

Whereas most of the effect of nozzle reaction is absorbed by the ground when handheld lines are worked at ground level, the effect can be dangerous when lines are worked from ladders. In such cases, it is important that firefighters be secured to the ladder and that the ladder itself be secured. Unsecured ladders can be pushed away from buildings by the nozzle reaction, placing firefighters in great danger (Figure 4.32).

NOZZLE REACTION ON ELEVATED MASTER-STREAM APPLIANCES

The nozzle reaction of large-tip streams at high pressures is quite high. For example, the nozzle reaction from a 2-inch tip at 100 psi nozzle pressure is approximately 600 pounds. This is a tremendous force acting on the leverage of an extended piece of equipment, and if improperly handled, might result in serious problems. It is particularly hazardous when ladder pipes are placed in operation. A ladder pipe or other large stream operated from an aerial ladder should always be operated in line with the main beams or trusses. The ladder turntable should be used if it becomes necessary to change the direction of the stream. Other precautions that should be taken are to never raise or lower the fly section of the ladder when the stream is in operation. Under no circumstances should the apparatus be moved while water is being discharged.

Particular care should be taken to ensure that streams worked from ladder pipes are shut down slowly. When the nozzle is discharging water, the nozzle reaction places a stress on the ladder perpendicular to the line of water discharge. A rapid shutdown of the nozzle results in an instant release of the pressure stress. This could result in a dangerous whipping action on the ladder. Under certain circumstances, this whipping action can be severe enough to cause the ladder to tip over. Such rapid shutdown could be caused by either the firefighter shutting off the nozzle or the pump operator suddenly shutting off the flow of water to the nozzle.

◆ WATER HAMMER

The sudden stopping of a moving body exerts a force, the amount of which depends upon the rate of deceleration and the weight of the moving body. The quicker a body is stopped, the greater is the force: maximum force is produced by instantaneous stoppage.

Sudden closing of nozzles results in shock waves that travel from the nozzle to the source of supply (pumper or hydrant) and then back to the nozzle. Pressures developed by this wave can be twice that of the hydrant or pump pressure. This sudden stoppage of water with its resultant shock is called **water hammer**. Water hammer has caused hose lines to burst during firefighting operations. This reaction can be prevented by closing nozzles slowly.

Occasionally, due to emergencies, it becomes necessary to shut off lines as quickly as possible. Water hammer can be prevented, or at least minimized, by partially shutting down the nozzle immediately and then slowly closing it until it is completely off. A quick, partial shutdown releases most of the nozzle reaction and avoids the stresses that would otherwise be placed on members.

Summary of Chapter Formulas

Horizontal Reach

$$S = \sqrt{(HF)(P)}$$

where S = distance

     HF = horizontal factor

     P = nozzle pressure

or

$$S = \tfrac{1}{2}P + 26$$

where S = distance

     P = nozzle pressure

Vertical Reach

$$S = \sqrt{(VF)(P)}$$

where S = distance

     VF = vertical factor

     P = nozzle pressure

Stream Penetration

$$\frac{A}{B} = \frac{C}{D}$$

where A = height of ceiling above window sill

     B = distance from outer wall to point of contact

     C = height of window sill above ground level

     D = distance from nozzle to the building

$$C^2 = A^2 + B^2$$

where C = hypotenuse of a right triangle

     A = one side of the right triangle

     B = other side of the right triangle

Velocity Flow

$$V = 8\sqrt{H}$$

where V = velocity flow

     H = head

and

$$V = 12.14\sqrt{P}$$

where V = velocity flow

     P = discharge pressure

Comparing the Velocity Flow in the Hose with the Velocity Flow at the Nozzle Tip

$$VD^2 = vd^2$$

where V = velocity flow through the hose

v = velocity flow from the nozzle tip

D = diameter of the hose

d = diameter of the nozzle tip

Nozzle Reaction

$$NR = 1.57D^2P$$

where NR = nozzle reaction

D = nozzle diameter

P = nozzle pressure

Momentary Nozzle Reaction

$$MNR = 1.88D^2P$$

where MNR = momentary nozzle reaction

D = nozzle diameter

P = nozzle pressure

Nozzle Reaction on Fog Nozzles

$$NR = (.0505)(\text{gpm})(\sqrt{P})$$

where NR = nozzle reaction

P = nozzle pressure

The formula for fog nozzles rated at 100 psi is

$$NR = \frac{\text{gpm}}{2}$$

■ ■

Review Questions

1. What is the formula for horizontal reach? For vertical reach?
2. Give the formula for (a) finding velocity flow when the head is known; (b) finding velocity flow when the pressure is known; (c) comparing the velocity flow in a hose with the velocity flow at the tip; (d) finding the velocity flow in a hose when the velocity flow at the tip is known; (e) finding the velocity flow at the tip when the velocity flow in the hose is known; (f) for determining nozzle reaction; (g) for determining momentary nozzle reaction; (h) for comparing nozzle reaction of various size tips.
3. What are the horizontal reach factors for tips from ½ inch to 2 inches?
4. What are the vertical reach factors for tips from ½ inch to 2 inches?
5. From an extinguishment standpoint, what is the ultimate objective of fire-fighting tactics?
6. What are some of the factors that affect a fire stream as it leaves the nozzle tip, and then after it leaves the nozzle tip?
7. When are solid streams normally used on fires?
8. What is necessary in order for a solid fire stream to be effective?
9. When is perfect use made of water as an extinguishing agent?
10. What precautions must be taken when working streams from ladders?
11. What care should be taken in operating ladder pipe streams?
12. Theoretically, when has a solid stream the greatest horizontal reach?
13. What are the recommended working nozzle pressures for solid streams?
14. Under actual conditions, when does a solid stream have the greatest horizontal reach?
15. What method can be used to make a quick shutdown of a nozzle and also restrict or eliminate water hammer?
16. What system is used for determining the theoretical stream penetration into a building?
17. How can one determine the unknown side of a right triangle when the lengths of the other two sides are known?
18. What is the normal nozzle pressure used on fog streams?
19. What is the basic foundation of fire department operations?
20. What has been the greatest impact on extinguishing operations made in recent years?
21. What is probably the next best thing to having a fire hydrant in front of every building?

22. What is the friction loss in 100 feet of 5-inch hose when moving 500 gpm?

23. What are some of the considerations to be given when using LDH?

24. What is the range of nozzle sizes for smooth-bore tips?

25. What is probably the maximum discharge for hand-held lines?

26. What are similar right triangles?

27. Define the fire problems of a community.

28. What are the four classifications of fog nozzles?

29. What are automatic nozzles?

30. What is the principle of the water triangle?

31. Define a constant-flow fog nozzle.

32. What is a high-pressure fog nozzle?

33. How does a manually adjustable fog nozzle work?

34. What is the field formula for the nozzle reaction of fog nozzles operating at 100 psi? At 75 psi?

■■

Test Four

1. What is the approximate horizontal reach of a 1½-inch tip when the discharge pressure is 60 psi?

2. What is the approximate vertical reach from a 2-inch tip when the discharge pressure is 80 psi?

3. Water is being discharged from a nozzle located 65 feet from a building. The water is just passing over the window sill on the fourth floor. How far in on the ceiling will the stream hit?

4. The nozzle on a ladder pipe is located 50 feet above ground level. Water from the ladder pipe is just passing over the window sill on the seventh floor and hitting the ceiling at a point 20 feet from the outside wall. What is the distance from the building to the tip on the ladder pipe?

5. Water from a deck gun is being directed through the fourth floor window of a building. The tip of the deck gun is 55 feet from the building. The stream is just passing over the window sill and hitting 15 feet in on the ceiling. How far above ground level is the tip of the deck gun located?

6. What is the velocity flow of the water issuing from a hole in the side of a tank if the hole is located 60 feet below the surface of the water?

7. A hydrant is being supplied from a reservoir located in the hills behind Centerville. The level of the water in the reservoir is 250 feet above the hydrant. Due to friction loss in the mains, the flow pressure on the hydrant is 75 psi. What is the velocity flow of water from the hydrant?

8. If the velocity flow from a 1¼-inch tip on a 2½-inch line is 125 fps, what is the velocity flow in the hose?

9. Water is flowing through a layout of 1½-inch hose at a rate of 25 fps. The line is equipped with a ⅝-inch tip. What is the velocity flow from the tip?

10. A handheld 2½-inch line with a 1⅛-inch tip is being used on a fire in a warehouse. Nozzle pressure is 60 psi. What is the nozzle reaction?

11. What is the momentary nozzle reaction on a 1¼-inch tip at 60 psi?

12. Engine 5 is using a ⅝-inch tip on a single 1½-inch line at a fire. Engine 3 is using a 1-inch tip on a single 2½-inch line at the same fire. The nozzle pressure on both tips is 50 psi. How many times greater is the nozzle reaction on the line used by Engine 3 than on the line used by Engine 5?

Discharge

5 CHAPTER

◆ DEVELOPMENT OF A DISCHARGE FORMULA

Fire streams are intended to discharge a sufficient amount of water to accomplish fire extinguishment. The selection of a garden hose, a booster line, a 1½-inch line, a 1¾-inch line, a 2-inch line, a 2½-inch line, or a master stream by the officer in charge at a fire is primarily based on the amount of water he or she believes is needed to properly attack the fire. Most officers operate on the **fail-safe principle**: If there is any doubt as to whether the amount of water that could be discharged from a hose line of a given size is adequate to do the job, then the next larger size is to be used. When the fire is so large and intense that there is doubt that it can be extinguished with the use of hand lines alone, then master streams are put into operation.

Fire personnel should be familiar with two primary factors involved in the discharge of water:

1. They should have some idea regarding the approximate amount of water that is discharged when various sizes of hose lines are placed in operation.
2. They should be familiar with the formulas and methods of estimating the amount of water discharged from various sizes and types of openings.

With the increasing use of constant-flow fog nozzles, the knowledge of the amount of water being discharged is more often readily available; however, other types of nozzles that do not provide this information are still in common use.

Although flows vary from different-size tips at the same nozzle pressure and from identical tips at different nozzle pressures, there are some generalities that can be developed to provide good guidelines for use in field operations. Many fire departments have standardized on the use of 50 psi nozzle pressure for solid streams on handheld lines, 80 psi for solid streams for heavy-stream appliances, and 100 psi for fog nozzles, regardless of whether used on booster lines, handheld heavier lines, or master-stream appliances.

1. The discharge from nozzle tips on 1-inch hose varies from about 13 gpm when a ¼-inch tip is used to approximately 29 gpm when a ⅜-inch tip is used. Discharge from tips used on 1-inch hose should be considered as varying from 15 to 30 gpm.
2. The discharge from nozzle tips on 1½-inch hose approximates 30 gpm when a ⅜-inch tip is used, 52 gpm when a ½-inch tip is used, and 81 gpm when a ⅝-inch tip is used. Discharge should be considered as 30, 50, and 80 gpm, respectively, for these three tips.
3. The advantage of using 1¾-inch hose and 2-inch hose is that the lines can still be maneuvered without too much difficulty and they can achieve a discharge approaching that from a 2½-inch line.
4. The most common sizes of tips used on 2½-inch hose are 1, 1⅛, and 1¼ inches. These tips should be considered as discharging 200, 250, and 325 gpm, respectively.
5. The 1½-, 1¾-, and 2-inch tips used on master streams should be considered as discharging 600, 800, and 1000 gpm, respectively.

Discharge may be defined as the amount of water issuing from an opening. It is normally calculated in gallons per minute (gpm). The most accurate and easiest method of determining the discharge from a nozzle tip at a known nozzle pressure is to look up the information in the discharge table; however, fire personnel should have some knowledge regarding the theory of discharge. In addition, they should be familiar with the formulas and methods of estimating the flow from various openings without referring to the discharge tables.

Theoretically, the amount of water discharged from an opening is a function of the area of the opening times the velocity of water issuing from the opening. This thought is expressed in the following basic discharge formula:

$$\text{Discharge} = AV$$

where
$$A = \text{area of the opening}$$
$$V = \text{velocity of flow}$$

The basic discharge formula considers the two chief factors needed to determine the amount of water issuing from an opening; from a practical standpoint, however, the formula has little value. Nevertheless, the basic discharge formula can be used as the basis for the development of a useful formula from which the discharge can be determined when the diameter of the opening and the discharge pressure are known. To develop a useful formula from the basic discharge formula, it is necessary to express the area of the opening in terms of the diameter and the velocity of flow in terms of the discharge pressure.

Nozzle tips, hoses, hydrant outlets, sprinkler heads, and other openings from which water is discharged are generally circular. The diameters of the openings are measured in inches. When the diameter of a circle is known, the area of the circle can be determined by the formula

$$\text{Area} = .7854D^2$$

Discharge pressure is expressed in pounds per square inch (psi). The formula $V = 12.14\sqrt{P}$ is used to determine the velocity of flow when the discharge pressure is known, where V in the formula is expressed in feet per second (fps).

As nozzle tips are expressed in inches, it is necessary that the velocity flow be expressed in inches in the development of a practical discharge formula. Feet per second (fps) can be changed to inches per minute by multiplying by 720 (12×60).

The number of cubic inches discharged from an opening in 1 minute can be determined by multiplying the area of the opening (in inches) by the linear inches per minute of the velocity flow. The number of gallons discharged in 1 minute can be determined by dividing the number of cubic inches discharged in 1 minute by 231, the number of cubic inches in 1 gallon.

Figure 5.1 shows the discharge of 1-inch cubes from a nozzle having a square 1-inch by 1-inch waterway. The purpose of the illustration is to provide a better understanding of the practical development of the discharge formula. The discharge formula can be developed by substituting knowns into the basic discharge formula for the area (A) and the velocity (V):

$$\text{Discharge} = \frac{AV}{231}$$

where
$$A = .7854D^2$$
$$V = 12.14\sqrt{P} \times 720$$

Then
$$\text{Discharge} = \frac{(.7854D^2)(12.14\sqrt{P})(720)}{231}$$
$$= \frac{6865.02D^2\sqrt{P}}{231}$$
$$= 29.72D^2\sqrt{P}$$

This book uses the formula:

$$\text{Discharge} = 29.7D^2\sqrt{P}$$

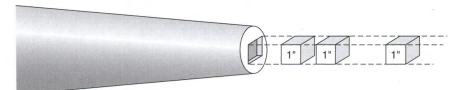

FIGURE 5.1 ◆ For illustration purposes, this shows 1-inch cubes being discharged from a nozzle tip with a square waterway.

COEFFICIENT OF DISCHARGE

A standard orifice is one with a sharp entrance edge. Water discharging from a standard orifice contracts to form a stream with a cross-sectional area less than that of the orifice. Contraction of the stream is complete at a point measured from the plane of the orifice equal to a distance approximately one-half the diameter of the jet (Figure 5.2).

Theoretically, the amount of water discharged from a standard orifice can be determined by the formula

$$Discharge = 29.72D^2\sqrt{P}$$

The actual discharge is less than the theoretical discharge due to the contraction of the stream. Actual discharge is based upon the diameter of the stream at the point where contraction is completed.

The percentage of water discharged from any opening that is not specifically streamlined, compared with the theoretical amount that should be discharged from the opening, is referred to as the **coefficient of discharge**. For example, if the amount of water discharged from an opening is 99 percent of the theoretical amount, the coefficient of discharge is .99. A coefficient of discharge of .985 means that the actual discharge from an opening is 98½ percent of the theoretical amount.

QUESTION The coefficient of discharge of an opening in the side of a tank is .71. By formula, the theoretical discharge is 350 gpm. What is the actual discharge?

ANSWER Actual discharge = theoretical discharge × coefficient of discharge

where

$$Theoretical\ discharge = 350\ gpm$$
$$Coefficient\ of\ discharge = .71$$

Then

$$Actual\ discharge = (350)(.71)$$
$$= 248.5\ gpm$$ ◼

Orifices can be designed to take advantage of stream contraction and therefore increase the coefficient of discharge. As an example, by rounding the entrance edge of the orifice, a coefficient approaching a value of 1.0 may be obtained. The effect of rounding the edges can be seen in Figure 5.3.

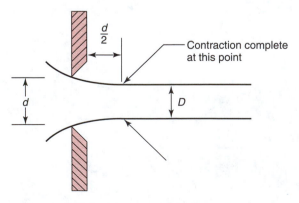

FIGURE 5.2 ◆ Contraction of the water from a standard orifice.

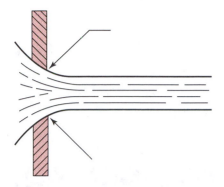

FIGURE 5.3 ◆ The edges of the entrance have been rounded to conform to the shape of the stream.

Nozzle tips are designed to secure maximum coefficients. In general, the larger the tip, the larger is the coefficient. The assumed coefficients of discharge for common tips are as follows:

¼″ and ⁵⁄₁₆″	.9825
⅜″, ⁷⁄₁₆″, and ½″	.985
⅝″, ¾″, and ⅞″	.9875
1″, 1⅛″, and 1¼″	.99
1⅜″	.9925
1½″, 1⅝″, and 1¾″	.995
1⅞″	.996
2″ and 2¼″	.997

DISCHARGE FROM NOZZLE TIPS

Substitutions are made in the basic discharge formula in order to develop a formula that can be used for solving practical problems:

Determining the Discharge from a Nozzle Tip

$$\text{Discharge} = 29.72D^2\sqrt{P}$$

where
$$D = \text{diameter}$$
$$P = \text{nozzle pressure}$$

The figure 29.7 is commonly used in place of 29.72 for convenience in solving problems. This introduces an error, but the error is small and insignificant for practical purposes.

QUESTION From the formula, what is the discharge from a 1½-inch tip at 50 psi?

ANSWER
$$\text{Discharge} = 29.7D^2\sqrt{P}$$
where
$$D = 1.5 \text{ inches}$$
$$P = 50 \text{ psi}$$

Then $\qquad$ Discharge $= (29.7)(1.5)(1.5)(\sqrt{50})$

$$= (29.7)(1.5)(1.5)(7.07)$$

$$= 472.45 \text{ gpm}$$ ■

The discharge table for smooth-bore nozzles (Table 5.1) shows a discharge of 473 gpm from a 1½-inch tip at 50 psi. In this example, the difference between the flow as determined by the formula and the flow given in the discharge table is 0.55 gpm. This difference is considered insignificant.

TABLE 5.1 ◆ Pump Operator Chart

Flow (gpm) at Various Nozzle Sizes (in.)

Nozzle Pressure (psi)	½	⅝	¾	⅞	1	1⅛	1¼	1⅜	1½	1⅝	1¾	1⅞	2	2¼	2½	3
30	41	64	92	125	163	206	254	308	366	430	498	572	651	824	1017	1464
35	44	69	99	135	176	222	275	332	395	464	538	618	703	890	1098	1581
40	47	73	106	144	188	238	294	355	423	496	575	660	751	951	1174	1691
45	50	78	112	153	199	252	311	377	448	525	610	700	797	1009	1245	1793
50	53	82	118	161	210	266	328	397	473	555	643	738	840	1063	1313	1890
55	55	86	124	169	220	279	344	417	496	582	675	774	881	1115	1377	1982
60	58	90	130	176	230	291	360	435	518	608	705	809	920	1165	1438	2071
62	58	91	132	179	234	296	366	442	526	618	716	822	935	1184	1462	2105
64	59	93	134	182	238	301	371	449	535	628	728	835	950	1203	1485	2138
66	60	94	136	185	241	305	377	456	543	637	739	848	965	1222	1508	2172
68	61	96	138	181	245	310	383	463	551	647	750	861	980	1240	1531	2204
70	62	97	140	190	248	315	388	470	559	656	761	874	994	1258	1553	2236
72	63	99	142	193	252	319	394	477	567	666	772	886	1008	1276	1575	2268
74	64	100	144	196	255	323	399	483	575	675	783	898	1022	1293	1597	2299
76	65	101	146	198	259	328	405	490	583	684	793	910	1036	1311	1618	2330
78	66	103	148	201	262	332	410	496	590	693	803	922	1049	1328	1639	2361
80	66	104	150	203	266	336	415	502	598	702	814	934	1063	1345	1660	2391
85	68	107	154	210	274	347	428	518	616	723	839	963	1095	1386	1711	2465
90	70	110	159	216	282	357	440	533	634	744	863	991	1127	1427	1761	2536
95	72	113	163	222	289	366	452	547	651	765	887	1018	1158	1466	1809	2605
100	74	116	167	228	297	376	464	562	668	784	910	1044	1188	1504	1856	2673
105	76	119	171	233	304	385	476	575	685	804	932	1070	1217	1541	1902	2739
110	78	122	175	239	311	394	487	589	701	823	954	1095	1246	1577	1947	2803
115	80	125	179	244	319	403	498	602	717	841	976	1120	1274	1613	1991	2867
120	81	127	183	249	325	412	509	615	732	859	997	1144	1301	1647	2034	2928

Courtesy of the Maryland Fire and Rescue Institute, University of Maryland, College Park, MD.

To solve discharge problems, the square root of the nozzle pressure must be known. The following are the square roots of some of the common nozzle pressures:

Nozzle Pressure	Square Root
40	6.32
45	6.71
50	7.07
55	7.42
60	7.75
70	8.37
80	8.94
90	9.49

QUESTION   What is the discharge from a ½-inch tip at 80 psi nozzle pressure?

ANSWER

$$\text{Discharge} = 29.7D^2\sqrt{P}$$

where

$$D = .5 \text{ inch}$$
$$P = 80 \text{ psi}$$

Then

$$\text{Discharge} = (29.7)(.5)(.5)(\sqrt{80})$$
$$= (29.7)(.5)(.5)(8.94)$$
$$= 66.38 \text{ gpm}$$

The flow according to the discharge table is 66 gpm.   ■

QUESTION   What is the discharge from a 1-inch tip at 50 psi nozzle pressure?

ANSWER

$$\text{Discharge} = 29.7D^2\sqrt{P}$$

where

$$D = 1 \text{ inch}$$
$$P = 50 \text{ psi}$$

Then

$$\text{Discharge} = (29.7)(1)(1)(\sqrt{50})$$
$$= (29.7)(1)(1)(7.07)$$
$$= 209.98 \text{ gpm}$$

The flow according to the discharge table is 210 gpm (Table 5.1).   ■

OPEN-BUTT DISCHARGE FORMULA

An **open butt** refers to a hose line without a nozzle attached or to the outlet from a hydrant. The coefficients of discharge of open butts are much less than those for well-designed nozzle tips. Figure 5.4 illustrates three different designs of hydrant outlets, each having a different coefficient of discharge. The coefficients shown are merely estimates and are used only for illustration. The generally accepted coefficient for a hydrant is 0.9.

If the actual coefficient of an open butt is known, it should be used to solve problems involving discharge from the opening. Prior to actually flow testing a hydrant, the cap is

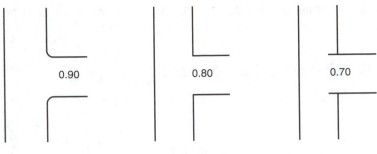

FIGURE 5.4 ◆

removed and the inside of the outlet checked to determine the construction of the outlet. In the absence of known coefficients, it is commonly assumed that the coefficient of an open butt is 0.9. The discharge from an open butt can then be determined by the formula

$$\text{Discharge} = 29.7D^2\sqrt{P} \times .9$$

Since both the 29.7 and 0.9 are constants, they can be combined, $(29.7)(.9) = 26.73$. Since the discharge is only an estimate, this result can be rounded off to 27 for the development of an open-butt formula:

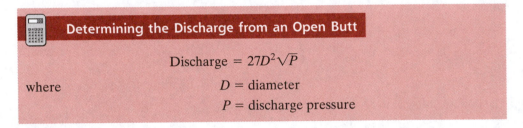

Determining the Discharge from an Open Butt

$$\text{Discharge} = 27D^2\sqrt{P}$$

where

D = diameter

P = discharge pressure

QUESTION What is the discharge from a 2½-inch hydrant when the discharge pressure is 65 psi?

ANSWER $\text{Discharge} = 27D^2\sqrt{P}$

where
D = 2.5 inches
P = 65 psi
$\sqrt{P}$ = 8.06

Then $\text{Discharge} = (27)(2.5)(2.5)(\sqrt{65})$
$= (27)(2.5)(2.5)(8.06)$
$= 1360.13 \text{ or } 1360 \text{ gpm}$ ■

QUESTION How much water is discharged from two 3-inch hose lines in 10 minutes if the discharge pressure is 40 psi?

ANSWER $\text{Discharge} = 27D^2\sqrt{P}$ (for one line)

where
D = 3 inches
P = 40 psi
$\sqrt{P}$ = 6.32

Then
$$\text{Discharge} = (27)(3)(3)(\sqrt{40})$$
$$= (27)(3)(3)(6.32)$$
$$= 1535.76 \text{ or } 1536 \text{ gpm}$$
$$\text{Total discharge} = (2 \text{ lines}) \times (1536)(10 \text{ min})$$
$$= (2)(1536)(10)$$
$$= 30,720 \text{ gallons}$$

◆ **ESTIMATING THE FLOW FROM A HYDRANT OUTLET**

For many years, apparatus operators have used various methods of estimating whether or not an additional working line could be supplied from the pump. Some of the methods have been objective; most have been based on intuition and experience. Professional operation of pumping apparatus requires knowledge, not intuition. Pump operators should be familiar with both the ability and the limitations of pump operations. The number and the size of lines that can be operated from a pump are restricted by both the pump capacity and the amount of water available from the hydrant. Pump capacity of an individual pump is a constant, but water availability varies from hydrant to hydrant.

Warren Kimball did considerable research on the problem of estimating water flow from a hydrant, and published his results in his book, *Operating Fire Department Pumpers*. Kimball proposed a method of estimating flow from a hydrant outlet that has proved to be quite helpful to pump operators. Here is how his system works.*

After the hydrant outlet is opened and water let into the pump, but before opening any pump discharge gate, the incoming pressure should be observed on the suction gage. The pressure reading on the suction gage is hydrant static pressure (Figure 5.5). After the static pressure is mentally recorded, water should be provided for a working line. Pump pressure should be increased to provide the desired nozzle pressure on the working line. When the desired pump pressure is reached, the incoming pressure should again be observed on the suction gage. This pressure is the residual (remaining) pressure at the hydrant outlet (Figure 5.6).

If the drop in pressure between the static reading and the residual reading on the **suction gage** is 10 percent or less of the static pressure, three additional lines with the same size

suction gage

pressure gage

FIGURE 5.5 ◆ On centrifugal pumps, before an outlet is opened on the pump, both the suction gage and the pressure gage will read *hydrant static pressure*.

*Reprinted with permission from *Operating Fire Department Pumpers*, copyright ©1974 National Fire Protection Association, Quincy, MA 02269. This reprinted material is not the complete and official position of the National Fire Protection Association on the referenced subject, which is represented only by the standard in its entirety.

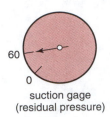

suction gage
(residual pressure)

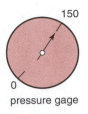

pressure gage

FIGURE 5.6 ◆ Once the discharge gate is opened and working pressure obtained, the suction gage will read *residual* (remaining) *pressure*.

tip at the same nozzle pressure as the first line can be supplied, provided the total discharge does not exceed pump capacity. If the drop in pressure is more than 10 percent but 15 percent or less, two additional lines can be supplied. If the drop is more than 15 percent but 25 percent or less, one additional line can be supplied. If the drop is more than 25 percent, an additional line with the same-size tip at the same nozzle pressure cannot be supplied.

It should be noted that this system is based on using the same-size tip at the same nozzle pressure. In reality, it is based upon the delivery of a given amount of water. For example, if 250 gpm is discharged from the first working line, additional available water is as follows:

Drop in Pressure	*Additional Water Available*
10% or less	(3)(250), or 750 gpm
More than 10%, but 15% or less	(2)(250), or 500 gpm
More than 15%, but 25% or less	(1)(250), or 250 gpm
More than 25%	Less than 250 gpm

It should also be noted that this system has reference to the amount of additional water available from a hydrant outlet. If the hydrant in use has more than one outlet and the maximum amount of water from one outlet is being used, it is possible that some additional water can be obtained from one of the other outlets.

There is a quick method for determining the amount of pressure drop. For a 10 percent drop in pressure, move the decimal point one place to the left. For example, if the static pressure is 80 psi, an 8-psi loss would be a 10 percent drop. If the residual pressure is below 72 psi, there is more than a 10 percent drop.

For a 15 percent drop in pressure, use the 10 percent drop, take one-half of it, and add to the 10 percent drop. In the example used, a 10 percent drop is 8 psi; one-half of 8 is 4; therefore, a 15 percent drop is 12 psi. If the residual pressure is below 68 psi, there is more than a 15 percent drop.

For a 25 percent drop in pressure, divide the static pressure by 4. In the example used, 80 psi divided by 4 is 20. If the residual pressure is below 60 psi, there is more than a 25 percent drop.

QUESTION After letting water into his pump, a pump operator notices that the hydrant static pressure is 90 psi. A discharge gate is opened and water supplied to a 1-inch tip at 50 psi. At this time the residual pressure is 66 psi. How many additional lines with 1-inch tips at 50 psi nozzle pressure can be supplied from the hydrant outlet?

ANSWER The pressure drop from a static pressure of 90 psi to a residual pressure of 66 psi is 24 psi. A 25 percent drop would be 22.5 psi (90 divided by 4). As the drop is greater than 25 percent, it is not possible to supply an additional line with a 1-inch tip at 50 psi nozzle pressure. It would probably be possible to supply a 1½-inch line with a ⅝-inch tip without too much trouble. ■

QUESTION The static pressure on a hydrant outlet is 60 psi. A pumper is connected to the hydrant. A single 2½-inch line with a 1¼-inch tip is taken off the pump and supplied at a nozzle pressure of 50 psi. The residual hydrant pressure is 52 psi. Approximately how much more water can be taken from the hydrant outlet?

ANSWER A 10 percent drop from the 60-psi static pressure is 6 psi. A 15 percent drop is 9 psi. The drop from a pressure of 60 psi to a residual pressure of 52 psi is an 8-psi drop. The 8 psi is between a 10 percent drop and a 15 percent drop, but close to the 15 percent drop. At least two more 1¼-inch tips at 50 psi nozzle pressure can be supplied.

The discharge from a 1¼-inch tip at 50 psi nozzle pressure is 328 gpm (Table 5.1). It is known that the hydrant outlet is capable of delivering at least an additional 656 gpm. Total available water from the hydrant outlet is at least 984 gpm (the first line plus the additional two available). ■

◆ **COMPARING DISCHARGE**

There are direct relationships between the discharge of tips of various sizes at the same nozzle pressure and between the discharge of identical tips at various nozzle pressures. Occasionally these relationships prove useful in the application of hydraulic principles. These relationships are examined in the following.

COMPARING DISCHARGE WHEN NOZZLE PRESSURES ARE IDENTICAL

At the same nozzle pressure, the increase in discharge from a larger tip compared with the discharge from a smaller tip varies directly as the square of the increase in tip size. For example, if the size of the tip is doubled while the nozzle pressure remains the same, the discharge will be four times as great. This relationship can be seen by comparing discharge formulas:

$$\text{Increase in discharge} = \frac{\text{discharge of larger tip}}{\text{discharge of smaller tip}}$$

$$= \frac{29.7D^2\sqrt{P}}{29.7d^2\sqrt{P}}$$

where

$$D = \text{larger tip}$$
$$d = \text{smaller tip}$$

Assume that the nozzle pressure is 50 psi on both tips.

Then

$$\text{Increase in discharge} = \frac{29.7D^2\sqrt{50}}{29.7d^2\sqrt{50}}$$

$$\text{Increase in discharge} = \frac{29.7D^2\sqrt{50}}{29.7d^2\sqrt{50}}$$

Then

$$\text{Increase in discharge} = \frac{D^2}{d^2}$$

$$= \left(\frac{D}{d}\right)^2$$

Thus, the final formula is as follows:

Determining the Increase in Discharge from the Larger of Two Tips When Their Nozzle Pressures Are the Same

$$\text{Increase in discharge} = \left(\frac{D}{d}\right)^2$$

where
$$D = \text{diameter of larger tip}$$
$$d = \text{diameter of smaller tip}$$

QUESTION How many times greater is the discharge from a 3-inch tip than from a 2-inch tip if the nozzle pressure is the same on both tips?

ANSWER
$$\text{Increase in discharge} = \left(\frac{D}{d}\right)^2$$

where
$$D = 3 \text{ inches}$$
$$d = 2 \text{ inches}$$

Then
$$\text{Increase in discharge} = \left(\frac{3}{2}\right)^2$$
$$= (1.5)^2$$
$$= 2.25$$

Therefore, there is 2.25 times as much water discharged from a 3-inch tip at a given nozzle pressure than from a 2-inch tip at the same nozzle pressure. ■

QUESTION The discharge from a 2½-inch hydrant outlet is 500 gpm. What is the discharge from a 4-inch hydrant outlet at the same discharge pressure?

ANSWER The same relationship exists for the open-butt discharge formula as for the nozzle formula. Therefore,

$$\text{Increase in discharge} = \left(\frac{D}{d}\right)^2$$

where
$$D = 4 \text{ inches}$$
$$d = 2\frac{1}{2} \text{ inches}$$

Then
$$\text{Increase in discharge} = \left(\frac{4}{2.5}\right)^2$$
$$= (1.6)^2$$
$$= 2.56$$
$$\text{Discharge from 4-inch outlet} = (2.56)(500 \text{ gpm})$$
$$= 1280 \text{ gpm}$$
■

Knowing that with the same nozzle pressure doubling the size of a tip gives a discharge four times greater is quite useful in solving problems. The reverse is also true: A tip that is half as large as another will produce one-fourth the discharge. The following example shows an application of this knowledge.

The discharge from a 1-inch tip at 50 psi nozzle pressure is 210 gpm. Thus the discharge from a 2-inch tip at 50 psi nozzle pressure is approximately 840 gpm, and from a ½-inch tip is approximately 52 gpm. Table 5.1 indicates a flow of 840 gpm from the 2-inch tip and 53 gpm from the ½-inch tip.

This application can be used with a 1⅛-inch tip to find the discharge from a 2¼-inch tip and with a 1¼-inch tip to find the discharge from a 2½-inch tip and from a ⅝-inch tip.

COMPARING DISCHARGE WHEN TIP SIZES ARE IDENTICAL

The increase in discharge from the same-size tip as the nozzle pressure increases can be found by comparing the square roots of the nozzle pressures. This relationship can be seen by comparing discharge formulas:

$$\text{Increase in discharge} = \frac{\text{discharge at higher nozzle pressure}}{\text{discharge at lower nozzle pressure}}$$

$$= \frac{29.7D^2\sqrt{P}}{29.7D^2\sqrt{p}}$$

where

P = higher nozzle pressure

p = lower nozzle pressure

D = diameter of tip

Assume that the tip size is 2 inches on both lines. Then

$$\text{Increase in discharge} = \frac{(29.7)(2)^2(\sqrt{P})}{(29.7)(2)^2(\sqrt{p})}$$

Then, by canceling,

$$\text{Increase in discharge} = \frac{\cancel{(29.7)}\,\cancel{(2)^2}(\sqrt{P})}{\cancel{(29.7)}\,\cancel{(2)^2}(\sqrt{p})}$$

$$= \frac{\sqrt{P}}{\sqrt{p}}$$

$$= \sqrt{\frac{P}{p}}$$

Thus, the final formula is as follows:

Determining the Increase in Discharge When Two Tips Are Identical but Their Nozzle Pressures Are Different

$$\text{Increase in Discharge} = \sqrt{\frac{P}{p}}$$

where

P = greater pressure

p = smaller pressure

QUESTION How much greater is the discharge from a 2-inch tip at 100 psi nozzle pressure than from a 2-inch tip at 50 psi nozzle pressure?

ANSWER
$$\text{Increase in discharge} = \sqrt{\frac{P}{p}}$$

where
$$P = 100 \text{ psi}$$
$$p = 50 \text{ psi}$$

Then
$$\text{Increase in discharge} = \sqrt{\frac{100}{50}}$$
$$= \sqrt{2}$$
$$= 1.41$$

Therefore, there is approximately 1.41 times as much water discharged from a 2-inch tip at 100 psi nozzle pressure than from a 2-inch tip at 50 psi nozzle pressure. Notice that in Table 5.1 the discharge from a 2-inch tip at 100 psi is 1188 gpm and the discharge from a 2-inch tip at 50 psi is 840 gpm, and that (840)(1.41) = 1184.4 gpm.

A 2-inch tip was used in the problem; however, it can be seen that there is a general application of the result: The discharge from a tip of any size is approximately 1.41 times greater when the discharge pressure is doubled. ■

QUESTION The discharge from a nozzle tip is 250 gpm with a nozzle pressure of 40 psi. What is the approximate discharge from this same tip at 50 psi nozzle pressure?

ANSWER
$$\text{Increase in discharge} = \sqrt{\frac{P}{p}}$$

where
$$P = 50 \text{ psi}$$
$$p = 40 \text{ psi}$$

Then
$$\text{Increase in discharge} = \sqrt{\frac{50}{40}}$$
$$= \sqrt{1.25}$$
$$= 1.12$$
$$\text{Discharge at 40 psi} = 250 \text{ gpm}$$
$$\text{Discharge at 50 psi} = (1.12)(250 \text{ gpm})$$
$$= 280 \text{ gpm (approx.)}$$ ■

◆ SPRINKLER DISCHARGE

The key to reducing fire losses to a minimum is early detection of fires and the placing of water on a fire as soon as possible after ignition. Both early detection and immediate placement of water on a fire can be achieved by a well-designed automatic sprinkler system.

An automatic sprinkler automatically distributes water on a fire in a quantity sufficient to extinguish the fire or help hold it in check. Water is fed to the sprinkler head through a system of piping that is usually suspended from the ceiling. Water is held in place within the head by a fusible link or a similar device that is designed to melt or break at a predetermined temperature, thereby releasing water onto the fire. At the same time that the water is released, an alarm is normally triggered that provides an audible warning to occupants of the building. An additional alarm is normally relayed to a central station, from which the fire department is notified.

Water from automatic sprinklers is discharged onto the fire in an umbrella-shaped spray. The pattern of the spray is relatively uniform, the circular area having a diameter of approximately 16 feet at a point 4 feet below the sprinkler head when the discharge is approximately 15 gpm. Each sprinkler head will protect approximately 100 square feet of floor space.

An automatic sprinkler system is one of the fire department's best friends. Fire department personnel should know the fundamental principles of sprinkler operation and be thoroughly familiar with sprinkler systems installed within their first-in districts.

Seven to 8 psi of working pressure at the head is considered the minimum for proper action of a sprinkler head with a ½-inch opening. For full effectiveness, a working pressure of at least 15 psi is required; at 15 psi, a sprinkler head will discharge approximately 22 gpm.

In some sprinkler systems, the minimum working pressure of 7 psi can be reached rather quickly if several heads are opened at the same time. Therefore, it is important that the pressure in the system be boosted by the fire department as soon as possible. Many fire departments have operational procedures requiring that either the first- or second-arriving company at a reported building fire where the building is protected by an automatic sprinkler system lay into the fire department sprinkler connection with two supply lines. It is good standard practice to pump into the system at a pressure of 150 psi, whether or not the fire is visible.

Fire department connections are a standard part of a sprinkler system. Piping from the fire department inlets is attached to the system side of the controlling valve on **wet-sprinkler systems** and between the dry-pipe valve and the controlling valve for dry-sprinkler systems. This makes it possible for the fire department to support the sprinkler system even if the gate valve has been closed.

A closed gate valve is a primary reason for a sprinkler system failing to extinguish or control a fire. The fire will open the heads, but only the water that is trapped in the system will be released. Under these conditions, the fire will probably have gained considerable headway prior to the arrival of the fire department; however, fire department officers should be cognizant of the fact that laying lines into the sprinkler system is probably the fastest and surest method of discharging water directly onto the fire. Heavy, dense smoke that might hamper firefighters who are trying to advance hose lines to the involved area will have no effect on the performance of the sprinkler system. A general rule to remember when pumping into the sprinkler system is that each opened sprinkler head will require approximately 20 gpm of water. This means that a 1000-gpm pumper, if attached to an adequate water supply, can support approximately 50 heads. This would provide coverage for approximately 5000 square feet of floor space.

There are two precautions that fire department officers should take when operating at fires where sprinkler systems are installed in the fire building.

1. Care must be taken to ensure that hose lines do not rob the sprinkler system. This is a possibility when the domestic system is weak and the sprinkler system is being supplied from the street mains. It is possible for a fire that is being held in check by the sprinkler system to accelerate rapidly if hose lines are placed into operation because this results in water being taken from the sprinkler system, which reduces the working pressure at the heads to below the operational minimum. This is most likely to happen when the officer in charge places hose streams in operation to protect exposures, feeling secure that the fire is being held in check by the sprinklers.

2. A sprinkler system should not be shut down until the officer in charge is certain that the fire has been extinguished or is under absolute control. Entire buildings have been lost due to sprinkler valves being closed too soon.

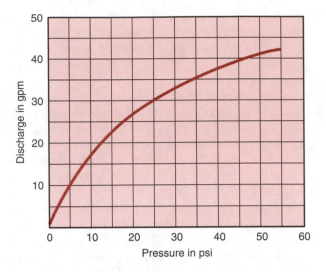

FIGURE 5.7 ◆ Sprinkler discharge.

The rate of discharge from sprinkler heads depends upon the size of the orifice and the discharge pressure. Standard sprinkler heads have a nominal ½-inch discharge orifice. The approximate rates of discharge can be obtained from Figure 5.7.

The coefficient of discharge for a ½-inch sprinkler opening is 0.75. Applying the coefficient to the discharge formula provides a formula for estimating the discharge from sprinkler heads. The development of the sprinkler discharge formula is as follows:

$$\text{Discharge} = 29.7D^2\sqrt{P} \times \text{coefficient of discharge}$$

where
$$D = .5 \text{ inches}$$
$$\text{coefficient} = .75$$

Then
$$\text{Discharge} = (29.7)(.5)(.5)(\sqrt{P})(.75)$$
$$= 5.57\sqrt{P}$$

While this formula will provide adequate results, a simpler formula has been developed for use in the fire service. The formula commonly used is as follows:

 Determining the Discharge from a Sprinkler Head

$$\text{Discharge} = \frac{P}{2} + 15$$

where
$$P = \text{discharge pressure}$$

QUESTION What is the discharge from a sprinkler head when the discharge pressure is 40 psi?

ANSWER
$$\text{Discharge} = \frac{P}{2} + 15$$

where
$$P = 40 \text{ psi}$$

Then
$$\text{Discharge} = \frac{40}{2} + 15$$

$$= 20 + 15$$
$$= 35 \text{ gpm}$$

For comparison, the answer from the first formula is

$$\text{Discharge} = (5.57)\sqrt{P}$$

where
$$P = 40 \text{ psi}$$
$$\sqrt{40} = 6.32$$

Then
$$\text{Discharge} = (5.57)(6.32)$$
$$= 35.2 \text{ gpm}$$

■

QUESTION What is the discharge from 10 sprinkler heads in 12 minutes if the discharge pressure at the heads is 50 psi?

ANSWER
$$\text{Discharge} = \frac{P}{2} + 15$$

where
$$P = 50$$

Then
$$\text{Discharge} = \frac{50}{2} + 15$$
$$= 25 + 15$$
$$= 40 \text{ gpm (for one head)}$$

Thus,

$$\text{Total discharge} = (\text{discharge of one head})(\text{number of heads})(\text{minutes})$$
$$= (40)(10)(12)$$
$$= 4800 \text{ gallons}$$

■

◆ **EQUIVALENT NOZZLE DIAMETER**

Occasionally, it is necessary to determine the size of nozzle that has the same discharge capacity as two or more nozzles combined, or to find the number of nozzles of a given size that have the same combined discharge capacity as a single larger nozzle. These types of problems are referred to as **equivalent-nozzle-diameter** problems.

ONE TIP REPLACING TWO OR MORE TIPS

The formula commonly used for determining the size of a single nozzle tip that has the same discharge capacity as two or more nozzle tips is as follows:

Determining the Size of a Single Tip That Has the Same Discharge as the Combined Discharge from Two or More Tips

$$END = \frac{D1^2 + D2^2 + D3^2 + \text{etc.}}{8}$$

where END = equivalent nozzle diameter
D_1 = number of eighths in the diameter of first nozzle tip (in.)
D_2 = number of eighths in diameter of second nozzle tip (in.)
D_3 = number of eighths in diameter of third nozzle tip (in.) etc.

TABLE 5.2 ◆ Nozzle Sizes Expressed in Eighths		
Tip Size (in.)	*Changed to Eighths (in.)*	*Square of the Number of Eighths*
½	4⁄8	16
5⁄8	5⁄8	25
¾	6⁄8	36
7⁄8	7⁄8	49
1	8⁄8	64
1⅛	9⁄8	81
1¼	10⁄8	100
1⅜	11⁄8	121
1½	12⁄8	144
1⅝	13⁄8	169
1¾	14⁄8	196
1⅞	15⁄8	225
2	16⁄8	256

To use this formula, the diameters of the tips are first expressed in eighths of an inch, and the number of eighths is then squared. As an example, a 1-inch tip equals 8⁄8 inch: There are 8 eighths, and 8 squared is (8)(8), or 64. A 1¼-inch tip equals 10⁄8 inches. There are 10 eighths, and 10 squared is (10)(10), or 100. Table 5.2 lists the eighths squared for the commonly used nozzle tips.

QUESTION In order to replace two 1¼-inch tips with a single tip having most nearly the same discharge as the two tips combined, what size of tip should be used (Figure 5.8)?

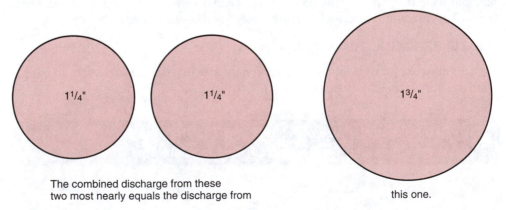

1¼"   1¼"   1¾"

The combined discharge from these
two most nearly equals the discharge from   this one.

FIGURE 5.8 ◆

ANSWER
$$END = \frac{\sqrt{D1^2 + D2^2}}{8}$$

where
$$D_1 = 10$$
$$D1^2 = (10)(10) \text{ or } 100$$
$$D_2 = 10$$
$$D2^2 = (10)(10) \text{ or } 100$$

Then
$$END = \frac{\sqrt{100 + 100}}{8}$$

$$= \frac{\sqrt{200}}{8}$$

Now $\sqrt{200} = 14.14$ (closest whole number is 14)

Then
$$END = \frac{14}{8}$$

$$= 1\tfrac{3}{4}\text{-inch tip}$$

■

QUESTION What is the smallest size of tip with a discharge capacity at least equal to the combined discharge capacities of a ½-inch tip, a ⅞-inch tip, and a 1-inch tip?

ANSWER
$$END = \frac{\sqrt{D1^2 + D2^2 + D3^2}}{8}$$

where
$$D_1 = 4$$
$$D1^2 = (4)(4) = 16$$
$$D_2 = 7$$
$$D2^2 = (7)(7) = 49$$
$$D_3 = 8$$
$$D3^2 = (8)(8) = 64$$

Then
$$END = \frac{\sqrt{16 + 49 + 64}}{8}$$

$$= \frac{\sqrt{129}}{8}$$

$$= \frac{11.36}{8}$$

The tip size exactly equaling the combined discharge of the three tips given in the problem is a fictional tip size of 11.36/8. It is then necessary to select the next larger tip in order to have a tip with a discharge capacity at least equal to an 11.36/8 tip. The next larger tip is a 12/8 inches, or a 1½-inch tip. If an individual prefers working with decimals, this formula can still be used for finding the equivalent nozzle diameter, as in the next example.

■

QUESTION In order to replace two 1¼-inch tips with a single tip having most nearly the same discharge as the two tips combined, what size of tip should be used?

ANSWER
$$END = \sqrt{D_1^2 + D_2^2}$$

where
$$D_1 = 1.25 \text{ inches}$$
$$D_2 = 1.25 \text{ inches}$$

Then
$$END = \sqrt{(1.25)^2 + (1.25)^2}$$
$$= \sqrt{1.56 + 1.56}$$

$$= \sqrt{3.12}$$
$$= 1.77, \text{ or } 1\tfrac{3}{4}\text{-inch tip} \qquad \blacksquare$$

SEVERAL TIPS REPLACING ONE TIP

The formula used to determine the number of smaller tips that have a combined discharge capacity of one larger tip is as follows:

Determining the Number of Smaller Tips That Have a Combined Discharge Capacity of One Larger Tip

$$N = \left(\frac{D}{d}\right)^2$$

where
    N = number of smaller tips required
    D = diameter of larger tip given in eighths (in.)
    d = diameter of smaller tip given in eighths (in.)

The diameters of the tips as used in this formula are in eighths of an inch.

QUESTION   How many ⅝-inch tips will it take to most nearly equal the discharge capacity of a 1¼-inch tip (Figure 5.9)?

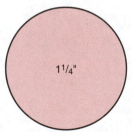

The discharge capacity from one of these

equals the discharge capacity of four of these.

FIGURE 5.9 ◆

ANSWER

$$N = \left(\frac{D}{d}\right)^2$$

where
$$D = 10$$
$$d = 5$$

Then
$$N = \left(\frac{10}{5}\right)^2$$

$$= (2)^2$$
$$= 4$$

QUESTION How many ¾-inch tips are required to provide a discharge capacity at least equal to that of a 1¾-inch tip?

ANSWER

$$N = \left(\frac{D}{d}\right)^2$$

where

$$D = 14$$
$$d = 6$$

Then

$$N = \left(\frac{14}{6}\right)^2$$
$$= (2.33)^2$$
$$= 5.43$$

It would take 5.43 three-quarter-inch tips to have a discharge capacity equal to that of a 1¾-inch tip; therefore, six ¾-inch tips would be required to at least equal the discharge capacity of the larger tip.

■ ■

Summary of Chapter Formulas

Discharge

$$\text{Discharge} = 29.7 D^2 \sqrt{P}$$

where D = diameter of opening
P = discharge pressure

Open-Butt Discharge

$$\text{Discharge} = 27 D^2 \sqrt{P}$$

where D = diameter of opening
P = discharge pressure

Comparing Discharge When Nozzle Pressures Are Identical

$$\text{Increase of discharge} = \left(\frac{D}{d}\right)^2$$

where D = diameter of larger tip
d = diameter of smaller tip

Comparing Discharge When Tip Sizes Are Identical

$$\text{Increase in discharge} = \sqrt{\frac{P}{p}}$$

where P = larger pressure
p = smaller pressure

Sprinkler Discharge

$$\text{Discharge} = \tfrac{1}{2}P + 15$$

where P = discharge pressure

Equivalent Nozzle Diameter

$$END = \frac{\sqrt{D1^2 + D2^2 + D3^2 + \text{etc.}}}{8}$$

where END = equivalent nozzle diameter

D_1 = number of eighths in diameter of first nozzle tip (in.)

D_2 = number of eighths in diameter of second nozzle tip (in.)

D_3 = number of eighths in diameter of third nozzle tip (in.) etc.

Several Tips Replacing One Tip

$$N = \left(\frac{D}{d}\right)^2$$

where N = number of smaller tips
required

D = diameter of larger tip
given in eighths (in.)

d = diameter of smaller tip
given in eighths (in.)

Review Questions

1. What is the purpose of a fire stream?
2. What is meant by the fail-safe principle?
3. What two primary factors should fire personnel be familiar with regarding the discharge of water?
4. What discharge should be expected on 1-inch lines? 1½-inch lines? 1¾-inch lines? 2½-inch lines?
5. What discharge should be expected from tips used on heavy-stream appliances?
6. Theoretically, what does the amount of water discharged from an opening depend upon?
7. What is the basic discharge formula?
8. What is the coefficient of discharge?
9. What formula is used to determine the actual discharge when the theoretical discharge and the coefficient of discharge are known?
10. In general, does the coefficient of discharge increase or decrease as tip size increases?
11. What is the discharge formula derived from the basic discharge formula?
12. What discharge formula is used in this book?
13. What is the open-butt discharge formula?
14. What rules are used to estimate the flow of water from a hydrant outlet?
15. Explain the suggested method for determining the pressure drop when estimating flow availability from a hydrant.
16. Give the formula for determining the increase in discharge from a smaller tip to a larger tip if the discharge pressure of the two tips is identical.
17. Give the formula for determining the increase in discharge as the pressure increases on a nozzle tip.
18. What is the key to reducing fire losses to a minimum?
19. What is the pattern of discharge from sprinkler heads?
20. How much floor space will each sprinkler head protect?
21. What is the minimum working pressure for the proper action of a sprinkler head?
22. What working pressure is required for full effectiveness of a sprinkler head?
23. What pressure should be supplied when supply lines are connected to sprinkler inlets?
24. Where are fire department connections attached to sprinkler systems?
25. What are the two precautions that fire department officers should take when operating at fires where sprinkler systems are installed in the fire building?
26. What is the commonly used sprinkler discharge formula?
27. What formula is used to determine the size of nozzle tip that can be used to replace two or more tips?
28. What formula is used to determine the number of smaller tips required to replace one larger tip?

Test Five

1. The coefficient of discharge of an opening in the side of a gravity tank is 0.82. By formula, the theoretical discharge is 425 gpm. What is the actual discharge?

2. Using the formula, find the discharge from a ¾-inch tip at 60 psi nozzle pressure.

3. Use the formula to find the discharge from a 1¼-inch tip at 40 psi nozzle pressure.

4. By formula, what is the discharge from a 2-inch tip at 90 psi nozzle pressure?

5. The discharge pressure on a single 4-inch hydrant is 70 psi. What is the flow from the hydrant?

6. How much water will be discharged from two 3½-inch lines without tips in 7 minutes if the discharge pressure is 35 psi?

7. After letting water into her pump, a pump operator notices that the hydrant static pressure is 55 psi. A discharge gate is opened and water is supplied to a 1¼-inch tip at 45 psi nozzle pressure. At this time, the residual pressure is 48 psi. How many additional lines with 1¼-inch tips at 45 psi nozzle pressure can be supplied from the hydrant outlet?

8. The static pressure on a hydrant outlet is 60 psi. A pump is connected to the hydrant outlet and a single 2½-inch line with a 1-inch tip is taken off the pump and supplied at a nozzle pressure of 40 psi. The residual pressure is then 51 psi. Approximately how much more water can be taken from the hydrant outlet?

9. How much greater is the discharge from a 2-inch tip than from a 1½-inch tip if the nozzle pressure on both tips is the same?

10. The discharge from a 1-inch tip is 209 gpm at 50 psi nozzle pressure. What would be the approximate discharge from a 1½-inch tip at the same nozzle pressure?

11. How much greater is the discharge from a 1¼-inch tip at 70 psi nozzle pressure than it is at 45 psi nozzle pressure?

12. The discharge from a tip at 35 psi nozzle pressure is 125 gpm. What is the approximate discharge from this same tip at 55 psi nozzle pressure?

13. What is the discharge from 35 sprinkler heads in 12 minutes if the discharge pressure at the heads is 45 psi?

14. In order to replace a 1-inch tip and a 1⅛-inch tip with a single tip having most nearly the same discharge as the two combined, which tip should be used?

15. What is the smallest tip with a discharge capacity at least equal to the combined capacities of a ¾-inch tip, a 1-inch tip, and a 1½-inch tip?

16. How many l-inch tips will it take to most nearly equal the discharge capacity of a 2½-inch tip?

CHAPTER 6

Friction Loss Principles and Applications

O f all the facets of fire department hydraulics, none has proven more controversial than the application of friction loss principles. For many years, there has been a firm understanding by most fire officials regarding friction loss principles, yet there has also been much disagreement in their application when applied to real-life situations.

The objective of this chapter is to provide a basic understanding of the friction loss principles involved together with a foundation for their application to hydraulic problems in accordance with formulas in use today.

◆ FRICTION LOSS PRINCIPLES

Friction can be defined as the resistance to relative motion between two bodies in contact. Friction is a part of our daily lives. Resistance due to friction is encountered whenever a person attempts to push an object across the floor or even move a pencil across a paper. In the fire service, friction is encountered as water moves through hose lines, pipes, fittings, and appliances.

Water moving through a fire hose encounters a resistance to flow in the form of friction as it rubs against the hose lining and as it rubs against itself. Eddy currents develop as a result of this friction. The faster the water moves, the more active these eddy currents become (Figure 6.1).

Friction loss in hose and piping may be defined as the loss of energy in the form of pressure due to friction. It follows that, whenever water moves through hose lines, sufficient pressure must be provided to overcome friction loss in addition to all other pressure requirements. If it were not for the problem of friction loss, a pump operator's task would be relatively simple. At ground level, the operator would simply have to set the discharge pressure of the pump to the desired nozzle pressure. In many cities, in fact, pumpers would not be required because the hydrant pressure would be sufficient to produce satisfactory hose streams. Only a reducing valve would be necessary, which would be set to the desired nozzle pressure. However, friction loss does exist, and it must be taken into consideration whenever hose streams are placed into operation.

Friction loss formulas and friction loss tables in use during the twentieth century were first developed around the turn of the century. As time progressed, a number of theorists criticized the earlier formulas and tables, stating that they were obsolete and inaccurate and claiming that the friction losses encountered in hose lines were actually less than those indicated in the tables. While it was true that substantial improvements had been made in fire hoses since the turn of the twentieth century and some tests on new hoses indicated friction loss figures less than those calculated by commonly used methods, the effectiveness of fire streams produced at emergencies by standard, accepted methods did not, for years, indicate a need for any radical changes. However, during the latter part of the twentieth century a change was made in both the friction loss formula and factors within it, which the majority of fire departments have accepted as the standard for use in solving problems. Despite this change, it is important that fire personnel be familiar not only with the friction loss formulas and

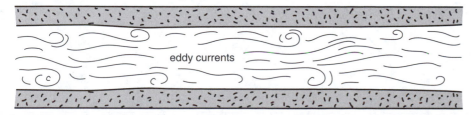

eddy currents

FIGURE 6.1 ◆

tables accepted as standard, but also with the principles that led to the development of the formulas and tables.

Understanding the principles of friction loss in hoses provides the basic foundation for understanding the method used for determining the pressure required at a pumper to develop efficient fire streams. The pressure required at a pumper when lines are laid at ground level is that necessary to overcome the friction loss in the hose and also provide for the required pressure at the nozzle tip. The formula is written as follows:

$$PDP = FL + NP$$

where
$$PDP = \text{pump discharge pressure}$$
$$FL = \text{friction loss in the hose}$$
$$NP = \text{nozzle pressure}$$

Rewritten for friction loss, the formula is

$$FL = PDP - NP$$

The friction loss in the hose is the difference between the pump pressure and the nozzle pressure whenever lines are laid at ground level. The standard practice when solving for pump pressure is to use the **friction loss coefficients** and formula for the layout being used to determine the friction loss in the layout and add it to the nozzle pressure. One of the objectives of this chapter is to present the principles of friction loss and their impact on friction loss problems. A formula for determining the friction loss in a hose is also presented, together with the application of this formula.

◆ FRICTION LOSS RULES

There are four fundamental rules governing friction loss in hose and pipe:

1. All other conditions being equal, friction loss varies directly as the length of the line.
2. In the same-size hose, friction loss varies directly as the square of the velocity flow.
3. For the same flow, friction loss varies inversely as the fifth power of the diameter of the hose.
4. For a given velocity of flow, the friction loss is independent of the pressure.

To understand these rules and their application to fire department hydraulics, it is first necessary to understand the terms "varies directly as" and "varies inversely as." Both of these terms indicate the relationship of change between two items. The term "varies directly as" means that when one item increases, the item used for comparison also increases. This principle is illustrated in Figure 6.2 with items *A* and *B*. As *A* increases, *B* also increases.

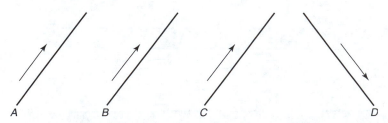

FIGURE 6.2 ◆ *A* and *B* illustrate the term "varies directly as"; *C* and *D* illustrate "varies inversely as."

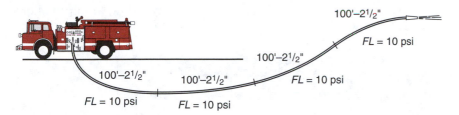

FIGURE 6.3 ◆

The term "varies inversely as" is illustrated in Figure 6.2 by items *C* and *D*. "Varies inversely as" means that when one item increases, the item used for comparison decreases. In Figure 6.2, as item *C* increases, item *D* decreases. These terms are used in Rules 2 and 3. Their application is discussed further as each of the rules is examined.

Rule 1. All other conditions being equal, friction loss varies directly as the length of the line. This rule basically means that as long as the size of the hose remains the same in a layout and the flow remains the same, the friction loss for each given length of hose is the same. This is true whether the line is laid uphill, downhill, or on level ground. For example, if the friction loss for a given flow in a layout of a single 2½-inch hose is 10 psi for the first 100 feet of hose in the layout, then as long as the size of hose and the flow remain the same, there is a 10-psi friction loss for each additional 100 feet of hose in the layout (see Figure 6.3).

QUESTION A single 2½-inch line 800 feet in length has been laid from a pumper. If the friction loss in the first 100 feet of the layout is 8 psi, what is the total friction loss in the hose?

ANSWER

$$TFL = (FL)(L)$$

where

TFL = total friction loss

FL = friction loss in 100 feet of hose

$$L = \frac{\text{total length of line}}{100}$$

Then

FL = 8 psi

$$L = \frac{800}{100} \text{ or } 8$$

$$TFL = (8)(8)$$
$$= 64 \text{ psi}$$ ■

Rule 2. In the same-size hose, friction loss varies directly as the square of the velocity flow. Friction loss in a fire hose is energy loss caused by the friction between the movement of the water and hose lining, together with friction caused by the rubbing of the water on itself. As the speed with which the water moves through the hose increases, the friction and consequently the friction loss increase. The rule states that the increase in the friction loss is proportional to the square of the increase in the speed of movement of the water. This means that if the speed of the water is doubled, the friction loss is four times as great (2×2). If the speed of the water is tripled, the friction loss is nine times as great (3×3). The principle of this increase in friction loss caused by an increase in the velocity flow within the hose can be expressed as follows:

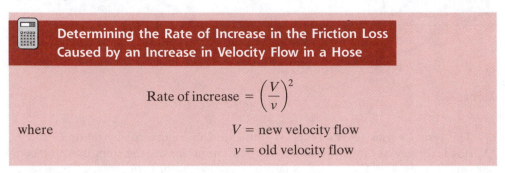

Determining the Rate of Increase in the Friction Loss Caused by an Increase in Velocity Flow in a Hose

$$\text{Rate of increase} = \left(\frac{V}{v}\right)^2$$

where

V = new velocity flow

v = old velocity flow

The application of this principle can best be illustrated by two examples.

QUESTION How much greater will the friction loss be if the velocity flow in a hose layout is increased from 120 fps to 200 fps?

ANSWER

$$\text{Rate of increase} = \left(\frac{V}{v}\right)^2$$

where

V = 200 fps

v = 120 fps

Then

$$\text{Rate of increase} = \left(\frac{200}{120}\right)^2$$

$$= (1.67)^2$$

$$= (1.67)(1.67)$$

$$= 2.79$$

Therefore, the friction loss will be 2.79 times as great. ■

QUESTION If the total friction loss in 500 feet of single $2\frac{1}{2}$-inch hose is 50 psi while the velocity flow is 150 fps, what will be the total friction loss in the 500 feet of hose if the velocity flow is increased to 200 fps (see Figure 6.4)?

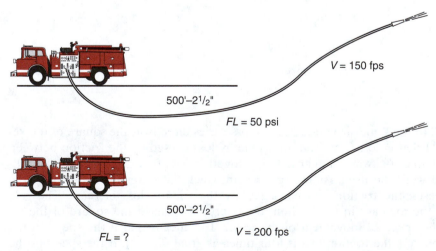

500'–$2\frac{1}{2}$" V = 150 fps

FL = 50 psi

500'–$2\frac{1}{2}$" V = 200 fps

FL = ?

FIGURE 6.4 ◆

ANSWER   New friction loss = (rate of increase)(old friction loss)

This can be shown in formula form as follows:

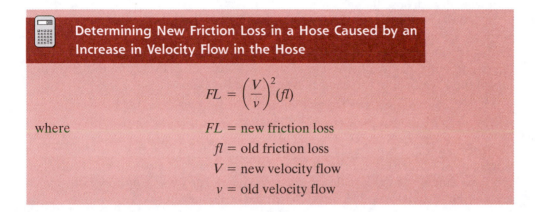

Determining New Friction Loss in a Hose Caused by an Increase in Velocity Flow in the Hose

$$FL = \left(\frac{V}{v}\right)^2 (fl)$$

where

FL = new friction loss

fl = old friction loss

V = new velocity flow

v = old velocity flow

For the problem given here

$$fl = 50 \text{ psi}$$
$$V = 200 \text{ fps}$$
$$v = 150 \text{ fps}$$

Then

$$FL = \left(\frac{200}{150}\right)^2 (50)$$
$$= (1.33)^2 (50)$$
$$= (1.33)(1.33)(50)$$
$$= 88.5 \text{ psi} \qquad \blacksquare$$

Rule 3. For the same flow, friction loss varies inversely as the fifth power of the diameter of the hose. This rule is by far the most important of the four rules governing friction loss in hose lines. It is the basis for the development of the friction loss coefficients that are used for determining the friction loss in a given layout. There are two points in the rule: (1) friction loss varies inversely and (2) to the fifth power.

As previously discussed, the term "varies inversely" means that as one item increases, the item being compared decreases. As this applies to friction loss in hose lines, if the flow is the same in both lines, the larger the hose, the less is the friction loss. Therefore, for the same flow, there is less friction loss in 3-inch hose than there is in 2½ -inch hose; similarly, there is less friction loss in 1½-inch hose than there is in 1-inch hose.

The rate of decrease is expressed by the term "to the fifth power." This means that the reduction of friction loss caused by substituting larger hose for smaller hose when the flow remains the same is to the fifth power of the increase in the size of the hose substituted. For example, if 4-inch hose is substituted for 2-inch hose and the flow remains the same, then the friction loss in the 4-inch hose will be only 1/32 that in a 2-inch hose (1/2 × 1/2 × 1/2 × 1/2 × 1/2 = 1/32).

The effect of this rule appears more dramatic when the size of the hose is tripled. For example, with the same amount of water flowing in a 3-inch hose as in a 1-inch hose, the friction loss in the 3-inch hose is only 1/243 that of the 1-inch hose (1/3 × 1/3 × 1/3 × 1/3 × 1/3 = 1/243). Expressed differently, the friction loss in 1-inch hose is 243 times as great as the friction loss in 3-inch hose when the same amount of water is flowing in each hose line.

A number of fire departments use $1^3/_4$-inch hose in lieu of $1^1/_2$-inch hose for inside work. A $1^3/_4$-inch line can be handled by one firefighter almost as readily as can a $1^1/_2$-inch line; yet because of the application of Rule 3, with the same amount of water flowing in both lines, the friction loss in the $1^1/_2$-inch line is twice that of the friction loss in the $1^3/_4$-inch line. This can be illustrated as follows:

$$\left(\frac{1.75}{1.5}\right)^5 = (1.17)^5 = 2.19$$

The following example illustrates the relationship between the friction loss in a single line of $3^1/_2$-inch hose compared with the loss in a single line of $2^1/_2$-inch hose when the same amount of water is flowing in each layout. The answer indicates how much greater the friction loss is in $2^1/_2$-inch hose than in $3^1/_2$-inch hose.

QUESTION How much greater is the friction loss in a given length of $2^1/_2$-inch hose than in the same length of $3^1/_2$-inch hose if the same amount of water is flowing through each line (see Figure 6.5)?

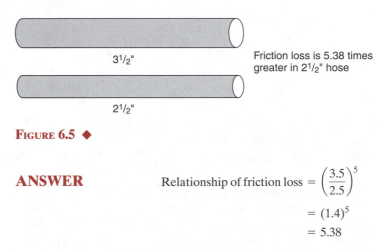

3½"

Friction loss is 5.38 times greater in 2½" hose

2½"

FIGURE 6.5 ◆

ANSWER   Relationship of friction loss $= \left(\dfrac{3.5}{2.5}\right)^5$

$$= (1.4)^5$$
$$= 5.38$$

According to Rule 3, the friction loss in a single line of 2½-inch hose is 5.38 times as great as the friction loss in an equal length of a single 3½-inch hose when the same amount of water is flowing in each line. This is just an approximation, as hose expansion has not been taken into consideration. Because of the effect of hose expansion, this rule is just an example of the relationship of the friction loss in various sizes of hose. The results will not be exactly the same as those determined by the friction loss formula.

Rule 4. For a given velocity of flow, the friction loss is independent of the pressure. In general, the amount of friction loss in a given size of hose depends upon the amount of water flowing in the hose and the speed with which it moves through the line. The pressure within the hose has no bearing on the friction loss. For example, if the loss in a layout of a single 2½-inch hose at ground level is 25 psi per 100 feet of hose and the nozzle pressure is 50 psi, gages placed in the line at 100-foot intervals will show the pressures illustrated in Figure 6.6.

Although the pressures at various points in the line are different, the friction loss in every 100 feet remains the same. This principle also applies to lines laid up or down grades.

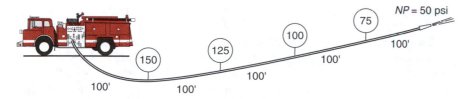

FIGURE 6.6 ◆

When lines are laid up grades, it is necessary for a pump to produce sufficient pressure to provide for back pressure, friction loss in the hose, and nozzle pressure. A pump providing the same amount of water at ground level, however, need only provide for friction loss in the hose and nozzle pressure. Although a pump providing water for lines laid up grades would have to pump at a higher pressure than one supplying the same amount of water to the same length of layout at ground level, the friction loss in the two layouts would be the same. *It is the amount of water flowing through the line, and not pressure, that determines the amount of friction loss in the hose.*

◆ THE FRICTION LOSS FORMULA

For many years there was a great difference of opinion among credible fire officials throughout the country regarding friction loss. Each of these individuals had arrived at a conclusion based upon the results of his or her experience. Most of this experience was based upon extensive tests using hose with which the individual conducting the tests was familiar. Using Rule 3, it is easy to see how different results could be obtained from the same-size hose. For example, all hoses expand a little when water is run through them and pressure applied. While technically the same, hoses manufactured by different companies may have a slightly different expansion ratio. However, it does not take too much of a difference to have a noticeable effect on friction loss.

If one hose expands from 2.5 inches to 2.52 inches whereas another expands from 2.5 inches to 2.55 inches, the one expanding to 2.52 inches would have a greater friction loss than the one expanding to 2.55 inches if the same amount of water were flowing through both. This can be illustrated as follows (Rule 3):

$$\frac{2.55}{2.52} = 1.0119 \text{ or } 1.01$$

Then

$$(1.01)(1.01)(1.01)(1.01)(1.01) = 1.05$$

If the friction loss in the hose expanding to 2.55 inches was 50 psi, then the friction loss in the hose expanding to 2.52 would be

$$(1.05)(50) = 52.5 \text{ psi}$$

The expansion difference of only .03 would result in a difference of 2.5 psi of friction loss. This makes it easy to see why different individuals could arrive at different figures for the friction loss in the same-size hose.

Although the controversy over friction loss formulas may not have entirely disappeared, two of the most respected organizations in the United States, the National Fire Protection Association and the International Fire Service Training Association, recommend the use of the same formula and friction coefficients for solving friction loss problems. A good portion of the fire departments in the country have adopted the formulas and methods recommended by the two organizations. Consequently, the methods recommended together with the friction coefficients are used in this book. It is recognized, however, that solutions determined by this method are not absolute, but merely estimates. Therefore, the results obtained, like those from previously used methods, produce results that are not absolute, but are satisfactory on the fire ground.

There are some fire departments that will continue to use other formulas and methods for solving hydraulic problems. There is nothing wrong with this. Any department that feels comfortable with the results obtained from their system should continue to use it. Remember the old saying, "If it isn't broken, don't fix it."

FRICTION LOSS APPLICATION

The objective of a friction loss formula is to determine the amount of loss due to friction in the hose that is required to be added to the formula for finding the required pump discharge pressure. The commonly used formula for determining the friction loss is as follows:

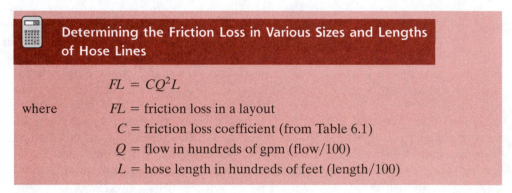

Determining the Friction Loss in Various Sizes and Lengths of Hose Lines

$$FL = CQ^2L$$

where
FL = friction loss in a layout
C = friction loss coefficient (from Table 6.1)
Q = flow in hundreds of gpm (flow/100)
L = hose length in hundreds of feet (length/100)

Note: A good point to remember is that whenever a number is being divided by 100, the answer can be determined by moving the decimal point two places to the left. For example, 565 divided by 100 = 5.65 and 300 divided by 100 = 3.

Although Table 6.1 was developed primarily for use in determining the friction loss in a hose, it can also be used to obtain the relationship between different-size hose lines. A formula for determining this relationship is as follows:

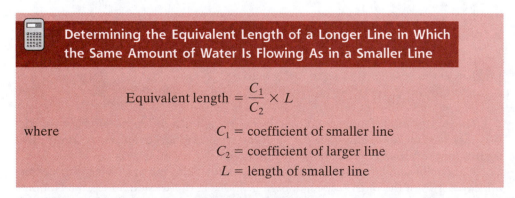

Determining the Equivalent Length of a Longer Line in Which the Same Amount of Water Is Flowing As in a Smaller Line

$$\text{Equivalent length} = \frac{C_1}{C_2} \times L$$

where
C_1 = coefficient of smaller line
C_2 = coefficient of larger line
L = length of smaller line

Note: The smaller the coefficient, the larger is the line.

TABLE 6.1 ◆ Friction Loss Coefficient for Single Hose Lines

Hose Diameter (in.)	Coefficient
¾ (booster)	1100
1 (booster) ·	150
1¼ (booster)	80
1½	24
1¾ (with 1½-in. couplings)	15.5
2	8
2½	2
3 (with 2½-in. couplings)	0.8
3 (with 3-in. couplings)	0.677
3½	0.34
4	0.2
4½	0.1
5	0.08
6	0.05
Standpipes	
4	0.374
5	0.126
6	0.052

QUESTION Water is flowing through a single 100-foot length of 1½-inch hose at 200 gpm. It is desired to replace this line and have the same amount of water flow in a 2½-inch line while maintaining the same friction loss. How long does the 2½-inch line have to be?

ANSWER

$$\text{Equivalent length} = \frac{C_1}{C_2} \times L$$

where

$$C_1 = 24$$
$$C_2 = 2$$
$$L = 100 \text{ feet}$$

Then

$$\text{Equivalent length} = \frac{24}{2} \times 100$$
$$= 12 \times 100$$
$$= 1200 \text{ feet}$$

■

Expressed another way, for friction loss purposes, when the flow is the same, a 2½-inch hose can be laid out 12 times as far as a 1½-inch hose.

Using the same logic, by dividing the coefficient of the smaller line by the coefficient of a larger line, one can find the equivalent length of layout for 100 feet of the smaller line when the flow is the same. For example, the coefficient for a 2-inch hose is 8 and the coefficient for a 5-inch hose is 0.08. Now, 8 divided by 0.08 is 100, and 100 × 100 (length of the 2-inch hose) is 10,000. This means that the friction loss in 10,000 feet of 5-inch hose is the same as the friction loss in 100 feet of 2-inch hose when the same amount of water is flowing in both lines.

Let us examine this principle a step further.

QUESTION What is the friction loss in 100 feet of 2-inch hose if the flow is 150 gpm (Figure 6.7)?

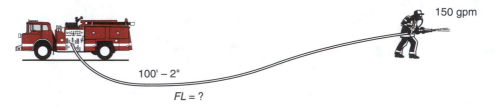

100' – 2"

FL = ?

FIGURE 6.7 ◆

ANSWER

$$FL = CQ^2L$$

where

$$C = 8$$

$$Q = 150/100 = 1.5 \text{ (moving the decimal point)}$$

$$L = 100/100 = 1$$

Then

$$FL = (8)(1.5)(1.5)(1)$$

$$= 18 \text{ psi}$$

QUESTION What is the friction loss in 10,000 feet of 5-inch hose if the flow is 150 gpm (Figure 6.8)?

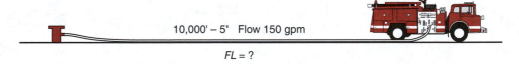

10,000' – 5"   Flow 150 gpm

FL = ?

FIGURE 6.8 ◆

ANSWER

$$FL = CQ^2L$$

where

$$C = 0.08$$

$$Q = 150/100 = 1.5$$

$$L = 10,000/100 = 100$$

Then

$$FL = (.08)(1.5)(1.5)(100)$$

$$= 18 \text{ psi}$$

USING THE FORMULA

In this chapter, an attempt is made to provide examples for the majority of pumping situations that might be encountered on the fire ground. This means that there is a large amount of repetition; however, repetition is the key to the establishment of a firm foundation.

Single Booster Lines

QUESTION What is the friction loss in 150 feet of 1-inch booster line when water is flowing at 30 gpm (Figure 6.9)?

FIGURE 6.9 ◆

ANSWER

$$FL = CQ^2L$$

where

$$C = 150$$
$$Q = 30/100 = .30$$
$$L = 150/100 = 1.5$$

Then

$$FL = (150)(.30)(.30)(1.5)$$
$$= 20.25 \text{ psi}$$

(handwritten margin notes:) 1100 245/100 (1100)(2.65)(2.65) 2.5

QUESTION Water is flowing through a ¾-inch booster line at 30 gpm. What is the friction loss in 200 feet of this line (Figure 6.10)?

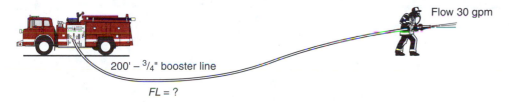

FIGURE 6.10 ◆

ANSWER

$$FL = CQ^2L$$

where

$$C = 1100$$
$$Q = 30/100 = .3$$
$$L = 200/100 = 2$$

Then

$$FL = (1100)(.3)(.3)(2)$$
$$= 198 \text{ psi}$$

Notice the big difference between the friction loss in a ¾-inch booster line as compared with that of a 1-inch booster line when the same amount of water is flowing in each. How much greater the friction loss is in the ¾-inch booster line compared with that of the 1-inch booster line can be determined by dividing the friction factor for a ¾-inch booster line (1100) by the friction factor for a 1-inch booster line (150): 1100/150 = 7.33.

Single Attack Lines

QUESTION A 350-foot, 1½-inch attack line is equipped with a fog nozzle through which water is flowing at 95 gpm. What should be allowed for the friction loss in the hose (Figure 6.11)?

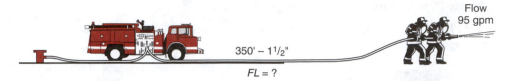

350' – 1½"

Flow 95 gpm

FL = ?

FIGURE 6.11 ◆

ANSWER

$$FL = CQ^2L$$

where

$C = 24$

$Q = 95/100 = .95$

$L = 350/100 = 3.5$

Then

$FL = (24)(.95)(.95)(3.5)$

$= 75.81$ psi

■

QUESTION A layout is 350 feet of single 2½-inch hose equipped with a 1¼-inch nozzle through which water is flowing at 325 gpm. What is the friction loss in the hose (Figure 6.12)?

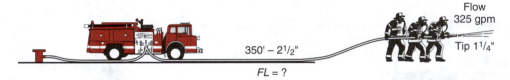

350' – 2½"

Flow 325 gpm

Tip 1¼"

FL = ?

FIGURE 6.12 ◆

ANSWER

$$FL = CQ^2L$$

where

$C = 2$

$Q = 325/100 = 3.25$

$L = 350/100 = 3.5$

Then

$FL = (2)(3.25)(3.25)(3.5)$

$= 73.94$ psi

■

Single Reduced Attack Lines

QUESTION Engine 3 has laid 400 feet of a single 2½-inch hose reduced to a single 1¾-inch line (1½-inch couplings) 200 feet long equipped with a 175-gpm constant-flow nozzle (Figure 6.13). What is the total friction loss in the layout?

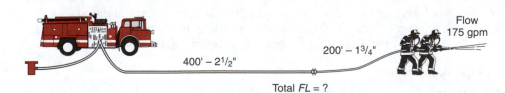

400' – 2½"

200' – 1¾"

Flow 175 gpm

Total FL = ?

FIGURE 6.13 ◆

ANSWER $TFL = FL$ in 2½-inch line $+ FL$ in 1¾-inch line

where $TFL =$ total friction loss in the layout

Note: This type of problem should be solved by determining the friction loss in the original line and adding it to the friction loss in the reduced line.

$$FL \text{ in 2½-inch line} = CQ^2L$$

where
$$C = 2$$
$$Q = 175/100 = 1.75$$
$$L = 400/100 = 4$$

Then
$$FL = (2)(1.75)(1.75)(4)$$
$$= 24.5 \text{ psi}$$

$$FL \text{ in 1¾-inch line} = CQ^2L$$

where
$$C = 15.5$$
$$Q = 175/100 = 1.75$$
$$L = 200/100 = 2$$

Then
$$FL = (15.5)(1.75)(1.75)(2)$$
$$= 94.94 \text{ psi}$$

$$\text{Total friction loss in the layout} = 24.5 + 94.94$$
$$= 119.44 \text{ psi} \quad ■$$

QUESTION What is the total friction loss for a layout of 500 feet of 3-inch hose (with 3-inch couplings) reduced to 150 feet of 2½-inch hose through which water is flowing at 325 gpm (Figure 6.14)?

Flow
325 gpm

150' – 2½"

500' – 3" (3" couplings)

Total FL = ?

FIGURE 6.14 ◆

ANSWER $TFL = FL$ in 3-inch hose $+ FL$ in 2½-inch hose
First,

$$FL \text{ in 3-inch hose} = CQ^2L$$

where
$$C = .677$$
$$Q = 325/100 = 3.25$$
$$L = 500/100 = 5$$

Then
$$FL = (.677)(3.25)(3.25)(5)$$
$$= 35.75$$

Then FL in 2½-inch hose $= CQ^2L$

where
$$C = 2$$
$$Q = 3.25$$
$$L = 150/100 = 1.5$$

Then $\qquad\qquad\qquad FL = (2)(3.25)(3.25)(1.5)$

$\qquad\qquad\qquad\qquad\qquad = 31.69$ psi

Finally,

$\qquad$ Total friction loss in the layout $= 35.75 + 31.69$

$\qquad\qquad\qquad\qquad\qquad\qquad\qquad = 67.44$ psi ∎

Single Large-Diameter Hose Lines (LDHs)

QUESTION What is the total friction loss in 500 feet of 3½-inch hose through which water is flowing at 500 gpm (Figure 6.15)?

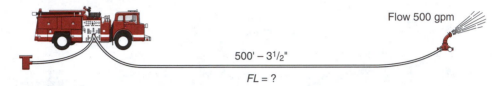

Flow 500 gpm

500' – 3¹/₂"

FL = ?

FIGURE 6.15 ◆

ANSWER $\qquad\qquad\qquad\qquad FL = CQ^2L$

where $\qquad\qquad\qquad\qquad\qquad C = .34$

$\qquad\qquad\qquad\qquad\qquad\qquad Q = 500/100 = 5$

$\qquad\qquad\qquad\qquad\qquad\qquad L = 500/100 = 5$

Then $\qquad\qquad\qquad\qquad\qquad FL = (.34)(5)(5)(5)$

$\qquad\qquad\qquad\qquad\qquad\qquad\quad = 42.5$ psi ∎

QUESTION What is the friction loss in 100 feet of 5-inch hose through which water is flowing at 1000 gpm (Figure 6.16)?

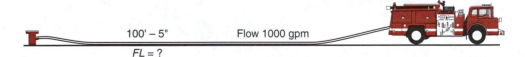

100' – 5" $\qquad\qquad$ Flow 1000 gpm

FL = ?

FIGURE 6.16 ◆

ANSWER $\qquad\qquad FL = CQ^2L$

where $\qquad\qquad\qquad C = .08$

$\qquad\qquad\qquad\qquad Q = 1000/100 = 10$

$\qquad\qquad\qquad\qquad L = 100/100 = 1$

Then $\qquad\qquad\qquad FL = (.08)(10)(10)$

$\qquad\qquad\qquad\qquad = 8$ psi (this is a good figure to remember) ∎

Multiple Lines: Same Size, Same Length, Same Flow

To determine the required friction loss for this type of layout, it is only necessary to solve the problem as if it were a single line.

QUESTION Two lines àre laid off a pumper. Each of them is a single 2½-inch line 350 feet in length equipped with a 1⅛-inch nozzle through which water is flowing at 265 gpm. What is the friction loss for this layout (Figure 6.17)?

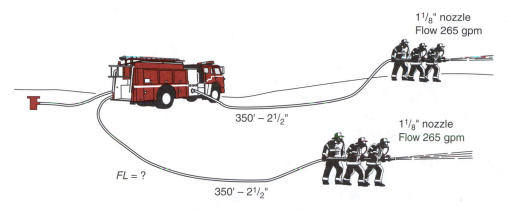

1¼₈" nozzle
Flow 265 gpm

350' − 2¹/₂"

1¼₈" nozzle
Flow 265 gpm

FL = ?

350' − 2¹/₂"

FIGURE 6.17 ◆

ANSWER

$$FL = CQ^2L$$

where

$$C = 2$$
$$Q = 265/100 = 2.65$$
$$L = 350/100 = 3.5$$

Then

$$FL = (2)(2.65)(2.65)(3.5)$$
$$= 49.16 \text{ psi}$$

Multiple Lines: Same Size, Same Length, Different Flows

With this type of problem it is only necessary to determine the required friction loss of the line having the most flow. The other line or lines have to be feathered (partially closed). There is usually a guess as to amount of feathering necessary. Of course, if the pumper is equipped with **flow gages** for each outlet, the outlet will be feathered to maintain the required flow at the nozzle. Feathering should commence at the point where the required flow is first encountered and continue until the desired pump pressure is reached.

QUESTION Engine 5 is pumping to two 450-foot lengths of 2-inch hose. One of the lines is equipped with a 175-gpm constant-flow nozzle. The other is equipped with a 200-gpm constant-flow nozzle. What is the friction loss for this setup (Figure 6.18)?

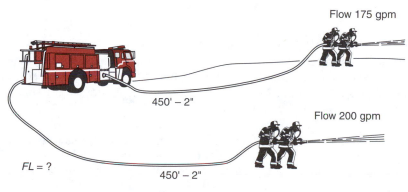

Flow 175 gpm

450' − 2"

Flow 200 gpm

FL = ?

450' − 2"

FIGURE 6.18 ◆

ANSWER

$$FL = CQ^2L$$

where

$$C = 8$$
$$Q = 200/100 = 2$$
$$L = 450/100 = 4.5$$

Then

$$FL = (8)(2)(2)(4.5)$$
$$= 144 \text{ psi}$$

■

Multiple Lines: Same Size, Different Lengths, Same Flow

With this type of problem, it is only necessary to solve for the friction loss of the longest line. The outlet to the other line or lines on the pumper has to be feathered.

QUESTION Two 1¾-inch lines equipped with 1½-inch couplings have been taken off a pumper. Both are equipped with 175-gpm constant-flow nozzles. One of the lines is 150 feet in length and the other is 250 feet in length. What is the friction loss for this layout (Figure 6.19)?

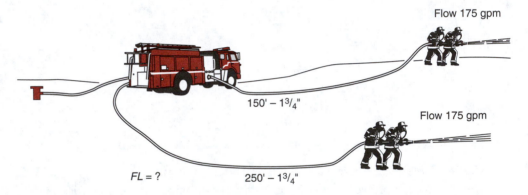

Flow 175 gpm

150' – 1³/₄"

Flow 175 gpm

FL = ?

250' – 1³/₄"

FIGURE 6.19 ◆

ANSWER

$$FL = CQ^2L$$

where

$$C = 15.5$$
$$Q = 175/100 = 1.75$$
$$L = 250/100 = 2.5$$

Then

$$FL = (15.5)(1.75)(1.75)(2.5)$$
$$= 118.67 \text{ psi}$$

■

Multiple Lines: Different Sizes, Same Length, Same Flow

When two lines are of different sizes but of the same length and with the same amount of water flowing, the friction loss in the largest line is the smallest. Therefore, with this type of problem, it is only necessary to solve for the friction loss in the smallest line. The outlet to the largest line has to be feathered.

QUESTION Engine 3 is pumping to two lines. Both of the lines are 300 feet in length and in both water is flowing at 125 gpm. One of the lines is 1½ inches in size and the other is 1¾ inches in size and equipped with 1½-inch couplings. What is the friction loss for this layout (Figure 6.20)?

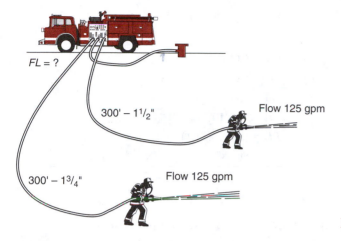

FL = ?

300' – 1½"

Flow 125 gpm

300' – 1¾"

Flow 125 gpm

FIGURE 6.20 ◆

ANSWER

$$FL = CQ^2L$$

where

$$C = 24$$
$$Q = 125/100 = 1.25$$
$$L = 300/100 = 3$$

Then

$$FL = (24)(1.25)(1.25)(3)$$
$$= 112.5 \text{ psi}$$ ■

Two Lines: Different Sizes, Different Lengths, Different Flows

With this type of problem, it is generally necessary to determine the friction loss for each line and then use the one with the largest friction loss to determine the required pump pressure. Gates on the other lines have to be feathered.

QUESTION   One line taken off Engine 5 is 150 feet of 2-inch hose equipped with a 250-gpm constant-flow nozzle. The other line is a 1½-inch line 100 feet in length equipped with a 200-gpm constant-flow nozzle. What is the friction loss for this layout (Figure 6.21)?

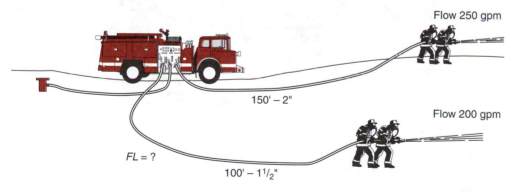

Flow 250 gpm

150' – 2"

Flow 200 gpm

FL = ?

100' – 1½"

FIGURE 6.21 ◆

ANSWER

$$FL \text{ for 2-inch line} = CQ^2L$$

where

$$C = 8$$
$$Q = 250/100 = 2.5$$
$$L = 150/100 = 1.5$$

Then
$$FL = (8)(2.5)(2.5)(1.5)$$
$$= 75 \text{ psi}$$

$$FL \text{ for } 1\frac{1}{2}\text{-inch line} = CQ^2L$$

where
$$C = 24$$
$$Q = 200/100 = 2$$
$$L = 100/100 = 1$$

Then
$$FL = (24)(2)(2)(1)$$
$$= 96 \text{ psi}$$

Pump to the 96 psi. ■

Multiple Lines: Different Sizes, Different Lengths, Different Flows

With this type of configuration, pump to the line requiring the greatest amount of pressure and feather the other two lines.

QUESTION Three lines taken off a pumper are (1) a 300-foot, 2½-inch line with a flow of 265 gpm; (2) a 450-foot, 2½-inch line with a flow of 210 gpm; (3) a 2-inch, 150-foot-long line with a flow of 195 gpm. Determine which line has the most friction loss (Figure 6.22).

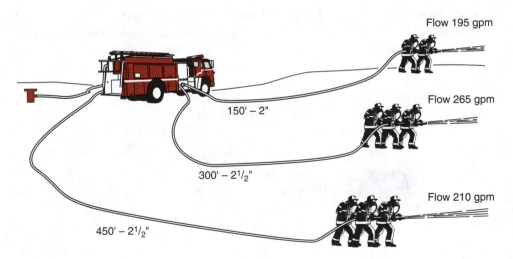

Flow 195 gpm
Flow 265 gpm
Flow 210 gpm
150' – 2"
300' – 2½"
450' – 2½"

FIGURE 6.22 ◆

ANSWER FL for 300 feet of 2½-inch line $= CQ^2L$

where
$$C = 2$$
$$Q = 265/100 = 2.65$$
$$L = 300/100 = 3$$

Then
$$FL = (2)(2.65)(2.65)(3)$$
$$= 42.14 \text{ psi}$$

$$FL \text{ for 450 feet of } 2\tfrac{1}{2}\text{-inch line} = CQ^2L$$

where
$$C = 2$$
$$Q = 210/100 = 2.1$$
$$L = 450/100 = 4.5$$

Then
$$FL = (2)(2.1)(2.1)(4.5)$$
$$= 39.69 \text{ psi}$$

$$FL \text{ for 150 feet of 2-inch line} = CQ^2L$$

where
$$C = 8$$
$$Q = 195/100 = 1.95$$
$$L = 150/100 = 1.5$$

Then
$$FL = (8)(1.95)(1.95)(1.5)$$
$$= 45.63 \text{ psi}$$

From a theoretical standpoint, pump to the 2-inch line and feather the gates on the two 2½-inch lines. From a practical standpoint, pump to the 2-inch line and do not worry about feathering the other two lines. ∎

Wyed Lines: Same Size, Same Length, Same Flow

The term **wyed lines** in this book refers to layouts where one line is divided into two or more lines (a one-into-two or a one-into-three situation). Unfortunately, this term is not standard throughout the country. Some sections of the country do not use the term at all. In those sections, the term siamesed lines refers both to two or more lines supplying a single line or appliance and one line supplying water to two or more lines.

QUESTION Engine 2 has laid 400 feet of a single 2½-inch hose, which has been connected to a preconnected wye assembly of two 1½-inch lines each 150 feet in length and with a water flow of 80 gpm. What is the total pressure loss in this layout due to friction (Figure 6.23)?

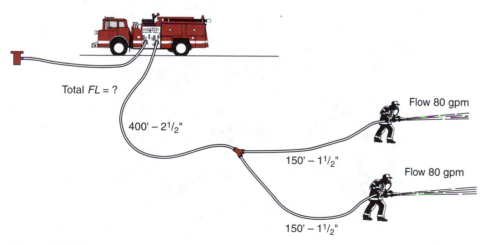

Total *FL* = ?

400' – 2½"

150' – 1½"   Flow 80 gpm

150' – 1½"   Flow 80 gpm

FIGURE 6.23 ◆

ANSWER This type of layout is typical of those used by fire companies throughout the United States. The supply line and attack lines may be different, but the concept and objective are the same. Because an equal split of the water occurs at the wye appliance, to solve the problem it is only necessary to determine the friction loss in one of the wyed lines and add the result to the friction loss in the supply line:

$$FL \text{ in } 1\frac{1}{2}\text{-inch line} = CQ^2L$$

where
$$C = 24$$
$$Q = 80/100 = .8$$
$$L = 150/100 = 1.5$$

Then
$$FL = (24)(.8)(.8)(1.5)$$
$$= 23.04 \text{ psi}$$

$$FL \text{ in } 2\frac{1}{2}\text{-inch line} = CQ^2L$$

where
$$C = 2$$
$$Q = 160/100 = 1.6 \text{ (total flow in the two wyed lines)}$$
$$L = 400/100 = 4$$

Then
$$FL = (2)(1.6)(1.6)(4)$$
$$= 20.48 \text{ psi}$$

Total friction loss = FL in $2\frac{1}{2}$-inch line + FL in $1\frac{1}{2}$-inch line
$$= 20.48 + 23.04$$
$$= 43.52 \text{ psi}$$

Wyed Lines: Same Size, Different Lengths, Different Flows

From a practical standpoint, this type of layout cannot be achieved without a gated wye, and so a gated wye is used in this problem. A three-way-gated wye is shown in Figure 6.24.

FIGURE 6.24 ◆ A three-way gated wye equipped with a Storz inlet. *Courtesy of Elkhart Brass Mfg. Company*

QUESTION A single 350-foot length of 3½-inch hose is supplying a gated wye. A single 2½-inch line 300 feet in length is attached to one of the gates. Water is flowing at 275 gpm. The other gate has a single 2½-inch line attached that is 400 feet in length and through which water is flowing at 210 gpm. What is the total friction loss in the hose for the layout (Figure 6.25)?

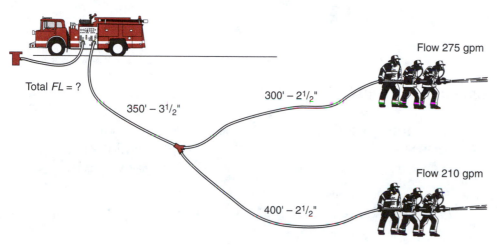

Flow 275 gpm

Total *FL* = ?

300' – 2½"

350' – 3½"

Flow 210 gpm

400' – 2½"

FIGURE 6.25 ◆

ANSWER To solve this type of problem it is necessary to determine the friction loss in each of the two 2½-inch lines and pump to the one having the most friction loss. It is necessary to feather the gate on the line with the least amount of friction loss. To complete the problem, determine the friction loss in the supply line and add it to the friction loss in the 2½-inch line having the greatest friction loss:

$$FL \text{ in 300-foot, 2½-inch line} = CQ^2L$$

where
$$C = 2$$
$$Q = 275/100 = 2.75$$
$$L = 300/100 = 3$$

Then
$$FL = (2)(2.75)(2.75)(3)$$
$$= 45.38 \text{ psi}$$

$$FL \text{ in 400-foot, 2½-inch line} = CQ^2L$$

where
$$C = 2$$
$$Q = 210/100 = 2.1$$
$$L = 400/100 = 4$$

Then
$$FL = (2)(2.1)(2.1)(4)$$
$$= 35.28 \text{ psi}$$

$$FL \text{ in 3½-inch supply line} = CQ^2L$$

where
$$C = .34$$
$$Q = 485/100 = 4.85 \text{ (total flow in}$$
$$\text{two 2½-inch lines)}$$
$$L = 350/100 = 3.5$$

Then
$$FL = (.34)(4.85)(4.85)(3.5)$$
$$= 27.99 \text{ psi}$$

Total $FL = FL$ in 3½-inch line $+ FL$ in 300-foot, 2½-inch line
$$= 27.99 + 45.38$$
$$= 73.37 \text{ psi}$$ ■

Summary of Chapter Formulas

Friction Loss in Relationship to the Pump Discharge Pressure

$$FL = PDP - NP$$

where FL = friction loss in hose
PDP = pump discharge pressure
NP = nozzle pressure

Total Friction Loss in the Hose

$$TFL = (FL)(L)$$

where TFL = total friction loss in hose
FL = friction loss in 100 feet of hose
L = total length of line/100

Rate of Friction Loss Increase Caused by an Increase in the Velocity Flow

$$\text{Rate of increase} = \left(\frac{V}{v}\right)^2$$

where V = new velocity flow
v = old velocity flow

New Friction Loss Resulting from an Increase in the Velocity Flow

$$FL = \left(\frac{V}{v}\right)^2 FL$$

where FL = new friction loss
fl = old friction loss
V = new velocity flow
v = old velocity flow

Friction Loss Formula

$$FL = CQ^2L$$

where FL = friction loss in layout
C = friction loss coefficient (from Table 6.1)
Q = flow in hundreds of gpm (flow/100)
L = hose length in hundreds of feet (length/100)

Relationship Between Different-Size Hose Lines

$$\text{Equivalent length} = \frac{C_1}{C_2} \times L$$

where C_1 = coefficient of smaller line
C_2 = coefficient of larger line
L = length of smaller line

Review Questions

1. Which facet of fire department hydraulics has been the most controversial?

2. What is the definition of friction?
3. What causes the friction in water moving through a hose line?

4. How can the friction loss in hose and piping be defined?

5. What formula is used to find the friction loss in a line when the pump discharge pressure and the nozzle pressure are known?

6. What are the four fundamental rules governing friction loss in a hose?

7. What is meant by "varies directly as"?

8. What is meant by "varies inversely as"?

9. If the speed of the water in a hose line is increased four times, how much will the friction loss be increased?

10. What is the formula for finding the rate of increase in relation to an increase in the velocity flow?

11. Which of the four friction loss rules is the most important?

12. What is the friction loss formula?

Test Six

1. A single 2½-inch line 500 feet in length has been laid from a pumper. If the friction loss in the first 100 feet of the layout is 7 psi, what is the total friction loss in the hose?

2. How much greater will the friction loss be if the velocity flow in a hose layout is increased from 100 to 140 fps?

3. If the total friction loss in 400 feet of a single 2½-inch line is 38 psi when the velocity flow is 120 fps, what is the total friction loss in 400 feet of hose if the velocity flow is increased to 180 fps?

4. How much greater is the friction loss in a given length of 1½-inch hose than in the same length of 2½-inch hose if the same amount of water is flowing through each line?

5. How much greater is the friction loss in 400 feet of 2½-inch hose than in 400 feet of 5-inch hose if the same amount of water is flowing in each line?

6. What is the friction loss in 250 feet of ¾-inch booster line when water is flowing at 25 gpm?

7. Water is flowing through a 1¼-inch booster line at 60 gpm. What is the friction loss in 150 feet of this line?

8. A 400-foot, 1¼-inch attack line is equipped with a fog nozzle through which water is flowing at 95 gpm. What should be allowed for the friction loss in the hose?

9. A layout is 550 feet of single 2½-inch hose equipped with a 1⅛-inch nozzle through which water is flowing at 265 gpm. What is the friction loss in the hose?

10. Engine 5 has laid 350 feet of a single 2½-inch hose reduced to a single 1½-inch line 150 feet long equipped with a 150-gpm constant-flow nozzle. What is the total friction loss in the layout?

11. What is the total friction loss for a layout of 400 feet of 3-inch hose with 3-inch couplings reduced to 200 feet of 2-inch hose with a flow of 175 gpm?

12. What is the total friction loss in 600 feet of 4-inch hose with a flow of 650 gpm?

13. Two lines are laid off a pumper. Each is a single 2-inch line 300 feet in length equipped with a nozzle with water flowing at 195 gpm. What is the friction loss for this layout?

14. Two 1½-inch lines have been taken off a pumper. Both are equipped with 80-gpm constant-flow nozzles. One line is 150 feet in length and the other is 250 feet in length. What is the friction loss for this layout?

15. What is the friction loss in 1200 feet of 5-inch hose with a flow of 1200 gpm?

16. Engine 6 is pumping to two lines. Both are 250 feet in length and the flow through both is 150 gpm. One of the lines is 1¾ inches in size and the other is 2 inches in size. What is the friction loss for this layout?

17. One line taken off Engine 2 is 200 feet of 2½-inch line equipped with a 275-gpm constant-flow nozzle. The other line is a 2-inch line 250 feet in length equipped with a 175-gpm constant-flow nozzle. What is the friction loss for this layout?

18. Three lines are taken off a pumper: a 300-foot, 1½-inch line with a flow of 85 gpm; a 450-foot, 2-inch line with a flow of 175 gpm; and a 300-foot, 2½-inch line with a flow of 210 gpm. Which line has the most friction loss and how much is it?

19. Engine 1 has laid 300 feet of single 3½-inch hose, which has been wyed to two 2-inch lines, each 150 feet in length and with a flow of 180 gpm. What is the total pressure loss in this layout due to friction?

20. A single 400-foot length of 3-inch hose with 3-inch couplings is supplying a gated wye. A single 2½-inch line 350 feet in length and with a flow of 210 gpm is attached to one of the gates. The other gate has a single 2½-inch line attached that is 300 feet in length and has a flow of 220 gpm. What is the total friction loss in the hose in this layout?

Required Pump Discharge Pressure

7 **CHAPTER**

Objectives

Upon completing this chapter, the reader should:

- Have a basic knowledge of the variables that must be considered in order to operate a pump at maximum efficiency on the fire ground.
- Understand the use of the required pump discharge pressure (RPDP) formula and its application to various pumping configurations.
- Understand the development of the RPDP formula.
- Be able to determine the RPDP for single lines at ground level.
- Be able to determine the RPDP for a single line wyed into two lines at ground level.
- Be able to determine the RPDP for layouts of siamesed lines into a single line at ground level.
- Be able to determine the RPDP for lines laid uphill.
- Be able to determine the RPDP for lines laid downhill.
- Be able to determine the RPDP for lines laid into standpipe systems.
- Be able to determine the RPDP for lines laid into portable monitors.
- Be able to determine the RPDP for lines laid into deck guns.
- Be able to determine the RPDP for lines laid into ladder pipes.
- Be able to determine the RPDP for lines laid into aerial platforms or other elevated master-stream devices.

A pump operator pulls up to a hydrant and spots his or her apparatus in the best position for making a supply connection from the hydrant to the pumper. Flames are billowing from a third-floor window half a block down the street. The pump operator connects a suction hose from the hydrant to the pumper, then turns water into the pump. In the meantime, lines have been laid and led into the building. Two of the lines are connected to the operator's pumper. The operator does a mental calculation of the various pressure requirements confronting him or her, then starts advancing the throttle until the pump is producing the pressure the operator considers necessary. The entire operation goes smoothly. The firefighters at the end of the lines open the nozzles and water is discharged at the proper pressure. The firefighters operating the lines start advancing on the fire. It is not long before the fire has been extinguished. The drills and the work in the classroom have paid off.

In the simplest pumping situation encountered by pump operators, a single line is laid at ground level and no appliance is involved. Complications arise when lines are laid up or down hills, into standpipes of multistory buildings, or where appliances are involved. The situation becomes even more critical when large tips are used on appliances, requiring the pump operator to balance his or her operations near the limited capability of the pump.

Standards exist for use on the fire ground that assist the pump operator in rapidly calculating the required pump discharge pressure while taking into consideration the several variables that are encountered in a pumping configuration. However, a pump operator must have a basic knowledge of the variables that must be considered in order to operate at maximum efficiency on the fire ground.

The friction loss problems in Chapter 6 considered only friction loss in the hose; therefore, the problems could be solved entirely by using the friction loss formula. This chapter considers hose layout configurations involving the need to solve for discharge and those involving elevation and appliances.

To produce effective streams on the fire ground, the pump operator must be capable of setting the correct discharge pressure at the pump to compensate for the various pressure requirements encountered in a particular situation. As with friction loss, there are various formulas available for calculating the so-called pump discharge pressure; however, only the most commonly used formula will be introduced in this book.

The formula introduced in this chapter is a pencil-and-paper formula. It is referred to as the **required pump discharge pressure** formula. It is much too complicated for use on the fire ground; however, a thorough understanding of the use of the formula and its application to various pumping configurations will provide a solid foundation for the understanding of fire ground hydraulics. The relationship of the methodology introduced in this chapter to fire ground operations is covered in another chapter.

The pump discharge pressure must be sufficient to provide for pressure losses due to:

1. The nozzle pressure.
2. The friction loss in the hose.
3. Back pressure.
4. **Appliance friction loss**. Friction loss in appliances includes the loss of pressure in the piping in **deck guns**, portable monitors, ladder pipes, elevated platforms, and standpipes and other appliances and apparatus.

Forward pressure is produced whenever the layout configuration is one in which the discharge of water at the nozzle tip or terminal outlet is below the level of the pump. Forward pressure assists the pumper; therefore, the amount of forward pressure can be subtracted from the required pump discharge pressure.

◆ DEVELOPMENT OF THE REQUIRED PUMP DISCHARGE FORMULA

The commonly accepted formula for determining the pump discharge pressure is

$$PDP = NP + TPL$$

where PDP = pump discharge pressure (psi)

NP = nozzle pressure (psi)

TPL = total pressure loss including friction loss in the hose, pressure loss in appliances, and pressure loss due to elevation

In this book, the total pressure loss portion of the formula is expanded for two reasons:

1. To provide a more thorough foundation for those items included in the total pressure loss.
2. To provide a check list to ensure that each of the items is considered when solving a problem.

The expanded formula, herein referred to as the required pump discharge pressure, can be illustrated as follows:

Determining the Required Pump Discharge Pressure

$$RPDP = NP + FL + BP + AFL - FP$$

where

NP = nozzle pressure

FL = friction loss in the hose

BP = back pressure

AFL = appliance friction loss

FP = forward pressure

Back pressure refers to those situations where the nozzle is being discharged at a point above the pump. When these situations are encountered, ½ psi is used for each foot of head. Head is the vertical distance above the pump where the nozzle is being used.

 Appliance friction loss refers to the loss of pressure as the water flows through fittings such as reducers, increasers, manifolds, siameses, wyes, standpipe systems, portable monitors, and aerial apparatus. For the purpose of solving problems, no loss will be considered whenever the flow through an appliance is less than 350 gpm. When the flow is 350 gpm or greater, a loss of 10 psi is used. An exception is the friction loss for master-stream appliances. In these appliances, a loss of 25 psi is considered as standard, regardless of the flow.

Forward pressure is a term used to identify a situation where the discharge from a nozzle takes place at a location below the pump. As with back pressure, ½ psi is used for each foot of head. Head is the vertical distance below the pump that the nozzle is being used.

Whenever lines are laid at ground level with no appliance involved, the required pump discharge pressure is only concerned with the nozzle pressure and the friction loss in the hose. It is best to consider the pump pressure as the pressure at a discharge outlet to which a hose line is attached. The outlet may be the discharge gate of a pumper, an outlet on a hydrant, an outlet on a standpipe, or any other outlet that provides a source of water for supplying a hose line.

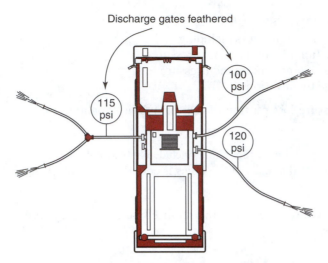

FIGURE 7.1 ◆ Multiple single lines from a pumper.

If a single line taken off a pumper is used to supply a single nozzle tip, then the pressure produced by the pump will need to be sufficient to provide for the nozzle pressure and the friction loss in the hose. If several lines are taken off a pumper and provided with individual tips, then the required pump discharge pressure for each of these lines must be calculated individually, which may result in the required pump discharge pressure on one or more of the lines being less than the required pump discharge pressure for another of the lines. This is accomplished by establishing a pump pressure equal to the highest required pump discharge pressure and feathering (partially closing) the discharge gate or gates on the hose line or lines that require less pressure than that being produced by the pump. In most cases, the individual discharge gates are equipped with separate pressure gages or flow gages, which permits the pump operator to properly set the required pump pressure or flow to each of the several separate hose lines. This principle is illustrated in Figure 7.1.

◆ REQUIRED PUMP DISCHARGE PRESSURE PROBLEMS

Required pump discharge pressure problems involve combining the information presented in Chapters 5 and 6 plus the addition of some other factors. If a constant-flow, constant-pressure nozzle is used in a problem, it is only necessary to solve for the friction loss in the hose layout and add the result to the nozzle pressure to solve the problem. However, if smooth-bore tips are used in the same type of problem, then it is necessary to determine the amount of water flowing before the friction loss in the hose can be determined. Examples of both types of situations are presented for the purpose of providing a firmer foundation in the simple use of the formula.

The standard established for handheld smooth-bore tips is 50 psi, while 80 psi has been established for master-stream smooth-bore tips. For fog nozzles, both hand-held and master-stream nozzles, nozzle pressures of 100, 75, and 50 psi are available. A nozzle pressure of 100 psi is used on most nozzles. Problems in this text involving required pump discharge pressure are confined to these standard pressures.

GROUND-LEVEL PROBLEMS

Required pump discharge pressure problems involving lines laid at ground level can be divided into three principal categories:

1. Single lines.
2. A single line wyed into two lines.
3. Siamesed lines into a single line.

Single Lines

The solution to required pump discharge pressure problems where single lines are in use at ground level involves the following:

1. Determine the nozzle pressure.
2. Solve for the discharge or use that given.
3. Solve for the friction loss in the hose.
4. Add the nozzle pressure to the friction loss in the hose.

The formula required is as follows:

Determining the Required Pump Discharge Pressure for Single Lines

$$RPDP = NP + FL$$

where

$RPDP$ = required pump discharge pressure

NP = nozzle pressure

FL = friction loss in the hose

QUESTION   Engine 6 is pumping through 400 feet of a single 1½-inch hose that is equipped with a ⅝-inch tip. What is the required pump discharge pressure if the nozzle pressure is 50 psi (Figure 7.2)?

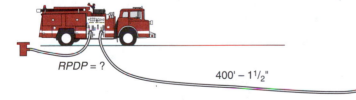

⅝" tip
NP 50 psi

RPDP = ?

400' – 1½"

FIGURE 7.2 ◆

ANSWER   First, determine the discharge from the ⅝-inch tip:

$$\text{Discharge} = 29.7D^2\sqrt{P}$$

where

D = ⅝ or .625

P = 50 psi

Then

$$\text{Discharge} = (29.7)(.625)^2(\sqrt{50})$$

$$= (29.7)(.625)(.625)(7.07)$$

$$= 82.02 \text{ gpm}$$

Next, solve for the friction loss in the hose:

$$FL = CQ^2L$$

where
$$C = 24$$
$$Q = 82.02/100 = .82$$
$$L = 400/100 = 4$$

Then
$$FL = (24)(.82)(.82)(4)$$
$$= 64.55 \text{ psi}$$

Next, solve for the required pump discharge pressure:

$$RPDP = NP + FL$$

where
$$NP = 50 \text{ psi}$$
$$FL = 64.55 \text{ psi}$$

Then
$$RPDP = 50 + 64.55$$
$$= 114.55 \text{ psi}$$
■

QUESTION Engine 1 is pumping through 250 feet of a single 1¾-inch hose equipped with a ¾-inch tip. The nozzle pressure is 50 psi. What is the required pump discharge pressure (Figure 7.3)?

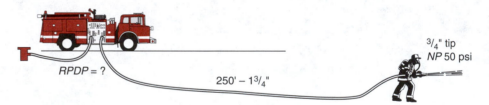

3/4" tip
NP 50 psi

RPDP = ?

250' – 1³/₄"

Figure 7.3 ◆

ANSWER First, determine the discharge from the ¾-inch tip:

$$\text{Discharge} = 29.7D^2\sqrt{P}$$

where
$$D = ¾ \text{ or } .75$$
$$P = 50 \text{ psi}$$

Then
$$\text{Discharge} = (29.7)(.75)^2\sqrt{50}$$
$$= (29.7)(.75)(.75)(7.07)$$
$$= 118.11 \text{ gpm}$$

Next, solve for the friction loss in the hose:

$$FL = CQ^2L$$

where
$$C = 15.5$$
$$Q = 118.11/100 = 1.18$$
$$L = 250/100 = 2.5$$

Then
$$FL = (15.5)(1.18)(1.18)(2.5)$$
$$= 53.96 \text{ psi}$$

Next solve for the required pump discharge pressure:

$$RPDP = NP + FL$$

where

$$NP = 50 \text{ psi}$$
$$FL = 53.96 \text{ psi}$$

Then

$$RPDP = 50 + 53.96$$
$$= 103.96 \text{ psi}$$ ■

QUESTION The layout from Engine 3 is 800 feet of single 3-inch hose equipped with 3-inch couplings reduced to 200 feet of single 2½-inch hose. The 2½-inch line is equipped with a 1⅛-inch tip at 50 psi nozzle pressure. What is the required pump discharge pressure (Figure 7.4)?

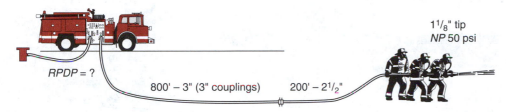

RPDP = ?

800' – 3" (3" couplings)   200' – 2½"

1⅛" tip
NP 50 psi

FIGURE 7.4 ◆

ANSWER First, determine the discharge from the 1⅛-inch nozzle tip:

$$\text{Discharge} = 29.7 D^2 \sqrt{P}$$

where

$$D = 1⅛ \text{ or } 1.125$$
$$P = 50 \text{ psi}$$

Then

$$\text{Discharge} = (29.7)(1.125)^2 \sqrt{50}$$
$$= (29.7)(1.125)(1.125)(7.07)$$
$$= 265.75 \text{ gpm}$$

Next solve for the friction loss in the 2½-inch line:

$$FL = CQ^2 L$$

where

$$C = 2$$
$$Q = 265.75/100 = 2.66$$
$$L = 200/100 = 2$$

Then

$$FL = (2)(2.66)(2.66)(2)$$
$$= 28.3 \text{ psi}$$

Next, solve for the friction loss in the 3-inch line:

$$FL = CQ^2 L$$

where

$$C = .677$$
$$Q = 265.75/100 = 2.66$$
$$L = 800/100 = 8$$

Then

$$FL = (.677)(2.66)(2.66)(8)$$
$$= 38.32 \text{ psi}$$

The required pump discharge pressure is given by

$$RPDP = NP + FL$$

where
$$NP = 50 \text{ psi}$$
$$FL = 28.3 + 38.32$$
$$= 66.62 \text{ psi}$$

Then
$$RPDP = 50 + 66.62$$
$$= 116.62 \text{ psi} \qquad ■$$

QUESTION A 300-foot, 1¾-inch attack line has been laid at a single-family house fire. The line is equipped with a constant-gallonage 175-gpm fog nozzle rated at 75 psi. What is the required pump discharge pressure for this layout (Figure 7.5)?

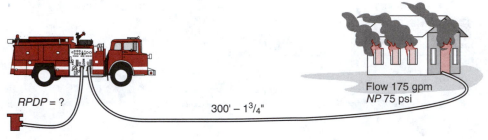

RPDP = ?

300' – 1³/₄"

Flow 175 gpm
NP 75 psi

FIGURE 7.5 ◆

ANSWER First, solve for the friction loss in the hose:

$$FL = CQ^2L$$

where
$$C = 15.5$$
$$Q = 175/100 = 1.75$$
$$L = 300/100 = 3$$

Then
$$FL = (15.5)(1.75)(1.75)(3)$$
$$= 142.41 \text{ psi}$$

The required pump discharge pressure is given by

$$RPDP = NP + FL$$

where
$$NP = 75 \text{ psi}$$
$$FL = 142.41$$

Then
$$RPDP = 75 + 142.41$$
$$= 217.41 \text{ psi} \qquad ■$$

A Single Line Wyed into Two Lines

The solution to problems where wyed lines are used is limited in this section to those situations where the wyed lines are of equal length and the same amount of water is flowing through each. Variations of this will be covered in another chapter. The solution to problems in this category involves the following steps:

1. Determine the nozzle pressure on one of the nozzles of the wyed lines (the other line will be the same).
2. Solve for the flow in one of the wyed lines, if not given.

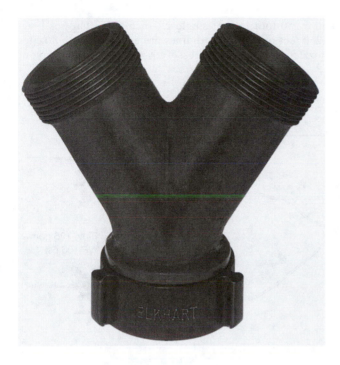

FIGURE 7.6 ◆ A plain wye. *Courtesy of Elkhart Brass Mfg. Company*

3. Double the flow in one of the wyed lines to determine the flow in the supply line.
4. Solve for the friction loss in one of the wyed lines. The reason that it is only necessary to determine the friction loss in one of the lines is that the flow and the pressure provided by the pump operator at the wye are sufficient to take care of both of the lines and their nozzle pressures.
5. Solve for the friction loss in the supply line.
6. Add the friction loss in one of the wyed lines to the friction loss in the supply line to determine the total friction loss in the hose layout.
7. If the total flow in the supply line is 350 gpm or more, add 10 psi to the problem for the appliance friction loss. If less than 350 gpm, consider the loss as zero.
8. Solve the problem for the required pump discharge pressure.

In a wye appliance the flow from a single line is divided into two or more lines. A plain one-into-two wye is shown in Figure 7.6.

The formula to use is the following:

Determining the Required Pump Discharge Pressure for a Single Line Wyed into Two Lines

$$RPDP = NP + FL + AFL$$

where $RPDP$ = required pump discharge pressure
NP = nozzle pressure
FL = friction loss in the hose
AFL = appliance friction loss

QUESTION Engine 10 has laid a 550-foot line of single 2½-inch hose that has been wyed into two 1¾-inch lines, each 250 feet in length. Both lines are equipped with nozzles with a flow of 125 gpm at 100 psi nozzle pressure. What pressure is required for this layout (Figure 7.7)?

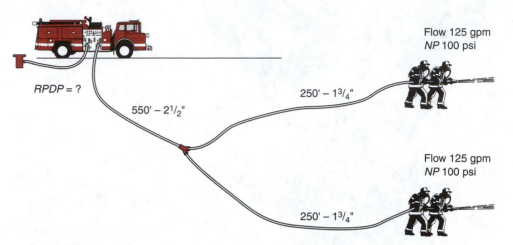

FIGURE 7.7 ◆

ANSWER Since both 1¾-inch lines are the same length and have the same amount of water flowing, it is only necessary to determine the friction loss in one of the lines to solve the problem. The total friction loss in the layout is the friction loss in the 2½-inch line plus the friction loss in one of the 1¾-inch lines.

The friction loss in the 1¾-inch line is given by

$$FL = CQ^2L$$

where
$$C = 15.5$$
$$Q = 125/100 = 1.25$$
$$L = 250/100 = 2.5$$

Then
$$FL = (15.5)(1.25)(1.25)(2.5)$$
$$= 60.55 \text{ psi}$$

Next, solve for the friction loss in the 2½-inch line. The flow in this line is twice that of the flow in the wyed line. Then

$$FL = CQ^2L$$

where
$$C = 2$$
$$Q = 250/100 = 2.5$$
$$L = 550/100 = 5.5$$

Then
$$FL = (2)(2.5)(2.5)(5.5)$$
$$= 68.75 \text{ psi}$$

The total friction loss in the hose is equal to the friction loss in the 2½-inch line plus the friction loss in one of the 1¾-inch lines, expressed as

$$TFL = 68.75 + 60.55$$
$$= 129.3 \text{ psi}$$

Next, solve for the required pump discharge pressure:

$$RPDP = NP + FL + AFL$$

where

$$NP = 100 \text{ psi}$$
$$FL = 129.3 \text{ psi}$$
$$AFL = 0 \text{ (total flow is less than 350 gpm)}$$

Then

$$RPDP = 100 + 129.3 + 0$$
$$= 229.3 \text{ psi}$$

■

QUESTION A single 3½-inch line 550 feet in length has been laid and reduced to two wyed 2½-inch lines. Each of the 2½-inch lines is 150 feet in length and equipped with a 1-inch tip at 50 psi nozzle pressure. What is the required pump discharge pressure for this layout (Figure 7.8)?

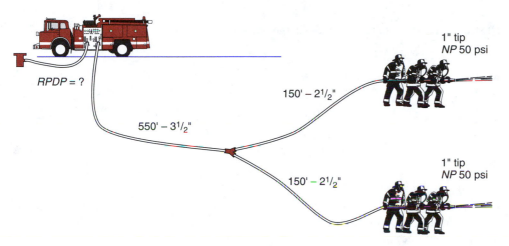

1" tip
NP 50 psi

RPDP = ?

150' – 2½"

550' – 3½"

150' – 2½"

1" tip
NP 50 psi

FIGURE 7.8 ◆

ANSWER First, solve for the discharge from the 1-inch tip:

$$\text{Discharge} = 29.7D^2\sqrt{P}$$

where

$$D = 1.0 \text{ inches}$$
$$P = 50 \text{ psi}$$

Then

$$\text{Discharge} = (29.7)(1)^2(\sqrt{50})$$
$$= (29.7)(1)(1)(7.07)$$
$$= 209.98 \text{ gpm}$$

Next, solve for the friction loss in a 2½-inch line:

$$FL = CQ^2L$$

where

$$C = 2$$
$$Q = 209.98/100 = 2.1$$
$$L = 150/100 = 1.5$$

Then

$$FL = (2)(2.1)(2.1)(1.5)$$
$$= 13.23 \text{ psi}$$

Next, solve for the friction loss in the 3½-inch line. The flow in the 3½-inch line is twice that in the 2½-inch line, or $2 \times 210 = 420$ gpm:

$$FL = CQ^2L$$

where

$$C = .34$$
$$Q = 420/100 = 4.2$$
$$L = 550/100 = 5.5$$

Then

$$FL = (.34)(4.2)(4.2)(5.5)$$
$$= 32.99 \text{ psi}$$

Next, solve for the total friction loss in the hose. The total friction loss in the hose is equal to the friction loss in the supply line plus the friction loss in one of the 2½-inch lines:

$$FL = 32.99 + 13.23$$
$$= 46.22 \text{ psi}$$

Next, solve for the required pump discharge pressure:

$$RPDP = NP + FL + AFL$$

where

$$NP = 50 \text{ psi}$$
$$FL = 46.22 \text{ psi}$$
$$AFL = 10 \text{ psi (flow is greater than 350 gpm)}$$

Then

$$RPDP = 50 + 46.22 + 10$$
$$= 106.22 \text{ psi}$$

Siamesed Lines into a Single Line

In Chapter 6, friction loss coefficients for single hose lines were introduced and used in solving problems. The coefficients in Table 7.1 are used for problems involving siamesed lines of equal length.

TABLE 7.1 ◆ Friction Loss Coefficients for Siamesed Lines of Equal Length	
Number of Hose Lines and their Diameter	*Coefficient*
Two 2½″	0.5
Three 2½″	0.22
Two 3″ with 2½″ couplings	0.2
One 3″ with 2½″ couplings, one 2½″	0.3
One 3″ with 3″ couplings, one 2½″	0.27
Two 2½″, one 3″ with 2½″ couplings	0.16
Two 3″ with 2½″ couplings, one 2½″	0.12
Standpipes	
4″	0.374
5″	0.126
6″	0.052

The solution to problems in this category is limited to those situations where the siamesed lines are of equal length. This type of layout is normally used where the friction loss in a single line on a long layout is too large. The friction loss in siamesed lines is approximately one-fourth that in a single line of the same size. The solution to problems in this category involves the following steps:

1. Determine the nozzle pressure on the attack line.
2. Solve for the flow in the attack line, if not given.
3. Determine the friction loss in the single line.
4. Determine the friction loss in the siamesed lines.
5. Determine the total friction loss in the hose in the layout.
6. Add the appliance friction loss if the total flow is 350 gpm or more.
7. Solve the problem for the required pump discharge pressure.

The formula to use is as follows:

Determining the Required Pump Discharge Pressure for Siamesed Lines Laid into a Single Line

$$RPDP = NP + FL + AFL$$

where
$RPDP$ = required pump discharge pressure
NP = nozzle pressure
FL = friction loss in the hose
AFL = appliance friction loss

QUESTION Two 700-foot lengths of 3-inch hose with 2½-inch couplings have been laid from Engine 5. The two lines have been siamesed into a single 2½-inch line 250 feet in length. The 2½-inch line is equipped with a selectable-gallonage nozzle set at 175 gpm with a 100 psi nozzle pressure. What is the required pump discharge pressure (Figure 7.9)?

A siamesed layout is one where the flow from two or more lines converge into a flow for one line. A plain siamese fitting is shown in Figure 7.10.

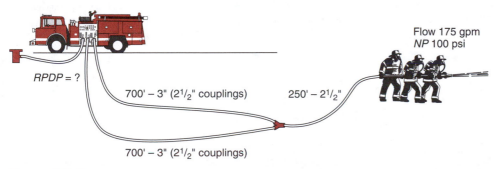

Flow 175 gpm
NP 100 psi

RPDP = ?

700' – 3" (2¹/₂" couplings)

250' – 2¹/₂"

700' – 3" (2¹/₂" couplings)

FIGURE 7.9 ◆

FIGURE 7.10 ◆ A plain 2-into-1 siamese. *Courtesy of Elkhart Brass Mfg. Company*

ANSWER First, determine the friction loss in the 2½-inch line:

$$FL = CQ^2L$$

where
$$C = 2$$
$$Q = 175/100 = 1.75$$
$$L = 250/100 = 2.5$$

Then
$$FL = (2)(1.75)(1.75)(2.5)$$
$$= 15.31 \text{ psi}$$

Next, solve for the friction loss in the siamesed lines:

$$FL = CQ^2L$$

where
$$C = .2$$
$$Q = 175/100 = 1.75$$
$$L = 700/100 = 7$$

Then
$$FL = (.2)(1.75)(1.75)(7)$$
$$= 4.29 \text{ psi}$$

Next, solve for the total friction loss in the hose. The total friction loss in the hose equals the friction loss in the siamese lines plus the friction loss in the attack line. Since the flow is less than 350 gpm, there is no friction loss added for the appliance friction loss:

$$TFL = 4.29 + 15.31$$
$$= 19.6 \text{ psi}$$

Now, solve for the required pump discharge pressure:

$$RPDP = NP + FL$$

where
$$NP = 100 \text{ psi}$$
$$FL = 19.6 \text{ psi}$$

Then $$RPDP = 100 + 19.6$$
$$= 119.6 \text{ psi}$$ ■

QUESTION What is the required pump discharge pressure if two 2½-inch lines, each 400 feet in length, are siamesed into a single 2-inch line 200 feet in length? The 2-inch line is equipped with an automatic nozzle with a flow of 200 gpm at a nozzle pressure of 75 psi (Figure 7.11)?

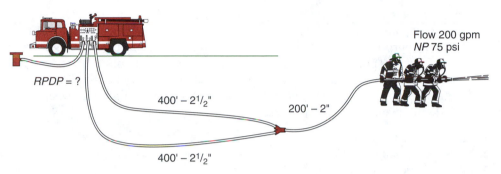

Flow 200 gpm
NP 75 psi

RPDP = ?

400' – 2½"

200' – 2"

400' – 2½"

FIGURE 7.11 ◆

ANSWER First, solve for the friction loss in the 2-inch line:

$$FL = CQ^2L$$

where $$C = 8$$
$$Q = 200/100 = 2$$
$$L = 200/100 = 2$$
Then $$FL = (8)(2)(2)(2)$$
$$= 64 \text{ psi}$$

Next, solve for the friction loss in the siamesed lines:

$$FL = CQ^2L$$

where $$C = .5$$
$$Q = 200/100 = 2$$
$$L = 400/100 = 4$$
Then $$FL = (.5)(2)(2)(4)$$
$$= 8 \text{ psi}$$

Next, solve for the total friction loss in the hose. The total friction loss equals the friction loss in the siamese lines plus the friction loss in the 2-inch line:

$$TFL = 8 + 64$$
$$= 72 \text{ psi}$$

Now, solve for the required pump discharge pressure. The required pump discharge pressure is equal to the nozzle pressure plus the friction loss in the hose. Since the flow is less than 350 gpm, there is no loss calculated for the appliance:

$$RPDP = NP + FL$$

where
$$NP = 75 \text{ psi}$$
$$FL = 72 \text{ psi}$$

Then
$$RPDP = 75 + 72$$
$$= 147 \text{ psi}$$

ELEVATION AND MASTER-STREAM PROBLEMS

Elevation problems were introduced in Chapter 1, where they were discussed from a technical standpoint. The pressure created by 1 foot of head was introduced as $P = .434H$. For the purpose of simplifying the solution to elevation problems, a slight error is injected: From this point on, the formula $P = .5H$ is used in the solution to problems. In addition, 5 pounds per floor is used for problems in multistory buildings.

Required pump discharge problems involving elevation and master streams can be divided into seven principal categories:

1. Lines laid uphill.
2. Lines laid downhill.
3. Lines laid into standpipe systems.
4. Lines laid into portable monitors.
5. Lines laid into deck guns.
6. Lines laid into ladder pipes.
7. Lines laid into aerial platforms or other elevated master-stream devices.

This is a good place to review part of Chapter 1.

Lines laid uphill or downhill are generally said to be laid up or down a given grade. For example, a line may be laid up a 10 percent grade or down a 12 percent grade. Grade is the vertical rise or decline of elevation for every 100 feet of hose laid out. "Up a 10 percent grade" refers to an elevation rise of 10 feet in 100 feet. "Down a 15 percent grade" means a decline of 15 feet in 100 feet. This concept is illustrated in Figure 1.30 on page 18.

Lines laid up or down grades result in back pressure against the pump or forward pressure aiding the pump. The amount of back pressure or forward pressure is determined by the resultant head. Head is the vertical distance from the pump to the place where the water is discharged. Head can be determined by the following formula:

$$H = GL$$

where
$$H = \text{resultant head}$$
$$G = \text{percent of grade}$$
$$L = \frac{\text{length of line}}{100}$$

Lines Laid Uphill

The solution to required pump discharge problems when lines are laid uphill involves the following:

1. Determine the nozzle pressure.
2. Solve for the discharge, if not given.
3. Solve for the head, if not given.
4. Solve for the friction loss in the hose.
5. Solve for the back pressure.

Add the nozzle pressure, the friction loss in the hose, and the back pressure to deter-mine the required pump discharge pressure. Use the following formula:

> ### Determining the Required Pump Discharge Pressure for Lines Laid up a Hill
>
> $$RPDP = NP + FL + BP$$
>
> where $RPDP$ = required pump discharge pressure
> NP = nozzle pressure
> FL = friction loss
> BP = back pressure

QUESTION A single 2½-inch line 300 feet in length has been taken off a pumper and ex-tended uphill to a point 30 feet above the pumper, where it is put into operation. The line is equipped with a 1-inch tip operating at 50 psi nozzle pressure. What is the required pump dis-charge pressure for this layout (Figure 7.12)?

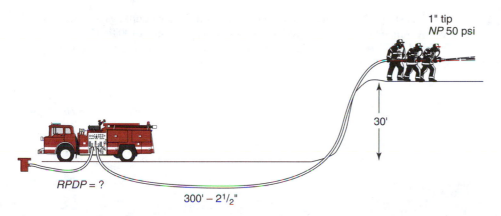

1" tip
NP 50 psi

30'

RPDP = ?

300' − 2½"

FIGURE 7.12 ◆

ANSWER First, solve for the discharge from the 1-inch tip:

$$\text{Discharge} = 29.7D^2\sqrt{P}$$

where $D = 1$
$P = 50 \text{ psi}$

Then $\text{Discharge} = (29.7)(1)^2\sqrt{50}$
$= (29.7)(1)(1)(7.07)$
$= 209.98 \text{ gpm}$

Next solve for the friction loss in the hose:

$$FL = CQ^2L$$

where $C = 2$
$Q = 209.98/100 = 2.10$
$L = 300/100 = 3$

Then
$$FL = (2)(2.1)(2.1)(3)$$
$$= 26.46 \text{ psi}$$

Next solve for the back pressure:
$$BP = .5H$$

where
$$H = 30$$
Then
$$BP = (.5)(30)$$
$$= 15 \text{ psi}$$

Now, solve for the required pump discharge pressure:
$$RPDP = NP + FL + BP$$

where
$$NP = 50 \text{ psi}$$
$$FL = 26.46 \text{ psi}$$
$$BP = 15 \text{ psi}$$
Then
$$RPDP = 50 + 26.46 + 15$$
$$= 91.46 \text{ psi}$$

QUESTION A single 3-inch line, 400 feet in length, with 3-inch couplings has been laid up a 12% grade. The 3-inch line has been wyed into two 1½-inch lines, each 100 feet in length. Both lines are equipped with ⅝-inch tips at 50 psi nozzle pressure. Solve for the required pump discharge pressure (Figure 7.13).

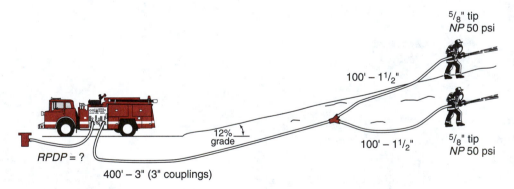

FIGURE 7.13 ◆

ANSWER First, determine the discharge from a ⅝-inch tip:
$$\text{Discharge} = 29.7D^2\sqrt{P}$$

where
$$D = \text{⅝-inch or .625}$$
$$P = 50 \text{ psi}$$
Then
$$\text{Discharge} = (29.7)(.625)^2(\sqrt{50})$$
$$= (29.7)(.625)(.625)(7.07)$$
$$= 82.02 \text{ gpm}$$

Next, solve for the friction loss in 100 feet of 1½-inch line:
$$FL = CQ^2L$$

where
$$C = 24$$
$$Q = 82.02/100 = .82$$
$$L = 100/100 = 1$$

Then
$$FL = (24)(.82)(.82)(1)$$
$$= 16.14 \text{ psi}$$

Next, solve for the friction loss in the 3-inch line:

$$FL = CQ^2L$$

where
$$C = .677$$
$$Q = 164.04/100 = 1.64$$
$$L = 400/100 = 4$$

Then
$$FL = (.677)(1.64)(1.64)(4)$$
$$= 7.28 \text{ psi}$$

Next, determine the total friction loss in the hose. The total friction loss equals the friction loss in the 3-inch line plus the friction loss in one of the 1½-inch lines:

$$TFL = 7.28 + 16.14$$
$$= 23.42 \text{ psi}$$

Next, solve for the back pressure. To solve for back pressure, it is first necessary to determine the head. The head can be determined by the formula

$$H = GL$$

where
$$G = 12$$
$$L = 500/100 = 5 \text{ (3-in. line + one wyed line)}$$

Then
$$H = (12)(5)$$
$$= 60 \text{ feet}$$

Next, solve for the back pressure:

$$BP = .5H$$
$$= (.5)(60)$$
$$= 30 \text{ psi}$$

Now, solve for the required pump discharge pressure:

$$RPDP = NP + FL + BP$$

where
$$NP = 50 \text{ psi}$$
$$FL = 23.42 \text{ psi}$$
$$BP = 30 \text{ psi}$$

Then
$$RPDP = 50 + 23.42 + 30$$
$$= 103.42 \text{ psi}$$

Lines Laid Downhill

The solution to required pump discharge problems when lines are laid downhill involves the following:

1. Determine the nozzle pressure.
2. Solve for the discharge, if not given.

3. Solve for the head, if not given.
4. Solve for the friction loss in the hose.
5. Solve for the forward pressure.

Add the nozzle pressure to the friction loss in the hose, then subtract the forward pressure to determine the required pump discharge pressure. Use the following formula:

Determining the Required Pump Discharge for Lines Laid Down a Hill

$$RPDP = NP + FL - FP$$

where $RPDP$ = required pump discharge pressure

NP = nozzle pressure

FL = friction in hose line

FP = forward pressure

QUESTION A 550-foot length of 1½-inch hose has been laid downhill from Engine 9. The line is equipped with a ⅝-inch tip at 50 psi nozzle pressure. What is the required pump discharge pressure (Figure 7.14)?

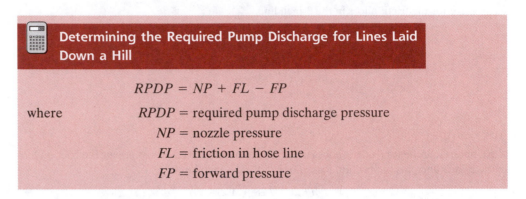

RPDP = ?

550' – 1¹⁄₂"

⁵⁄₈" tip
NP 50 psi

FIGURE 7.14 ◆

Note: When the grade is not given in a problem, consider it to be 10 percent.

ANSWER First, solve for the discharge:

$$Discharge = 29.7D^2\sqrt{P}$$

where D = ⅝ or .625

P = 50 psi

Then $Discharge = (29.7)(.625)^2\sqrt{50}$

$$= (29.7)(.625)(.625)(7.07)$$

$$= 82.02 \text{ gpm}$$

Next, solve for the friction loss in the hose:

$$FL = CQ^2L$$

where

$$C = 24$$
$$Q = 82.02/100 = .82$$
$$L = 550/100 = 5.5$$

Then

$$FL = (24)(.82)(.82)(5.5)$$
$$= 88.76 \text{ psi}$$

Next, solve for the forward pressure. First solve for the head:

$$H = GL$$

where

$$G = 10 \text{ percent}$$
$$L = 550/100 = 5.5$$

Then

$$H = (10)(5.5)$$
$$= 55 \text{ feet}$$

Next, solve for forward pressure:

$$FP = .5H$$
$$= (.5)(55)$$
$$= 27.5 \text{ psi}$$

Now solve for the required pump discharge pressure:

$$RPDP = NP + FL - FP$$

where

$$NP = 50 \text{ psi}$$
$$FL = 88.76 \text{ psi}$$
$$FP = 27.5 \text{ psi}$$

Then

$$RPDP = 50 + 88.76 - 27.5$$
$$= 111.26 \text{ psi}$$

QUESTION   The layout from Engine 1 is down a 15 percent grade. An 800-foot length of single 3-inch hose equipped with 2½-inch couplings has been reduced to 500 feet of 2½-inch hose. The 2½-inch line is equipped with a 1⅛-inch tip at 50 psi nozzle pressure. What is the required pump discharge pressure for this layout (Figure 7.15)?

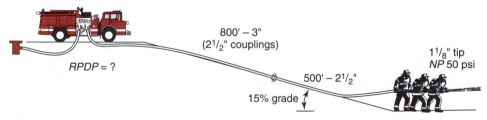

FIGURE 7.15 ◆

ANSWER   First, solve for the discharge from the 1⅛-inch tip:

$$\text{Discharge} = 29.7D^2\sqrt{P}$$

where $\qquad\qquad\qquad\qquad\qquad\qquad D = 1\tfrac{1}{8}$ or 1.125 inches

$$P = 50 \text{ psi}$$

Then $\qquad\qquad$ Discharge $= (29.7)(1.125)^2\sqrt{50}$

$$= (29.7)(1.125)(1.125)(7.07)$$

$$= 265.75 \text{ gpm}$$

Next, solve for the friction loss in the 2½-inch line:

$$FL = CQ^2L$$

where $\qquad\qquad\qquad\qquad\qquad\qquad C = 2$

$$Q = 265.75/100 = 2.66$$

$$L = 500/100 = 5$$

Then $\qquad\qquad\qquad\qquad\qquad FL = (2)(2.66)(2.66)(5)$

$$= 70.76 \text{ psi}$$

Next, solve for the friction loss in the 3-inch line:

$$FL = CQ^2L$$

where $\qquad\qquad\qquad\qquad\qquad\qquad C = .8$

$$Q = 265.75/100 = 2.66$$

$$L = 800/100 = 8$$

Then $\qquad\qquad\qquad\qquad\qquad FL = (.8)(2.66)(2.66)(8)$

$$= 45.28 \text{ psi}$$

Now, solve for the total friction loss in the hose:

$$TFL = FL \text{ in 3-inch line} + FL \text{ in 2½-inch line}$$

$$= 45.28 + 70.76$$

$$= 116.04 \text{ psi}$$

Next, solve for the forward pressure. First, determine the head:

$$H = GL$$

where $\qquad\qquad\qquad\qquad\qquad\qquad G = 15 \text{ percent}$

$$L = (800 + 500)/100$$

$$= 1300/100$$

$$= 13$$

Then $\qquad\qquad\qquad\qquad\qquad H = (15)(13)$

$$= 195 \text{ feet}$$

Then

$$\text{Forward pressure} = (.5)(H)$$

$$= (.5)(195)$$

$$= 97.5 \text{ psi}$$

Now, solve for the required pump discharge pressure:

$$RPDP = NP + FL - FP$$

$$= 50 + 116.04 - 97.5$$

$$= 68.54 \text{ psi}$$

Lines Laid into Standpipe Systems

Normally one of the factors that needs to be considered when lines are laid into standpipe systems is whether an allowance needs to be added for friction loss in the standpipe systems. This factor should be reviewed prior to the presentation of problems in standpipe systems.

QUESTION   Lines have been laid into a standpipe with water flowing at a rate of 500 gpm. The standpipe system is made of 6-inch piping. Several lines have been taken off and are working on the ninth and tenth floors. What pump discharge pressure is needed for the friction loss in the standpipe (Figure 7.16)?

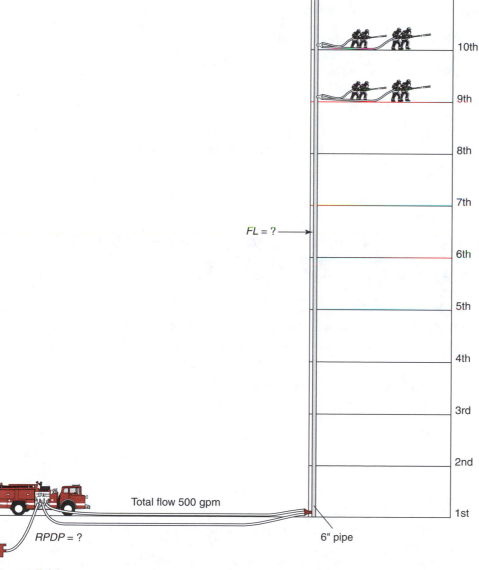

FIGURE 7.16 ◆

ANSWER

$$FL = CQ^2L$$

where

$$C = .052$$
$$Q = 500/100 = 5$$
$$L = 80/100 = .8$$

Then

$$FL = (.052)(5)(5)(.8)$$
$$= 1.04 \text{ psi}$$

The friction loss pressure of approximately 1 pound is insignificant to the overall problem. Therefore, the friction loss in standpipe piping where 6-inch piping is involved is not considered for problems in this book. However, the friction loss of 10 psi for the appliance is still added if the total flow of water in the standpipe is 350 gpm or more. ■

The solution to required pump discharge problems when lines are laid into standpipe systems involves the following:

1. Determine the nozzle pressure.
2. Solve for the discharge, if not given.
3. Solve for the back pressure.
4. Solve for the friction loss in the hose.
5. Consider whether appliance friction loss is to be added.

Add the nozzle pressure, the friction loss in the hose, the back pressure, and perhaps the appliance friction loss to determine the required pump discharge pressure. Use the following formula:

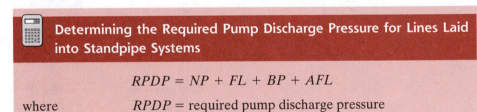

Determining the Required Pump Discharge Pressure for Lines Laid into Standpipe Systems

$$RPDP = NP + FL + BP + AFL$$

where

$$RPDP = \text{required pump discharge pressure}$$
$$NP = \text{nozzle pressure}$$
$$FL = \text{friction loss in the hose}$$
$$BP = \text{back pressure}$$
$$AFL = \text{appliance friction loss}$$

QUESTION A 1¼-inch tip attached to 100 feet of 2½-inch hose is working on the sixth floor of a ten-story building. A single 3-inch line with 2½-inch couplings, 200 feet in length, has been taken off Engine 7 and is supplying the standpipe. What is the required pump discharge pressure if the nozzle pressure is 50 psi (Figure 7.17)?

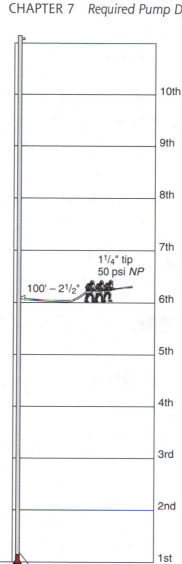

10th

9th

8th

7th

1¼" tip
50 psi *NP*

100' – 2½"   6th

5th

4th

3rd

2nd

200' – 3" (2½" couplings)

RPDP = ?   6" pipe   1st

FIGURE 7.17 ◆

ANSWER First, solve for the discharge from the 1¼-inch tip:

$$\text{Discharge} = 29.7D^2\sqrt{P}$$

where
$$D = 1¼ \text{ or } 1.25 \text{ inches}$$
$$P = 50 \text{ psi}$$

Then
$$\text{Discharge} = (29.7)(1.25)^2\sqrt{50}$$
$$= (29.7)(1.25)(1.25)(7.07)$$
$$= 328.09 \text{ gpm}$$

Next, solve for the back pressure. The line is working on the sixth floor, which is five floors above ground level. The head in the problem is then given by

$$H = (5)(10)$$
$$= 50 \text{ feet}$$

The back pressure is given by

$$BP = .5H$$
$$= (.5)(50)$$
$$= 25 \text{ psi}$$

Next, solve for the friction loss in the hose. The friction loss in the 100 feet of 2½-inch hose is given by

$$FL = CQ^2L$$

where
$$C = 2$$
$$Q = 328.09/100 = 3.28$$
$$L = 100/100 = 1$$

Then
$$FL = (2)(3.28)(3.28)(1)$$
$$= 21.52 \text{ psi}$$

The friction loss in the 3-inch line is given by

$$FL = CQ^2L$$

where
$$C = .8$$
$$Q = 328.09/100 = 3.28$$
$$L = 200/100 = 2$$

Then
$$FL = (.8)(3.28)(3.28)(2)$$
$$= 17.21 \text{ psi}$$

Next, solve for the total friction loss in the hose:

$$TFL = FL \text{ in 2½-inch line} + FL \text{ in 3-inch line}$$
$$= 21.52 + 17.21$$
$$= 38.73 \text{ psi}$$

Since the flow is less than 350 psi, there is no appliance friction loss.
  Next, determine the required pump discharge pressure:

$$RPDP = NP + FL + BP$$
$$= 50 + 38.73 + 25$$
$$= 113.73 \text{ psi}$$

■

QUESTION A 500-foot length of 3½-inch hose is laid into the standpipe of a twelve-story building. A 1¾-inch line 150 feet in length is taken off the standpipe on the sixth floor and advanced up the stairway to the eighth floor, where it is working on a well-involved fire in an office. The line is equipped with a dual-pressure automatic nozzle set to provide a nozzle pressure of 50 psi. Water is flowing at 225 gpm. What is the required pump discharge pressure for this layout (Figure 7.18)?

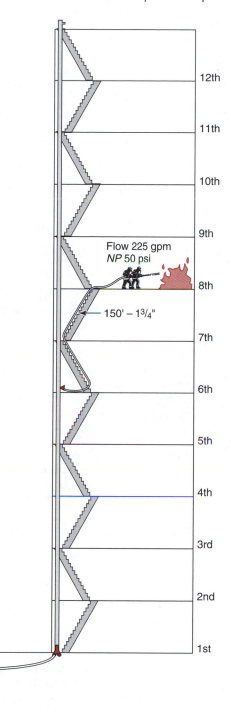

Flow 225 gpm
NP 50 psi

150' – 1³/₄"

500' – 3¹/₂"

RPDP = ?

FIGURE 7.18 ◆

ANSWER First, determine the back pressure. The attack line is working on the eighth floor, which is seven floors above the street level. As there is 10 feet per floor, the height above ground level is (10)(7) = 70 feet. The back pressure is given by

$$BP = .5H$$

where $\qquad H = 70$ feet

Then $\qquad BP = (.5)(70)$

$$= 35 \text{ psi}$$

Next, solve for the friction loss in the hose. First, determine the friction loss in the 1¾-inch line:

$$FL = CQ^2L$$

where $\qquad C = 15.5$

$$Q = 225/100 = 2.25$$

$$L = 150/100 = 1.5$$

Then $\qquad FL = (15.5)(2.25)(2.25)(1.5)$

$$= 117.70 \text{ psi}$$

Next, determine the friction loss in the 3½-inch hose:

$$FL = CQ^2L$$

where $\qquad C = .34$

$$Q = 225/100 = 2.25$$

$$L = 500/100 = 5$$

Then $\qquad FL = (.34)(2.25)(2.25)(5)$

$$= 8.61 \text{ psi}$$

Next, determine the total friction loss in the hose:

$$TFL = FL \text{ in } 1\text{¾-inch line} + FL \text{ in } 3\text{½-inch line}$$

$$= 117.7 + 8.61$$

$$= 126.31 \text{ psi}$$

Now, solve for the required pump discharge pressure. Since the flow is less than 350 gpm, no loss for friction loss in the appliance is required:

$$TFL = NP + FL + BP$$

$$= 50 + 126.31 + 35$$

$$= 211.31 \text{ psi} \qquad ■$$

Lines Laid into Portable Monitors

A portable monitor is one of several different types of appliances used to provide master streams. A master stream is defined as any stream that is too large to be hand held, or a stream producing 350 gpm or more. Either definition conveys the fact that this is a stream producing a large quantity of water. Master streams are recognized as the most potent weapon of any fire department.

The portable monitor is the only heavy-stream appliance capable of providing a heavy stream inside a building. Portable monitors are produced by many different manufacturers and are available in various shapes and sizes. Most are designed to be light in weight due to the necessity of carrying them to the location where they will be set up and used. Some are light enough to be carried by one firefighter. Some monitors are available with smooth-bore tips, generally of the stacked-tip variety; most, however, are equipped with fog nozzles of various sizes and types. The monitor may be designed with a single inlet, a dual inlet, or a triple inlet. Basically, the hydraulics for all types is the same (Figure 7.19).

FIGURE 7.19 ◆ A portable monitor manned by a single firefighter. *Courtesy of Task Force Tips*

One of the disadvantages of portable monitors is that they are limited in the amount of water that can be supplied. The maximum amount is generally set at approximately 1000 gpm. This is due to the tremendous back pressure exerted by the stream when large flows are used. The back pressure from a smooth-bore tip is greater than that produced by a fog stream with the same amount of water flowing.

The solution to required pump discharge problems when lines are laid to supply portable monitors involves the following:

1. Determine the nozzle pressure.
2. Solve for the discharge, if not given.
3. Solve for the friction loss in the hose.
4. Since all portable monitors are designed as master-stream appliances, add 25 psi for the appliance friction loss.

Add the nozzle pressure, the friction loss in the hose, and 25 psi for the appliance friction loss to determine the required pump discharge pressure. Use the following formula:

Determining the Required Pump Discharge Pressure for Lines Laid into Portable Monitors

$$RPDP = NP + FL + AFL$$

where

$RPDP$ = required pump discharge pressure

NP = nozzle pressure

FL = friction loss in the hose

AFL = appliance friction loss

QUESTION A portable monitor is being supplied by three 2½-inch lines each 350 feet in length. The monitor is equipped with a 1¾-inch smooth-bore tip operating at 80 psi nozzle pressure. What is the required pump discharge pressure for this configuration (Figure 7.20)?

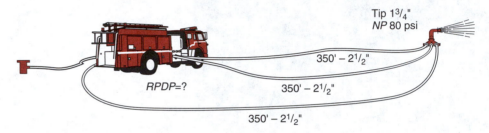

Tip 1³/₄"
NP 80 psi

350' – 2¹/₂"

RPDP=?

350' – 2¹/₂"

350' – 2¹/₂"

FIGURE 7.20 ◆

ANSWER First, solve for the discharge from the 1¾-inch tip:

$$\text{Discharge} = 29.7 D^2 \sqrt{P}$$

where

D = 1¾ or 1.75 inches

P = 80 psi

Then

$$\text{Discharge} = (29.7)(1.75)^2 \sqrt{80}$$
$$= (29.7)(1.75)(1.75)(8.94)$$
$$= 813.15 \text{ gpm}$$

Next, solve for the friction loss in the hose:

$$FL = CQ^2 L$$

where

C = .22

Q = 813.15/100 = 8.13

Then

$$FL = (.22)(8.13)(8.13)(3.5)$$
$$= 50.89 \text{ psi}$$

Now, solve for the required pump discharge pressure:

$$RPDP = NP + FL + AFL$$
$$= 80 + 50.89 + 25$$
$$= 155.89 \text{ psi}$$

■

QUESTION A dual-inlet portable monitor is supplied by one 3-inch line with 2½-inch couplings and one 2½-inch line. Each line is 250 feet in length. The monitor is equipped with an automatic nozzle with water flowing at 700 gpm. The nozzle pressure is 100 psi. What is the required pump discharge pressure (Figure 7.21)?

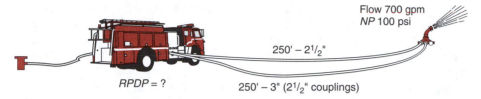

Flow 700 gpm
NP 100 psi

250' – 2¹/₂"

RPDP = ?   250' – 3" (2¹/₂" couplings)

FIGURE 7.21 ◆

ANSWER First, solve for the friction loss in the hose:

$$FL = CQ^2L$$

where
$$C = .3$$
$$Q = 700/100 = 7$$
$$L = 250/100 = 2.5$$

Then
$$FL = (.3)(7)(7)(2.5)$$
$$= 36.75 \text{ psi}$$

Now, solve for the required pump discharge pressure:

$$RPDP = NP + FL + AFL$$
$$= 100 + 36.75 + 25$$
$$= 161.75 \text{ psi}$$

FIGURE 7.22 ◆ An electric deck gun in the typical stowed position. *Courtesy of Akron Brass Company*

Lines Laid into Deck Guns

Deck guns are similar to portable monitors in designs and styles (Figures 7.22 and 7.23). The primary difference is that deck guns are secured in place. This gives them the ability to withstand a much greater back pressure than portable monitors. In fact, some deck guns can provide approximately twice the flow of portable units. In addition, some deck guns can be removed from their mounting and used as a portable monitor, although such units have a limited flow capacity when removed from their secure position.

Deck guns are supplied by two different methods. One variety uses some type of preconnected appliance. These may be supplied directly by the pumper to which they are attached or by a separate apparatus. Another type is supplied directly from the pump through a prepiped arrangement on the apparatus (Figure 7.24).

Although deck guns are capable of supplying a greater flow than portable monitors, there are some safety and operational concerns associated with their use. The guns are usually set 9 to 10 feet off the ground. If manually operated, this requires that a firefighter climb up to that level over a number of potential hazards along the way and operate from a relatively small platform. This is always hazardous,

FIGURE 7.23 ◆ An electric deck gun in the operational position. *Courtesy of Akron Brass Company*

FIGURE 7.24 ◆ A deck gun operating by remote control. This remote operation eliminates the need to have a firefighter on the deck to operate the monitor. *Courtesy of Akron Brass Company*

particularly if the firefighter is equipped with breathing apparatus. The hazard becomes greater in weather where ice has formed on various parts of the apparatus. To alleviate this problem, some deck guns are equipped with electric or other remote control devices.

The solution to required pump discharge problems when lines are laid to supply deck guns involves the following:

1. Determine the nozzle pressure.
2. Solve for the discharge, if not given.
3. Solve for the friction loss in the hose.
4. Allow 5 psi for the back pressure.
5. Since a deck gun is a master-stream appliance, add 25 psi for the appliance friction loss.

Add the nozzle pressure, the friction loss in the hose, 5 psi for the back pressure, and 25 psi for the appliance friction loss to determine the required pump discharge pressure. Use the following formula:

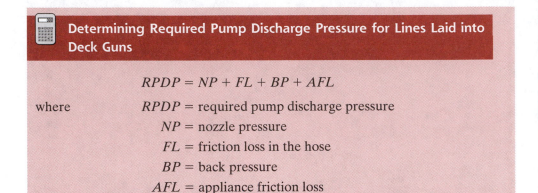

Determining Required Pump Discharge Pressure for Lines Laid into Deck Guns

$$RPDP = NP + FL + BP + AFL$$

where
$$RPDP = \text{required pump discharge pressure}$$
$$NP = \text{nozzle pressure}$$
$$FL = \text{friction loss in the hose}$$
$$BP = \text{back pressure}$$
$$AFL = \text{appliance friction loss}$$

QUESTION A deck gun mounted 9 feet off the ground is supplied by the pump to which it is attached through two 2½-inch preconnected lines each 25 feet in length. The gun is equipped with a 2-inch smooth-bore tip at 80 psi nozzle pressure. At what pressure should the pump operate to supply the deck gun (Figure 7.25)?

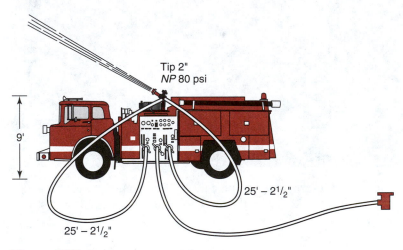

Tip 2"
NP 80 psi

9'

25' − 2¹/₂"

25' − 2¹/₂"

FIGURE 7.25 ◆

ANSWER First, solve for the discharge from the 2-inch tip:

$$\text{Discharge} = 29.7D^2\sqrt{P}$$

where
$$D = 2 \text{ inches}$$
$$P = 80 \text{ psi}$$

Then
$$\text{Discharge} = (29.7)(2)^2\sqrt{80}$$
$$= (29.7)(2)(2)(8.94)$$
$$= 1062.07 \text{ gpm}$$

Next, solve for the friction loss in the hose:

$$FL = CQ^2L$$

where

$$C = .5$$
$$Q = 1062.01/100 = 10.62$$
$$L = 25/100 = .25$$

Then

$$FL = (.5)(10.62)(10.62)(.25)$$
$$= 14.1 \text{ psi}$$

The back pressure is given by

$$BP = .5H$$

where

$$H = 9 \text{ feet}$$

Then

$$BP = (.5)(9)$$
$$= 4.5 \text{ psi}$$

Now, solve for the required pump discharge pressure:

$$RPDP = NP + FL + BP + AFL$$
$$= 80 + 14.1 + 4.5 + 25$$
$$= 123.6 \text{ psi}$$ ■

QUESTION A deck gun on Engine 7 is equipped with an automatic nozzle rated at a flow range of 150 to 1250 gpm at a constant nozzle pressure of 100 psi. The deck gun is 10 feet off the ground. The gun is supplied through two 2½-inch lines and one 3-inch line with 2½-inch couplings. All the supply lines are 300 feet in length. What is the required pump discharge pressure for this layout (Figure 7.26)?

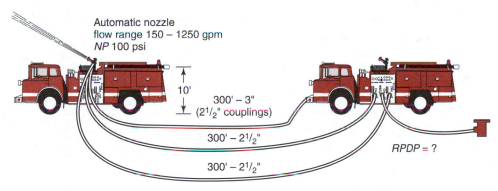

FIGURE 7.26 ◆

Note: When pumping to an automatic nozzle with a given flow range, if the flow is available from the water source, it is good practice to pump to the highest gpm flow. This provides some flexibility in the event handlines are later taken off the pumper.

ANSWER First, solve for the friction loss in the supply hose assembly:

$$FL = CQ^2L$$

where

$$C = .16$$
$$Q = 1250/100 = 12.5$$
$$L = 300/100 = 3$$

Then

$$FL = (.16)(12.5)(12.5)(3)$$
$$= 75 \text{ psi}$$

Back pressure is given by

$$BP = .5H$$

where $H = 10$ feet

Then $BP = (.5)(10)$

$$= 5 \text{ psi}$$

Now, solve for the required pump discharge pressure:

$$RPDP = NP + FL + BP + AFL$$

$$= 100 + 75 + 5 + 25$$

$$= 205 \text{ psi} \qquad ■$$

Lines Laid into Ladder Pipes

There are two general types of ladder pipes in common use. One type is permanently attached to the ladder and is prepiped with the fittings for supplying the unit located at or near the base of the ladder. The second type is detachable and is supplied by fire hose, usually a single 3-inch line (Figure 7.27). The prepiped units are usually operated from the turntable or pump panel area by some type of remote control unit. Some detachable units may also be operated by a remote control unit. Others are controlled from the ground by ropes. For safety, it is good practice to never operate a ladder pipe manually by a firefighter working on the ladder. Regardless of the type of ladder pipe, the hydraulic calculations necessary to supply the unit are the same.

The solution to required pump discharge problems when lines are laid to supply ladder pipes involves the following:

1. Determine the nozzle pressure.
2. Solve for the discharge, if not given.
3. Solve for the friction loss in the hose.
4. Solve for the back pressure.
5. Add 25 psi for the appliance friction loss, which is standard for master-stream appliances.
6. Prepiped units also receive an allowance for the friction loss in the piping. The amount of friction loss is usually provided by the factory that made the apparatus.

Use the following formula:

Determining the Required Pump Discharge Pressure for Lines Laid into Ladder Pipes

$$RPDP = NP + FL + BP + AFL + \text{piping allowance}$$

where $RPDP$ = required pump discharge pressure

NP = nozzle pressure

FL = friction loss in the hose

BP = back pressure

AFL = appliance friction loss

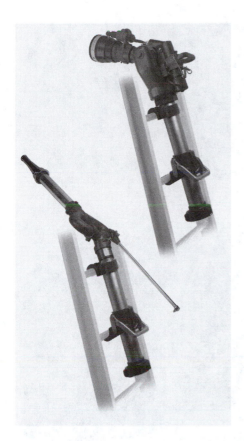

FIGURE 7.27 ◆ Two types of ladder pipes. The one on the left is controlled by ropes from the ground. The one on the right is electronically remote controlled. *Courtesy of Akron Brass Company*

QUESTION A prepiped ladder pipe is operating at a height of 75 feet above ground level. It is equipped with a smooth-bore 1½-inch nozzle operating at 80 psi nozzle pressure. The unit is being supplied by two 3-inch lines equipped with 2½-inch couplings. Each line is 450 feet in length. The friction loss in the piping is 35 psi. What pressure is required at the pump supplying the hose lines (Figure 7.28)?

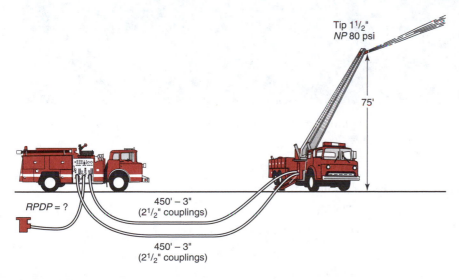

FIGURE 7.28 ◆

FIGURE 7.29 ◆ Elevated platforms are very effective on large fires. *Courtesy of Pierce Manufacturing, Inc.*

ANSWER First, solve for the discharge from the 1½-inch tip:

$$\text{Discharge} = 29.7D^2\sqrt{P}$$

where

$$D = 1½ \text{ or } 1.5 \text{ inches}$$
$$P = 80 \text{ psi}$$

Then

$$\text{Discharge} = (29.7)(1.5)^2\sqrt{80}$$

$$= (29.7)(1.5)(1.5)(8.94)$$
$$= 597.42 \text{ gpm}$$

Next, solve for the friction loss in the hose:

$$FL = CQ^2L$$

where
$$C = .2$$
$$Q = 597.42/100 = 5.97$$
$$L = 450/100 = 4.5$$

Then
$$FL = (.2)(5.97)(5.97)(4.5)$$
$$= 32.08 \text{ psi}$$

Next, solve for the back pressure:

$$BP = .5H$$

where
$$H = 75 \text{ feet}$$
Then
$$BP = (.5)(75)$$
$$= 37.5 \text{ psi}$$

Now, solve for the required pump discharge pressure:

$$RPDP = NP + FL + BP + AFL + \text{piping allowance}$$
$$= 80 + 32.08 + 37.5 + 25 + 35$$
$$= 209.58 \text{ psi}$$

Lines Laid into Aerial Platforms

The solution to required pump discharge problems when lines are laid to supply aerial platforms involves the following (Figure 7.29):

1. Determine the nozzle pressure.
2. Solve for the discharge, if not given.
3. Solve for the friction loss in the hose.
4. Solve for the back pressure.
5. Add 25 psi for the appliance friction loss, which is standard for master-stream appliances.
6. Add for prepiping.

Use the following formula:

Determining the Required Pump Discharge Pressure for Lines Laid into Aerial Platforms

$$RPDP = NP + FL + BP + AFL + \text{prepiping}$$

where
$$RPDP = \text{required pump discharge pressure}$$
$$NP = \text{nozzle pressure}$$
$$FL = \text{friction loss in the hose}$$
$$BP = \text{back pressure}$$
$$AFL = \text{appliance friction loss}$$

QUESTION An aerial platform is equipped with an automatic nozzle having a flow range of 150 to 1200 gpm and is operating at 100 psi nozzle pressure. It is a prepiped system raised to a height of 80 feet. The friction loss in the prepiping is 40 psi. The plan is to supply the unit by one 5-inch line 300 feet in length. What is the required pump discharge pressure for this layout (Figure 7.30)?

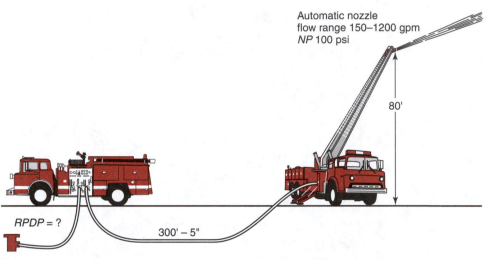

Automatic nozzle
flow range 150–1200 gpm
NP 100 psi

80'

RPDP = ?

300' – 5"

FIGURE 7.30 ◆

Note: The aerial platform is perhaps the most stable of all the master-stream appliances. A secure platform is provided for the firefighter operating the unit. However, some units are also designed to be operated remotely from the turntable or pump panel area. The stability of the unit allows greater amounts of water to be discharged than in other master-stream appliances. Some platforms are equipped with two master-stream turrets (see Figure 4.1), doubling the amount of water that can be discharged onto the fire.

ANSWER First, solve for the friction loss in the hose layout.

$$FL = CQ^2L$$

where

$$C = .08$$
$$Q = 1200/100 = 12 \text{ (pump to highest setting)}$$
$$L = 300/100 = 3$$

Then

$$FL = (.08)(12)(12)(3)$$
$$= 34.56 \text{ psi}$$

Next, solve for the back pressure:

$$BP = .5H$$

where

$$H = 80 \text{ feet}$$

Then

$$BP = (.5)(80)$$
$$= 40 \text{ psi}$$

Now, solve for the required pump discharge pressure:

$$RPDP = NP + FL + BP + AFL + \text{piping allowance}$$

$$= 100 + 34.56 + 40 + 25 + 40$$
$$= 239.56 \text{ psi}$$

Note: This pressure requirement is above the 185-psi maximum for pumping to 5-inch lines. A second pumper should be provided to assist with the supply, or rubber-lined supply lines can be used in lieu of the 5-inch line. Two 3-inch lines would do the job.   ■

QUESTION An elevated master-stream appliance is being supplied by 450 feet of 4-inch hose. It is raised to a height of 40 feet. The appliance is equipped with a 1⅜-inch smooth-bore tip at 80 psi nozzle pressure (Figure 7.31). What is the required pump discharge pressure for this configuration (do not consider any friction loss in the piping for this problem)?

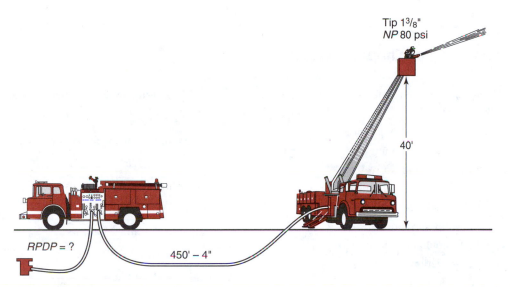

Tip 1³/₈"
NP 80 psi

40'

RPDP = ?

450' – 4"

FIGURE 7.31 ◆

ANSWER First, solve for the discharge in the 1⅜-inch nozzle:

$$\text{Discharge} = 29.7 D^2 \sqrt{P}$$

where
$$D = 1\tfrac{3}{8} \text{ or } 1.375 \text{ inches}$$
$$P = 80 \text{ psi}$$

Then
$$\text{Discharge} = (29.7)(1.375)(1.375)(8.94)$$
$$= 502 \text{ gpm}$$

Next, solve for the friction loss in the 4-inch line:

$$FL = CQ^2 L$$

where
$$C = .2$$
$$Q = 502/100 = 5.02$$
$$L = 450/100 = 4.5$$

Then
$$FL = (.2)(5.02)(5.02)(4.5)$$
$$= 22.68 \text{ psi}$$

Next, solve for the back pressure:

$$BP = .5H$$

where $\qquad$ $H = 40$ feet

Then $\qquad$ $BP = (.5)(40)$

$$= 20 \text{ psi}$$

Now, solve for the required pump discharge pressure:

$$RPDP = NP + FL + BP + AFL$$
$$= 80 + 22.68 + 20 + 25$$
$$= 147.68 \text{ psi} \qquad \blacksquare$$

■■■

Summary of Chapter Formulas

Terms used in the formulas have the following meanings:

PDP = pump discharge pressure

$RPDP$ = required pump discharge pressure

NP = nozzle pressure

FL = friction loss in the hose

BP = back pressure

FP = forward pressure

AFL = appliance friction loss

C = friction loss constant

Q = flow/100

L = length of line/100

Single Lines Laid at Ground Level (no appliances involved)

$$RPDP = NP + FL$$

A Single Line Wyed into Two Lines

$$RPDP = NP + FL + AFL$$

Siamesed Lines into a Single Line

$$RPDP = NP + FL + AFL$$

Lines Laid Uphill (no appliances involved)

$$RPDP = NP + FL + BP$$

Lines Laid Downhill (no appliances involved)

$$RPDP = NP + FL - FP$$

Lines Laid into Standpipe Systems

$$RPDP = NP + FL + BP + AFL$$

Lines Laid into Portable Monitors

$$RPDP = NP + FL + AFL$$

Lines Laid into Deck Guns

$$RPDP = NP + FL + BP + AFL$$

Lines Laid into Ladder Pipes

$$RPDP = NP + FL + BP$$
$$+ AFL + \text{piping allowance}$$

Lines Laid into Aerial Platforms

$$RPDP = NP + FL + BP$$
$$+ AFL + \text{prepiping}$$

■■■

Review Questions

1. What is the simplest pumping situation?
2. What factors must be considered when solving problems for the required pump discharge pressure?
3. How much should be allowed for the friction loss in appliances?
4. When is forward pressure produced?

5. What is included in the *TPL* in the formula *PDP = NP + TPL*?

6. What are the two reasons for expanding the *TPL* for use in this book?

7. What is the formula for solving the required pump discharge pressure?

8. How much loss is allowed for each foot of back pressure?

9. What loss is allowed for the appliance friction loss?

10. To what does appliance friction loss refer?

11. What is the appliance friction loss for master-stream appliances?

12. What factors are considered when lines are laid at ground level with no appliances involved?

13. How is the required pump discharge pressure handled when several lines are taken off a pumper?

14. What are the standard nozzle pressures for smooth-bore tips?

15. What are the standard nozzle pressures for fog nozzles?

16. What are the three principal categories for lines laid at ground level?

17. What is the formula for required pump discharge pressure when single lines are laid at ground level with no appliances involved?

18. What formula should be used when a single line is wyed into two lines at ground level with no appliances involved?

19. How much pressure loss is allowed per floor for back pressure?

20. What formula should be used when single lines are laid uphill if no appliances are involved?

21. What pressure loss should be allowed for the standpipe piping when lines are laid into the standpipe system?

22. What formula should be used to solve for the pump discharge pressure when lines are laid into portable monitors at ground level?

23. What pump discharge pressure formula should be used when lines are laid into deck guns?

24. What are the two types of ladder pipes?

25. How are ladder pipes controlled from the ground level?

Test Seven

For all of the following problems, solve for the required pump discharge pressure. When pumping to an automatic nozzle, pump to the highest gallonage of the flow range.

1. A 150 foot, ¾-inch booster line has been taken off Engine 3 for the purpose of extinguishing a car fire. Water through the nozzle is flowing at 20 gpm at a pressure of 100 psi.

2. Firefighters from Engine 2 have advanced a 150-foot section of 1-inch booster line to attack a small shed fire. The line is equipped with a 30-gpm fixed-gallonage fog nozzle rated at 50 psi nozzle pressure.

3. A 200-foot, 1¾-inch attack line with a fixed-gallonage nozzle of 200 gpm at 75 psi has been advanced from Engine 1 for use on a house fire.

4. Consider a layout of 400 feet of 2½-inch hose reduced to 150 feet of 1½-inch hose. The 1½-inch line is equipped with a ¾-inch smooth-bore nozzle at 50 psi nozzle pressure.

5. The 350-foot length of 2½-inch hose taken off Engine 6 has been advanced uphill. The 1⅛-inch smooth-bore tip is being worked at a point 85 feet above the pump. The nozzle pressure is 50 psi.

6. A 300-foot length of 1½-inch hose has been advanced up a 12 percent grade. It is equipped with an automatic nozzle having a flow range from 20 to 125 gpm at 100 psi nozzle pressure.

7. Two firefighters from Engine 5 have laid 400 feet of 2-inch hose down a 15 percent grade to work on a garage fire. The line is equipped with a 1-inch smooth-bore tip at 50 psi nozzle pressure.

8. A 400-foot length of 3-inch hose with 2½-inch couplings has been taken downhill and reduced to 100 feet of 2½-inch hose with a 250-gpm fixed-gallonage

nozzle working at 100 psi nozzle pressure. The nozzle is being used at a location that is 120 feet below the pump.

9. The crew from Engine 8 has taken off 500 feet of 3-inch hose with 3-inch couplings. They have wyed it into two 1½-inch attack lines each 150 feet in length. The attack lines are equipped with ½-inch tips at 50 psi nozzle pressure.

10. Two 600-foot lengths of 2½-inch hose have been siamesed into a single 2½-inch line 150 feet in length. The line is equipped with a 325-gpm fixed-gallonage 100-psi nozzle.

11. A crew member from Engine 1 has carried and set up a portable monitor with a 1½-inch smooth-bore tip. Other members have supplied the monitor with siamesed lines. One of the lines is a 3-inch line with 2½-inch couplings. The other is a 2½-inch line. Both lines are 650 feet in length. The nozzle pressure on the 1½-inch tip is 80 psi.

12. The deck gun on Engine 4 is located 10 feet above the ground. It is equipped with an automatic nozzle having a flow range of 150 to 1250 gpm at 100 psi nozzle pressure. The deck gun is supplied by two 3-inch lines with 2½-inch couplings and one 2½-inch line. All lines are 400 feet in length.

13. The deck gun on Engine 2 is prepiped from the pump. The gun is located 10 feet off the ground. Water is flowing at 1500 gpm at 100 psi nozzle pressure.

14. A detachable ladder pipe is attached to 100 feet of 3-inch hose with 2½-inch couplings. The ladder pipe has been raised to a height of 60 feet. The nozzle on the ladder pipe is a 1⅜-inch smooth-bore tip working at 80 psi nozzle pressure. Three 2½-inch lines each 500 feet in length have been attached to the 3-inch feeder line.

15. The 2-inch tip on an elevated platform has a nozzle pressure of 80 psi. The platform is working at a height of 80 feet. The fixed piping has been fed by 600 feet of 5-inch hose. Add 20 psi for the fixed piping system.

Unusual and Complex Problems

8 CHAPTER

Objectives

Upon completing this chapter, the reader should:

- Understand the functions of an attack pumper.
- Understand how some appliances can assume the role of an attack pumper.
- Be able to analyze and solve a variety of types of unusual hydraulic problems.
- Be able to analyze and solve a variety of types of complex hydraulic problems.
- Be able to solve complex siamesed problems.
- Be able to change hose other than 2½-inch hose into 2½-inch-hose equivalents.
- Be able to change lines other than 2½-inch siamesed lines into 2½-inch-siamesed equivalents.
- Be able to set up equal-length 2½-inch siamese layouts.

Almost every fire officer is aware of the premise that approximately 90 percent of the fires cause 10 percent of the fire loss, and approximately 10 percent of the fires cause 90 percent of the loss. However, a number of fire officers will disagree, taking the stand that in some cities the number of small fires is greater than 90 percent. In fact, in some cities, there are firefighters who have been on the department for four or five years who have never seen a 2½-inch or larger line laid at a fire. No doubt, there are a number of seasoned pump operators who will some day retire without having once pushed their pump to its maximum capability. Despite this, fire chiefs will continue to purchase 1500-gpm pumpers and master-stream appliances for use on that "big one."

When looking at the big picture, it is easy to see why some pump operators become complacent in their job. They are used to the fact that small fires are the norm. It is difficult for them to study and train for what they think will never happen. What a mistake! Preparing to handle unusual pumping situations is why they were hired or became a volunteer pump operator and why they are getting paid, if they are. It is the wise pump operator, or an individual who desires to become a pump operator, who thinks about and prepares for the unusual situations.

The objective of this chapter is to combine the principles introduced in Chapters 5 through 7 and build a foundation for unusual pumping situations. Several new principles are introduced that will enable the student to solve some problems that cannot be solved by the more generally used formulas and coefficients. The principles are introduced through problems, with their solution explained. However, some of the problems presented will probably never be found on the fire ground. They are presented for the objective of introducing a new concept and establishing a foundation for their use.

Tables 6.1 and 7.1 should be consulted when reviewing problems.

◆ UNDERSTANDING THE ATTACK PUMPER

Establishing a foundation for solving unusual and complex hydraulic problems requires a basic understanding of an attack pumper. An attack pumper may be defined as the water source at a pressure needed to properly supply the nozzles used at a fire. It can be thought of as a large pressurized storage tank equipped with a number of supply outlets (Figure 8.1).

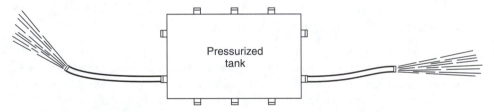

Pressurized tank

FIGURE 8.1 ◆

Whether the source is a pumper or a pressurized storage tank, the solution to adequately supplying each of the lines is to establish a pressure of sufficient magnitude to provide the needs of the line requiring the most pressure, then feathering (partially closing) the outlets to the other lines.

Regardless of whether the source is a pumper or a pressurized storage tank, a constant source of water supply must be maintained in order to supply the requirements for all of the attack lines. This supply to an attack pumper can be provided in several ways.

The most common method is to connect the pumper directly to a hydrant through a standard-size suction hose (Figure 8.2).

Standard soft suction

FIGURE 8.2 ◆ Most common method of supplying an attack pumper.

Used more frequently than in the past, a pumper may be supplied through a long supply line of large-diameter hose (LDH) from a hydrant (Figure 8.3). Alternatively, a **supply pumper** may be stationed at the supply source and provide water to the attack pumper through a long lay of LDH (Figure 8.4).

FIGURE 8.3 ◆ Using LDH as a long suction hose.

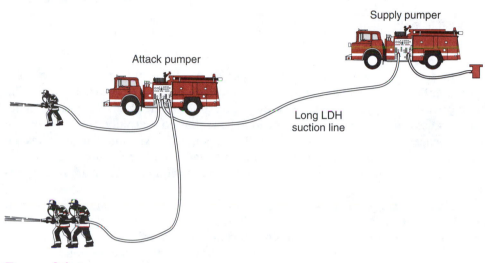

FIGURE 8.4 ◆

When supplied from a hydrant, either through a short, soft suction or a long LDH, the hydrant can be considered as a supply pumper (Figure 8.5).

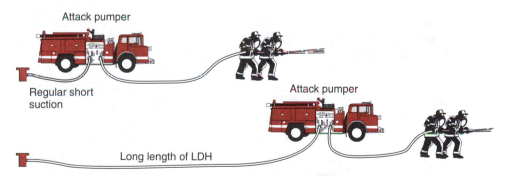

FIGURE 8.5 ◆ In both of these situations, the hydrant assumes the position of a supply pumper.

On some layouts, for the purpose of solving the hydraulics of a problem, the attack pumper assumes the role of the supply pumper, and appliances such as wyes, water thiefs, and manifolds assume the role of the attack pumper (Figure 8.6).

In each of these instances, the lines coming off the appliance can be solved in the same manner as if the lines were coming directly off the outlets of an attack pumper. Problems will be used to illustrate this point.

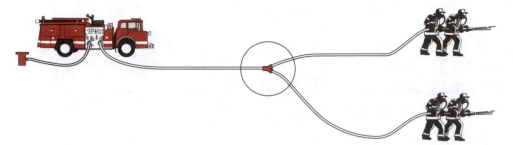

FIGURE 8.6 ◆ Problems worked off a wye are the same as if the lines were coming off a pumper.

When several lines are taken off an attack pumper, the general procedure for solving hydraulic problems is to determine the pressure requirement for each line. The pump operator then establishes a discharge pressure to that of the line requiring the highest pressure. The remaining lines need to be feathered (partially closed). Separate pressure gages or flow meters for each outlet are generally available, which makes the task simpler. However, if an appliance such as a water thief or monitor assumes the role of an attack pumper, then the feathering process requires making a professional estimate.

If a supply pumper is involved in the overall layout or an appliance has assumed the role of the attack pumper and the attack pumper has assumed the role of the supply pumper, the hydraulic calculations for the pump operator involve adding the pressure requirement for the line requiring the highest pressure to the friction loss in the supply line. In addition, the pump operator should consider whether additional factors should be included by using the formula

$$RPDP = NP + FL + BP + AFL - FP$$

as a check list.

QUESTION Three lines have been taken off an attack pumper. Attack Line 1 is a single 2½-inch line 300 feet in length equipped with a fixed-gallonage fog nozzle rated at 225 gpm at 100 psi. Attack Line 2 is a 2-inch line 250 feet in length equipped with a 1-inch smooth-bore tip with a nozzle pressure of 50 psi. Attack Line 3 is 150 feet of 1½-inch hose equipped with a dual-gallonage nozzle set to provide 60 gpm at 100 psi. What is the pump discharge pressure for this situation and how should the pump operator handle it (Figure 8.7)?

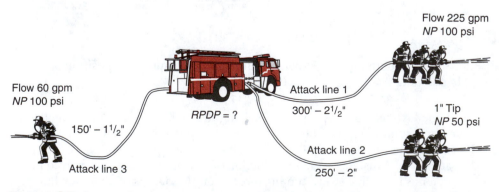

Flow 225 gpm
NP 100 psi

Flow 60 gpm
NP 100 psi

Attack line 1

RPDP = ?

300' – 2½"

1" Tip
NP 50 psi

150' – 1½"

Attack line 2

Attack line 3

250' – 2"

FIGURE 8.7 ◆

ANSWER   This type of problem is actually three separate problems combined into one. Each problem has to be solved separately. The three problems in this question are identified respectively as problem one, problem two, and problem three.

Problem one. First, solve the pressure requirement for Attack Line 1. This problem can be solved by using the formula

$$RPDP = NP + FL$$

The nozzle pressure is given and the friction loss can be solved by using the formula

$$FL = CQ^2L$$

where
$$C = 2$$
$$Q = 225/100 = 2.25$$
$$L = 300/100 = 3$$

Then
$$FL = (2)(2.25)(2.25)(3)$$
$$= 30.38 \text{ psi}$$

Now

$$RPDP = 100 + 30.38$$
$$= 130.38 \text{ psi}$$

Problem two. Next, solve for the required pump discharge pressure on Attack Line 2. This can be determined by using the formula

$$RPDP = NP + FL$$

The nozzle pressure is given; however, the flow from the 1-inch tip has to be determined in order to solve for the friction loss in the hose. The flow can be determined by using the formula

$$\text{Discharge} = 29.7D^2\sqrt{P}$$

where
$$D = 1$$
$$P = 50$$

Then
$$\text{Discharge} = (29.7)(1)^2\sqrt{50}$$
$$= (29.7)(1)(1)(7.07)$$
$$= 209.98 \text{ gpm}$$

Next, solve for the friction loss in the 2-inch hose, using the formula

$$FL = CQ^2L$$

where
$$C = 8$$
$$Q = (209.98/100) = 2.1$$
$$L = 250/100 = 2.5$$

Then
$$FL = (8)(2.1)(2.1)(2.5)$$
$$= 88.2 \text{ psi}$$

Now, solve for the required pump discharge pressure, using the formula:

$$RPDP = NP + FL$$

The nozzle pressure was given and the friction loss was determined to be 88.2 psi. Then

$$RPDP = 50 + 88.2$$
$$= 138.2 \text{ psi}$$

Problem three. Next, solve for the required pump discharge pressure for Attack Line 3. This can be determined by using the formula

$$RPDP = NP + FL$$

The nozzle pressure and the flow have been given. The friction loss in the 1½-inch line can be found by using the formula

$$FL = CQ^2L$$

where
$$C = 24$$
$$Q = 60/100 = .6$$
$$L = 150/100 = 1.5$$

Then
$$FL = (24)(.6)(.6)(1.5)$$
$$= 12.96 \text{ psi}$$

Now, solve for the required pump discharge pressure:

$$RPDP = NP + FL$$

The nozzle pressure is given and the friction loss in the hose has been determined to be 12.96 psi. Then

$$RPDP = 100 + 12.96$$
$$= 112.96 \text{ psi}$$

Summary. The pressure requirements for the three lines are: Attack Line 1, 130.38 psi; Attack Line 2, 138.2 psi; and Attack Line 3, 112.96 psi. The pump operator should set the pump pressure to supply line 2 at 138 psi (pump at 140 psi) and feather (partially close) the outlets to lines 1 and 3. With pressure gages or flow meters for each outlet, it is possible to use them to accurately feather the gate to provide the correct pressure for each line. ◼

QUESTION This is a problem where the manifold appliance assumes the role of an attack pumper. An attack pumper is supplying a portable manifold through a 500-foot length of 3½-inch hose. Three lines have been taken off the manifold for use in attacking the fire. Attack Line 1 is a single 2½-inch line 300 feet in length equipped with a fixed-gallonage fog nozzle rated at 225 gpm at 100 psi nozzle pressure. Attack Line 2 is a 2-inch line 250 feet in length equipped with a 1-inch smooth-bore tip with a nozzle pressure of 50 psi. Attack Line 3 is 150 feet of 1½-inch line equipped with a dual-gallonage automatic nozzle set to provide 60 gpm at 100 psi. What is the required pump discharge pressure for the attack pumper (Figure 8.8)?

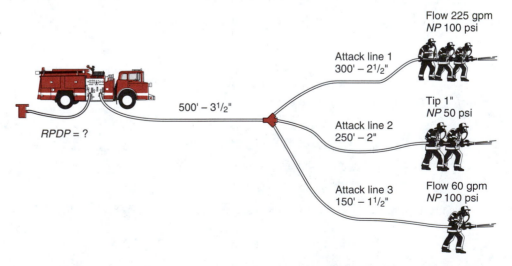

FIGURE 8.8 ◆

ANSWER This is a problem where the original attack pumper has become the supply pumper and the manifold has become the attack pumper. The assumed layout appears in Figure 8.9. The lines taken off the manifold are exactly the same as those taken off the pumper in the previous problem, and the problem at the manifold is handled in exactly the same manner. All that is necessary is to use the combined flows from the three lines to determine the friction loss in the 3½-inch supply line. This is added to the 138.2-psi pressure requirement for Attack Line 2 (the line requiring the greatest pressure) to determine the required pump discharge pressure at the original attack pumper. It is also necessary to add 10 psi for the friction loss in the manifold, as the total flow is more than 350 gpm.

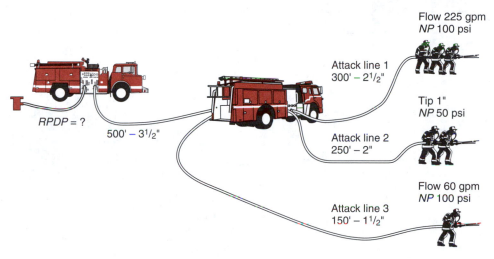

Flow 225 gpm
NP 100 psi

Attack line 1
300' − 2½"

Tip 1"
NP 50 psi

Attack line 2
250' − 2"

Flow 60 gpm
NP 100 psi

Attack line 3
150' − 1½"

RPDP = ?

500' − 3½"

FIGURE 8.9 ◆

Summary. The flows from the three attack lines are: Attack Line 1, 225 gpm; Attack Line 2, 210 gpm; and Attack Line 3, 60 gpm. The total amount of water flowing through the 3½-inch line is 225 + 210 + 60 = 495 gpm. The friction loss in the 3½-inch line can be determined by using the formula

$$FL = CQ^2L$$

where
$$C = .34$$
$$Q = 495/100 = 4.95$$
$$L = 500/100 = 5$$

Then
$$FL = (.34)(4.95)(4.95)(5)$$
$$= 41.65 \text{ psi}$$

Now the problem can be solved for the required pump discharge pressure. This is the total of the friction loss in the 3½-inch line plus the pressure requirement at the manifold (138.2 psi) plus 10 psi for the appliance friction loss. The solution is

$$RPDP = 41.65 + 138.2 + 10$$
$$= 184.85 \text{ psi}$$

■

QUESTION Engine 5 has responded to a fire in a hilly area of town to extinguish a small shed fire. A 400-foot length of 2½-inch hose has been laid into a water thief. A single 150-foot line of 2-inch hose equipped with a 150-gpm fixed-gallonage fog nozzle with a rated 75 psi nozzle pressure has been taken off one of the outlets on the water thief to extinguish the shed fire. A 200-foot length of 1½-inch hose equipped with a 90-gpm fixed-gallonage nozzle rated at 100 psi has been taken off a second outlet and advanced up a 12 percent grade to extinguish a spot fire. What is the required pump discharge pressure for Engine 5 with this layout (Figures 8.10 and 8.11)?

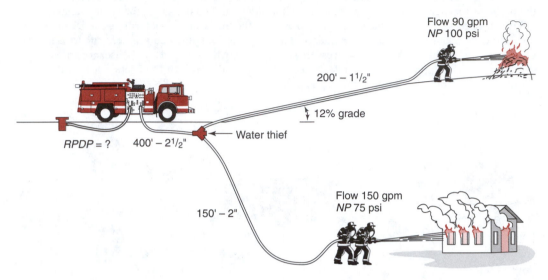

Flow 90 gpm
NP 100 psi

200' − 1¹/₂"

12% grade

RPDP = ? 400' − 2¹/₂"

Water thief

Flow 150 gpm
NP 75 psi

150' − 2"

FIGURE 8.10 ◆

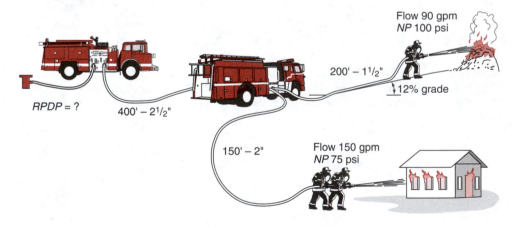

Flow 90 gpm
NP 100 psi

200' − 1¹/₂"

12% grade

RPDP = ? 400' − 2¹/₂"

150' − 2"

Flow 150 gpm
NP 75 psi

FIGURE 8.11 ◆

ANSWER This is a situation where the water thief has assumed the role of an attack pumper. The solution to the problem is the same as if the two lines had been taken off a regular pumper. The problem is divided into two parts.

Part one. It is first necessary to determine the pressure requirement for each of the lines. The 2-inch line can be solved by using the following formula:

$$RPDP = NP + FL$$

The nozzle pressure and the flow are given. The friction loss can be found by using the formula

$$FL = CQ^2L$$

where $C = 8$

$$Q = 150/100 = 1.5$$
$$L = 150/100 = 1.5$$

Then
$$FL = (8)(1.5)(1.5)(1.5)$$
$$= 27 \text{ psi}$$

Now, solve for the required pump discharge pressure:

$$RPDP = NP + FL$$
$$= 75 + 27$$
$$= 102 \text{ psi}$$

Part two. Next, solve for the required pump discharge pressure on the 1½-inch line. Use the formula

$$RPDP = NP + FL + BP$$

The nozzle pressure is given. The friction loss can be found by using the formula

$$FL = CQ^2L$$

where
$$C = 24$$
$$Q = 90/100 = .9$$
$$L = 200/100 = 2$$

Then
$$FL = (24)(.9)(.9)(2)$$
$$= 38.88 \text{ psi}$$

Next, solve for the back pressure. First, find the head:

$$H = GL$$
$$= (12)(2)$$
$$= 24 \text{ feet}$$

Then use

$$BP = .5H$$

and obtain

$$BP = (.5)(24)$$
$$= 12 \text{ psi}$$

Now, solve for the required pump discharge pressure:

$$RPDP = NP + FL + BP$$
$$= 100 + 38.88 + 12$$
$$= 150.88 \text{ psi}$$

Summary. Next, solve for the required pump discharge pressure on Engine 5. It is first necessary to determine the total flow in the 2½-inch line taken off Engine 5. The total flow is the combined flow of the two nozzles supplied by the water thief. The flow for the 2-inch line is 150 gpm and that for the 1½-inch line is 90 gpm, and 150 + 90 = 240 gpm. This flow can be used to determine the friction loss in the 2½-inch line. The friction loss can be determined by using the formula

$$FL = CQ^2L$$

where
$$C = 2$$
$$Q = 240/100 = 2.4$$
$$L = 400/100 = 4$$

Then
$$FL = (2)(2.4)(2.4)(4)$$
$$= 46.08 \text{ psi}$$

Now, solve for the required pump discharge pressure on Engine 5. The required pump discharge pressure is equal to the friction loss in the 2½-inch line plus the pressure requirement at the water thief (150.88 psi). This is the pressure required for the 1½-inch line. No appliance friction loss is added, as the total flow through the appliance is less than 350 gpm:

$$RPDP = 46.08 + 150.88$$
$$= 196.96 \text{ psi} \qquad \blacksquare$$

QUESTION A fire has occurred on the fifth floor of an old twelve-story building in the downtown section of the city. The building is equipped with an outside fire escape, with the standpipe system on the building adjacent to the fire escape. Engine 8, the first-in company, has laid 300 feet of 3-inch hose with 3-inch couplings into the standpipe system and commenced pumping. Upon Engine 4's arrival, the crew advanced two 2-inch attack lines up the fire escape. Attack Line 1, a 150-foot section of 2-inch hose with a 200-gpm fixed-gallonage nozzle rated at 75 psi, has been taken off the standpipe outlet on the fifth floor and advanced inside to attack the fire. Attack Line 2 is also 150 feet of 2-inch hose with a 200-gpm fixed-gallonage nozzle rated at 75 psi. It has been connected to the standpipe outlet on the seventh floor and taken inside to prevent the extension of the fire up the interior stairwell. Determine the required pump discharge pressure for Engine 8 (Figure 8.12).

ANSWER This is a situation where the standpipe system has assumed the role of an attack pumper. Each of the lines taken off the assumed attack pumper must be solved individually to determine its pressure requirement.

Problem one. First solve the pressure requirement for Attack Line 1. The formula needed for the solution is

$$RPDP = NP + FL$$

The nozzle pressure and the flow are given. The friction loss in the hose can be determined by using the formula

$$FL = CQ^2L$$

where
$$C = 8$$
$$Q = 200/100 = 2$$
$$L = 150/100 = 1.5$$

Then
$$FL = (8)(2)(2)(1.5)$$
$$= 48 \text{ psi}$$

Next, solve for the required pump discharge pressure:

$$RPDP = NP + FL$$
$$= 75 + 48$$
$$= 123 \text{ psi}$$

Problem two. The general procedure is to now solve for the pressure requirement for Attack Line 2. Since both lines are of the same length and have the same amount of water flowing, if this were a ground-level situation, the pressure requirement at the outlet would be the same. However, since Attack Line 2 is working two floors above Attack Line 1, 10 more pounds of pressure is required at the standpipe outlet to provide for the loss of pressure due to the back pressure. This is taken care of by the pump operator of Engine 8 when he or she solves the problem for the required pump pressure. In determining the required pump pressure, the pump operator utilizes the pressure requirement for either one of the lines, as it is the same for both. To

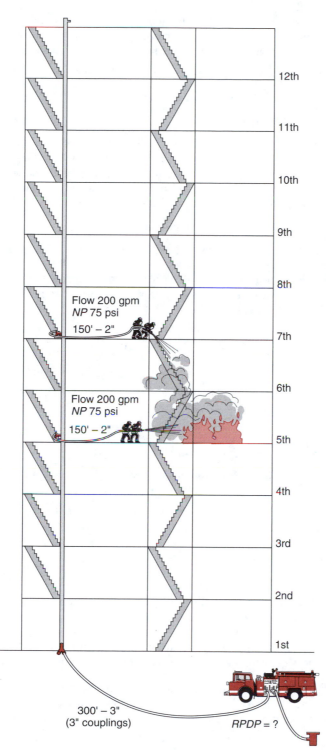

Flow 200 gpm
NP 75 psi
150' − 2"

Flow 200 gpm
NP 75 psi
150' − 2"

12th

11th

10th

9th

8th

7th

6th

5th

4th

3rd

2nd

1st

300' − 3"
(3" couplings)

RPDP = ?

FIGURE 8.12 ◆

add to the confusion, in a real-life situation, the pump operator of Engine 8 would be aware that the flow from Attack Line 2 is not continuous. However, he or she must consider that it is in continuous use. Anything less could put the members of both attack lines in jeopardy.

Summary. Now solve for the required pump discharge pressure for Engine 8:

$$RPDP = RPDP \text{ for one of the attack lines } + FL + BP + AFL$$

where $RPDP$(attack line) = 123 psi

FL = friction loss in 3-inch line

BP = 5 psi per floor to the highest floor where water is discharged (in this problem, the seventh floor, which is 6 floors above the ground level)

= 5 × 6 = 30 psi

AFL = appliance friction loss

Since the total flow from the two attack lines is 400 gpm, that is, greater than 350 gpm, AFL is taken to be 10 psi.

To complete the problem, it is first necessary to determine the friction loss in the 3-inch supply line. The total flow is the combined flow of the two nozzles. The friction loss can be determined by using the formula

$$FL = CQ^2L$$

where $C = .677$

Q = 400/100 = 4 (total flow in the two attack lines)

L = 300/100 = 3

Then $FL = (.677)(4)(4)(3)$

= 32.5 psi

Now solve for the required pump discharge pressure:

$$RPSP = \text{pressure requirement for one of the attack lines} + FL + BP + AFL$$

= 123 + 32.5 + 30 + 10

= 195.5 psi ◼

◆ UNUSUAL PROBLEMS

QUESTION Two 2½-inch lines each 800 feet in length have been taken off Engine 3 and laid downhill into the standpipe of a ten-story building. The downhill grade is 12 percent. A single 2½-inch line 200 feet in length has been taken off the standpipe on the eighth floor and is working on a fire on that floor. The line is equipped with a 1¼-inch tip at 50 psi nozzle pressure. What is the required pump discharge pressure for this problem (Figure 8.13)?

ANSWER In this problem, the standpipe does not assume the role of an attack pumper, as only one line has been taken off of it. The problem is unusual because it involves both back pressure and forward pressure. The formula required to solve this problem is

$$RPDP = NP + FL + BP - FP$$

First, determine the flow from the 1¼-inch tip:

$$\text{Discharge} = 29.7D^2\sqrt{P}$$

where D = 1¼ or 1.25 inches

P = 50 psi

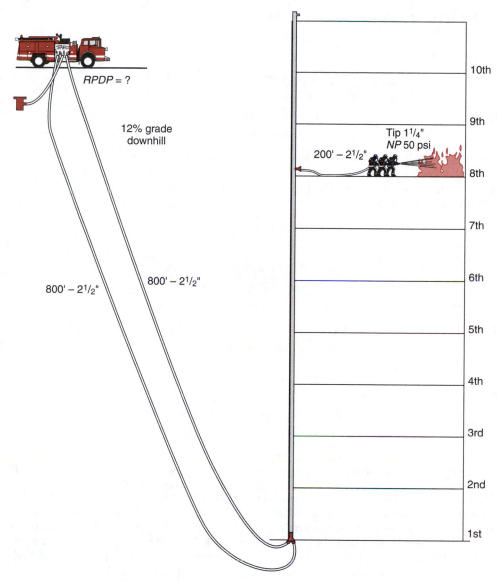

RPDP = ?

12% grade
downhill

Tip 1¼"
NP 50 psi

200' – 2½"

800' – 2½"

800' – 2½"

10th

9th

8th

7th

6th

5th

4th

3rd

2nd

1st

FIGURE 8.13 ◆

Then
$$\text{Discharge} = (29.7)(1.25)^2\sqrt{50}$$
$$= (29.7)(1.25)(1.25)(7.07)$$
$$= 328.09 \text{ gpm}$$

Next, solve for the friction loss in the hose. First solve for the friction loss in the single 2½-inch line, then for the friction loss in the siamesed lines:

$$FL = CQ^2L$$

where
$$C = 2$$
$$Q = 328.09/100 = 3.28 \text{ gpm}$$
$$L = 200/100 = 2$$

Then
$$FL = (2)(3.28)(3.28)(2)$$
$$= 43.03 \text{ psi}$$

The friction loss formula for the siamesed lines is

$$FL = CQ^2L$$

where

$$C = .5$$
$$Q = 3.28$$
$$L = 8$$

Then

$$FL = (.5)(3.28)(3.28)(8)$$
$$= 43.03 \text{ psi}$$

The total friction loss in the hose is $43.03 + 43.03 = 86.06$ psi.

Next, solve for the back pressure. The line is working on the eighth floor, which is seven floors above ground level. Back pressure is 5 psi per floor, or $5 \times 7 = 35$ psi.

Next, solve for the forward pressure:

$$H = GL$$

where

$$G = 12$$
$$L = 800/100 = 8$$

Then

$$H = (12)(8)$$
$$= 96 \text{ feet}$$

The forward pressure is ½ psi per foot, or

$$FP = (.5)(96)$$
$$= 48 \text{ psi}$$

Now solve for the required pump discharge pressure:

$$RPDP = NP + FL + BP - FP$$
$$= 50 + 86.06 + 35 - 48$$
$$= 123.06 \text{ psi}$$

Note: No loss was allowed for the appliance, as the flow was less than 350 gpm. ◼

QUESTION A 600-foot length of 5-inch hose has been laid from a pumper into a portable manifold. Three lines have been taken off the manifold. One is 200 feet of 2-inch hose equipped with a fixed-gallonage nozzle with water flowing at 175 gpm at 100 psi nozzle pressure. The other two lines are 3-inch lines with 2½-inch couplings, each 200 feet in length. They have been siamesed into a portable monitor with water flowing at 600 gpm at a nozzle pressure of 100 psi. What is the required pump discharge pressure for this setup (Figure 8.14)?

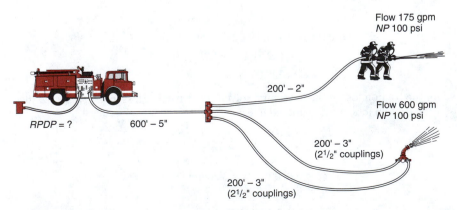

Flow 175 gpm
NP 100 psi

200' – 2"

Flow 600 gpm
NP 100 psi

RPDP = ?   600' – 5"

200' – 3"
(2¹/2" couplings)

200' – 3"
(2¹/2" couplings)

FIGURE 8.14 ◆

ANSWER This is a problem where the portable manifold assumes the role of an attack pumper. It is considered unusual because the manifold supplies both a portable monitor and a single attack line. It is also unusual because at first glance it would appear that the pressure requirement for the portable monitor would be much greater than that for the single attack line. That does not turn out to be the case. The assumed layout appears as in Figure 8.15.

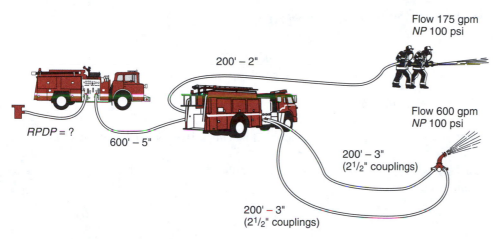

Flow 175 gpm
NP 100 psi

200' – 2"

Flow 600 gpm
NP 100 psi

RPDP = ?

600' – 5"

200' – 3"
(2¹/₂" couplings)

200' – 3"
(2¹/₂" couplings)

FIGURE 8.15 ◆

The solution to the problem is divided into three parts. The first two involve the portable manifold and the third involves the original attack pumper.

Part one. First, solve for the pressure requirement for the attack line. It can be determined by using the formula

$$RPDP = NP + FL$$

The nozzle pressure and the flow are given. The friction loss can be determined by using the formula

$$FL = CQ^2L$$

where
$$C = 8$$
$$Q = 175/100 = 1.75$$
$$L = 200/100 = 2$$
Then
$$FL = (8)(1.75)(1.75)(2)$$
$$= 49 \text{ psi}$$

Now

$$RPDP = 100 + 49$$
$$= 149 \text{ psi}$$

Part two. Next, solve for the required pump discharge pressure for the portable monitor. It can be determined by using the formula

$$RPDP = NP + FL + AFL$$

The nozzle pressure and the flow are given and the appliance friction loss for a master stream is 25 psi. The friction loss can be determined by using the formula

$$FL = CQ^2L$$

where
$$C = .2$$
$$Q = 600/100 = 6$$
$$L = 200/100 = 2$$

Then
$$FL = (.2)(6)(6)(2)$$
$$= 14.4 \text{ psi}$$

Now solve for the required pump discharge pressure:

$$RPDP = NP + FL + AFL$$
$$= 100 + 14.4 + 25$$
$$= 139.4 \text{ psi}$$

Part three. Next, solve for the required pump discharge pressure for the original attack pumper. It can be determined by using the formula

$RPDP$ = pressure requirement at the manifold (149 psi) + friction loss in 5-inch line + 10 psi for appliance friction loss

To solve the problem, it is necessary to determine the friction loss in the 5-inch line. It can be determined by using the formula

$$FL = CQ^2L$$

where
$$C = .08$$
Q = combined flow of attack line and
portable monitor divided by 100
$$= (175 + 600)/100$$
$$= 775/100$$
$$= 7.75$$
$$L = 600/100 = 6$$

Then
$$FL = (.08)(7.75)(7.75)(6)$$
$$= 28.83 \text{ psi}$$

Now solve for the required pump discharge pressure:

$RPDP$ = pressure requirement at the manifold + friction loss in 5-inch line + AFL
$$= 149 + 28.83 + 10$$
$$= 187.83 \text{ psi}$$

■

◆ COMPLEX PROBLEMS

Some problems that at first appear to be complex turn out not to be too difficult. As long as the problem can be divided into its component parts, the parts are not too difficult to solve if the layouts are those for which there are coefficients in Table 6.1 or Table 7.2. For example, consider the following problem.

QUESTION A portable monitor equipped with a 600-gpm fixed-gallonage nozzle at 100 psi is working on the roof of an eight-story building. Three 500-foot lines of hose have been taken off Engine 6 and laid into the standpipe of the building. Two of the lines are 3-inch lines with 2½-inch couplings and the other is a 2½-inch line. Two 3-inch lines with 2½-inch couplings each 200 feet in length have been taken off the standpipe outlets on the roof and laid into the inlets of the portable monitor. What is the required pump discharge pressure for Engine 6 (Figure 8.16)?

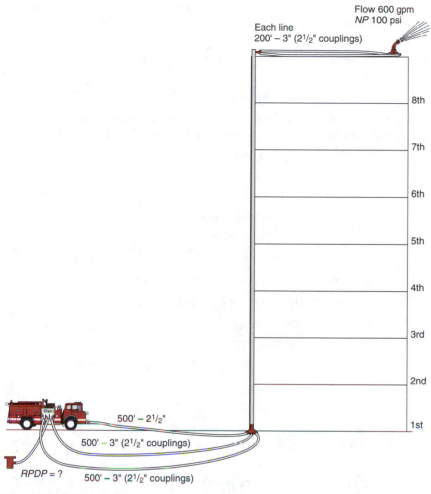

Flow 600 gpm
NP 100 psi

Each line
200' – 3" (2½" couplings)

8th

7th

6th

5th

4th

3rd

2nd

1st

500' – 2½"

500' – 3" (2½" couplings)

RPDP = ?

500' – 3" (2½" couplings)

FIGURE 8.16 ◆

ANSWER It must first be recognized that in this problem the standpipe assumes the role of an attack pumper. The reason for this is that more than one line has been taken off the standpipe. The assumed layout of the problem is shown in (Figure 8.17).

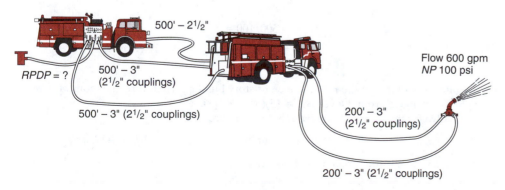

500' – 2½"

Flow 600 gpm
NP 100 psi

RPDP = ?

500' – 3"
(2½" couplings)

500' – 3" (2½" couplings)

200' – 3"
(2½" couplings)

200' – 3" (2½" couplings)

FIGURE 8.17 ◆

The problem is broken down into two parts. The first part involves solving for the pressure requirement at the standpipe outlet on the roof that is necessary to adequately supply the portable monitor.

Part one. The required pressure at the roof standpipe outlet can be found by using the formula

$$RPDP = NP + FL + AFL$$

The nozzle pressure and the flow have been given. The appliance friction loss for a master stream is 25 psi. Then the friction loss in the hose can be solved by using the formula

$$FL = CQ^2L$$

where
$$C = .2$$
$$Q = 600/100 = 6$$
$$L = 200/100 = 2$$
Then
$$FL = (.2)(6)(6)(2)$$
$$= 14.4 \text{ psi}$$

Now solve for the required pump discharge pressure at the roof outlet:

$$RPDP = NP + FL + AFL$$

where
$$NP = 100 \text{ psi}$$
$$FL = 14.4 \text{ psi}$$
$$AFL = 25 \text{ psi}$$
Then
$$RPDP = 100 + 14.4 + 25$$
$$= 139.4 \text{ psi}$$

Of course, the back pressure needs to be added to this to determine the actual pressure required at the outlet. In this problem, the back pressure is taken care of by the supplying pumper.

Part two. The second part consists in adding the pressure required at the roof outlet plus the friction loss in the hose laid into the standpipe plus the back pressure plus the appliance friction loss in the standpipe. The pressure requirement at the roof is known and the appliance friction loss in the standpipe is 10 psi since the flow is greater than 350 gpm. The friction loss for the lines laid into the standpipe can be determined by using the formula

$$FL = CQ^2L$$

where
$$C = .12$$
$$Q = 6$$
$$L = 500/100 = 5$$
Then
$$FL = (.12)(6)(6)(5)$$
$$= 21.6 \text{ psi}$$

The monitor is working on the roof of an eight-story building. Therefore, the monitor is working eight floors above ground level: $BP = (5)(8) = 40$ psi.

Now solve for the required pump discharge pressure:

$$RPDP = \text{roof outlet pressure} + FL + BP + AFL$$

where
$$\text{Roof outlet pressure} = 139.4 \text{ psi}$$
$$FL = 21.6 \text{ psi}$$
$$BP = 40 \text{ psi}$$

$$AFL = 10 \text{ psi (flow is over 350 gpm)}$$

Then

$$RPDP = 139.4 + 21.6 + 40 + 10$$
$$= 211 \text{ psi}$$ ■

COMPLEX SIAMESED PROBLEMS

The foregoing problem, which included siamesed assemblies on the roof and leading into the standpipe, could be solved relatively easily. The reason was because there were coefficients for the siamesed assemblies in Table 7.1. However, there are occasions when the siamesed assembly does not have a coefficient in Table 7.1. These are assemblies where the lines may be of the same or different sizes but of different lengths.

Two methods have been developed for use in this book to compensate for this deficiency. Both methods are designed to adjust the siamese assembly to one that complies with a siamese assembly of two 2½-inch lines or three 2½-inch lines as shown in Table 7.1. This involves changing hose other than 2½-inch lines into **2½-inch equivalents**. The coefficients required for this process are given in Table 8.1.

The formula to use in the application of the coefficients is the following

Determining How to Change Hose Other Than 2½-inch Hose to Equivalent Lengths of 2½-inch Hose

$$E\ 2\tfrac{1}{2}\text{-inch} = L/C$$

where

$E\ 2\tfrac{1}{2}$-inch = equivalent length of 2½-inch hose for the line being considered

L = length of line being considered

C = coefficient from Table 8.1

TABLE 8.1 ◆ Friction Loss Coefficients for Determining Equivalent Length of 2½-inch Hose

Hose Diameter (in.)	Coefficient
1½	0.08
1¾	0.13
2	0.25
3 (2½-in. couplings)	2.5
3 (3-in. couplings)	2.95
3½	5.88
4	10.0
4½	20.0
5	25.0
6	40.0

This table was developed by Gene Mahoney.

QUESTION What is the equivalent length of 2½-inch hose for 550 feet of single 2-inch hose?

ANSWER To solve, use the formula

$$E \; 2\tfrac{1}{2}\text{-inch} = L/C$$

where

$$L = 550$$
$$C = .25$$

Then

$$E \; 2\tfrac{1}{2}\text{-inch} = 550/.25$$
$$= 2200 \text{ feet}$$

This means that 2200 feet of 2½-inch hose has the same friction loss as 550 feet of 2-inch hose when the same amount of water is flowing in each line (Figure 8.18). ◼

500' – 2"

2200' – 2¹/₂"

FIGURE 8.18 ◆ The friction loss is the same when the flow is the same.

QUESTION What is the equivalent length of 2½-inch hose for 800 feet of 3-inch hose with 3-inch couplings if the same amount of water is flowing in each line?

ANSWER To solve, use the equivalent-length formula

$$E \; 2\tfrac{1}{2}\text{-inch} = L/C$$

where

$$L = 800$$
$$C = 2.95$$

Then

$$E \; 2\tfrac{1}{2}\text{-inch} = 800/2.95$$
$$= 271.19 \text{ or } 271 \text{ feet}$$

This means that the friction loss in 271 feet of 2½-inch hose is the same as that in 800 feet of 3-inch hose with 3-inch couplings if the same amount of water is flowing in both lines (Figure 8.19). ◼

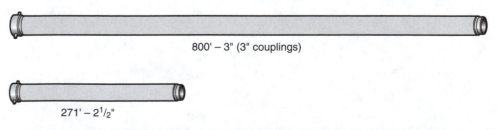

800' – 3" (3" couplings)

271' – 2¹/₂"

FIGURE 8.19 ◆ The friction loss is the same when the flow is the same.

◆ **DETERMINING EQUIVALENT LENGTHS OF HOSE FOR 100 FEET OF 2½-INCH HOSE**

By using the coefficients in Table 8.1, the lengths of lines having the same friction loss as that in 100 feet of 2½-inch hose can be determined. The process involves dividing 100 by the coefficient for that size as given in Table 8.1. For example:

QUESTION How many feet of single 3½-inch hose has the same friction loss as 100 feet of single 2½-inch hose if the same amount of water is flowing in each line?

ANSWER To solve this problem, divide 100 by the coefficient for a single 3½-inch line given in Table 8.1:

$$100/5.88 = 17.01 \text{ or } 17 \text{ feet}$$

This means that the friction loss in 100 feet of a single 2½-inch line is the same as that in 17 feet of a single 3½-inch hose if the same amount of water is flowing in each line (Figure 8.20). ■

100' – 3½"

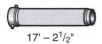

17' – 2½"

FIGURE 8.20 ◆ The friction loss is the same when the flow is the same.

Using the same process for each of the single commonly used hose lines from 1½ inches to 6 inches produces the results shown in Table 8.2.

TABLE 8.2 ◆ Equivalent Lengths for 100 Feet of 2½-inch Hose

Hose Diameter (in.)	Equivalent 2½-inch Length (ft)
1½	1250
1¾	769
2	400
3 (with 2½-in. couplings)	40
3 (with 3-in. couplings)	34
3½	17
4	10
4½	5
5	4
6	2.5

This table was developed by Gene Mahoney.

CHANGING SIAMESED LINES INTO 2½-INCH SIAMESED LINES

There are two methods of solving problems in this area. One method uses Table 8.1 and the other uses Table 8.2. Using Table 8.1 involves a division process. The use of Table 8.2 requires a multiplication process. Both methods produce the same, or nearly the same, result. The choice of which of the two systems to use is in the hands of the user. Problems are used to illustrate the principles involved in both methods.

QUESTION Three lines have been taken off Engine 4. The lines have been siamesed into a single 2½-inch line. One is 400 feet of 3-inch line with 2½-inch couplings. The second is 500 feet of 2½-inch line. The third is 300 feet of 2-inch line. Change this siamese setup into one having all 2½-inch hoses (Figure 8.21).

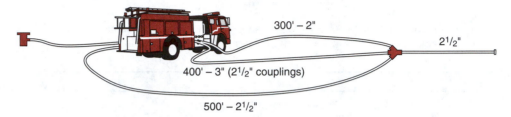

300' – 2"

400' – 3" (2½" couplings)

500' – 2½"

2½"

FIGURE 8.21 ◆

ANSWER First note that there is no coefficient in Table 7.1 that can be used to solve for the friction loss in this siamese layout. It is then necessary to change the layout into one that can be used. Since it is a three-line siamese, it will be changed into three lines of 2½-inch hose, and then the coefficient in Table 7.1 can be used.

First, change the 400 feet of 3-inch line by using the formula

$$E \text{ 2½-inch} = L/C$$

where
$$L = 400$$
$$C = 2.5$$

Then
$$E \text{ 2½-inch} = 400/2.5$$
$$= 160 \text{ feet}$$

Next, change the 2-inch line:

$$E \text{ 2½-inch} = L/C$$

where
$$L = 300$$
$$C = .25$$

Then
$$E \text{ 2½-inch} = 300/.25$$
$$= 1200 \text{ feet}$$

The layout now appears as shown in Figure 8.22.

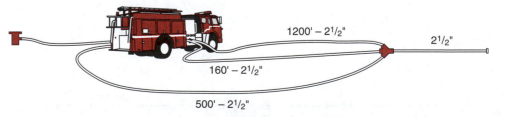

1200' – 2½"

160' – 2½"

500' – 2½"

2½"

FIGURE 8.22 ◆

QUESTION Three lines have been taken off Engine 4. The lines have been siamesed into a single 2½-inch line. One is 400 feet of 3-inch line with 2½-inch couplings. The second is 500 feet of 2½-inch line. The third is 300 feet of 2-inch line. Change this siamese setup into one having all 2½-inch hoses.

ANSWER First note that this is the same question previously asked. However, this time it will be solved by the use of Table 8.2.

First, change the 3-inch line. The change can be made by using the following formula:

$$E\,2\tfrac{1}{2}\text{-inch} = (L/100) \times C2$$

where

$$L = \text{length of line}$$
$$C2 = \text{equivalent from Table 8.2}$$

Then

$$E\,2\tfrac{1}{2}\text{-inch} = (400/100) \times 40$$
$$= 4 \times 40$$
$$= 160 \text{ feet}$$

Next, change the 2-inch hose:

$$E\,2\tfrac{1}{2}\text{-inch} = (300/100) \times 400$$
$$= 3 \times 400$$
$$= 1200 \text{ feet}$$

Notice that the results by both methods are the same. ■

SETTING UP EQUAL-LENGTH 2½-INCH SIAMESE LAYOUTS

QUESTION Convert a siamese layout consisting of a single 3½-inch line 600 feet in length and 450 feet of 2-inch hose into a siamese of two 2½-inch lines of equal length (Figure 8.23).

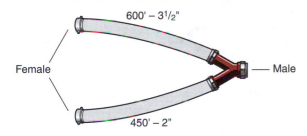

600' – 3½"

Female

Male

450' – 2"

FIGURE 8.23 ◆

ANSWER First, convert each of the lines to 2½-inch equivalents using either Table 8.1 or Table 8.2. For demonstrative purposes, Table 8.1 is used to convert the 3½-inch line and Table 8.2 is used to convert the 2-inch line.

First, convert the 3½-inch line by using Table 8.1:

$$E\,2\tfrac{1}{2}\text{-inch} = L/C$$

where

$$L = 600$$
$$C = 5.88$$

Then

$$E\,2\tfrac{1}{2}\text{-inch} = 600/5.88$$
$$= 102 \text{ feet}$$

Next, convert the 2-inch line using Table 8.2:

$$E\,2\tfrac{1}{2}\text{-inch} = (L/100) \times C2$$

where
$$L = 450$$
$$C2 = 400$$

Then
$$E\ 2\tfrac{1}{2}\text{-inch} = (450/100) \times 400$$
$$= 4.5 \times 400$$
$$= 1800\ \text{feet}$$

The layout now appears as shown in Figure 8.24.

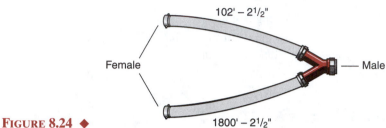

FIGURE 8.24 ◆

At this point it is necessary to change a siamese assembly consisting of the same-size hose but of different lengths into one having equal lengths. This is accomplished by adding the lengths of hose in each line and dividing by the number of lines in the siamese assembly. In this problem, 102 + 1800 = 1902 feet. Then 1902/2 = 951 feet. The layout now appears as shown in Figure 8.25. With this configuration, the friction loss in the siamese assembly can be determined by using the coefficient for two 2½-inch lines from Table 7.1.   ■

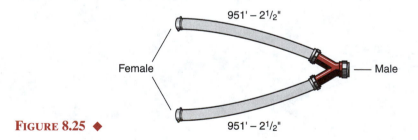

FIGURE 8.25 ◆

QUESTION   Three lines have been taken off Engine 1 and siamesed into a single 3-inch line. One of the lines is 500 feet of 5-inch hose; one is 300 feet of 4-inch hose; and one is 350 feet of 3½-inch hose. Convert this layout into three siamesed 2½-inch hoses of equal length (Figure 8.26).

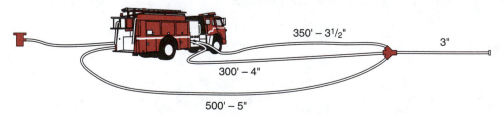

FIGURE 8.26 ◆

ANSWER Use Table 8.1 for this problem. First, convert the 5-inch line:

$$E\ 2\tfrac{1}{2}\text{-inch} = L/C$$

where
$$L = 500$$
$$C = 25$$

Then
$$E\ 2\tfrac{1}{2}\text{-inch} = 500/25$$
$$= 20 \text{ feet}$$

Next, convert the 4-inch hose:

$$E\ 2\tfrac{1}{2}\text{-inch} = L/C$$

where
$$L = 300$$
$$C = 10$$

Then
$$E\ 2\tfrac{1}{2}\text{-inch} = 300/10$$
$$= 30 \text{ feet}$$

Now, convert the 3½-inch hose:

$$E\ 2\tfrac{1}{2}\text{-inch} = L/C$$

where
$$L = 350$$
$$C = 5.88$$

Then
$$E\ 2\tfrac{1}{2}\text{-inch} = 350/5.88$$
$$= 59.52 \text{ or } 60 \text{ feet}$$

The layout now appears as shown in Figure 8.27.

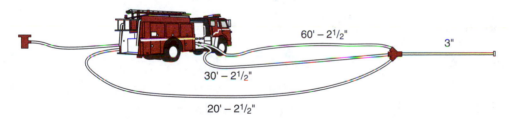

FIGURE 8.27 ◆

The next step is to average the converted three lines. This is done by adding the three lines and dividing by three:

$$\text{Average length} = \frac{\text{Line 1 + Line 2 + Line 3}}{3}$$
$$= \frac{20 + 30 + 60}{3}$$
$$= \frac{110}{3}$$
$$= 36.67 \text{ or } 37 \text{ feet}$$

The layout now appears as shown in Figure 8.28. The friction loss in this layout can now be solved by using the coefficient for three 2½-inch lines from Table 7.1. ■

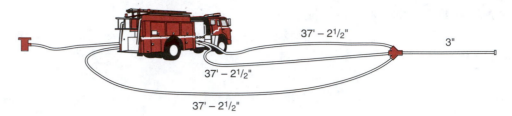

FIGURE 8.28 ◆

TESTING THE SYSTEM

To test the system, a problem is presented that can be solved directly by the use of Table 7.1. The hose in the problem is then converted into 2½-inch equivalents and the problem solved by the method being tested. The answers to the two problems are compared to determine the accuracy of the method.

QUESTION What is the friction loss in two 500-foot siamesed lines of 3-inch hose with 2½-inch couplings if the flow is 300 gpm (Figure 8.29)?

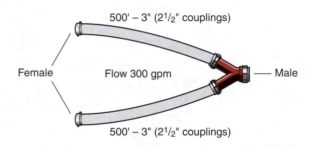

FIGURE 8.29 ◆   Friction loss = ?

ANSWER The friction loss can be found from

$$FL = CQ^2L$$

where

$$C = .2 \text{ (from Table 7.1)}$$
$$Q = 300/100 = 3$$
$$L = 500/100 = 5$$

Then

$$FL = (.2)(3)(3)(5)$$
$$= 9 \text{ psi}$$

QUESTION Now, solve the problem by first converting the layout into 2½-inch-equivalent hose. Since both lines are the same length and size, we can write

$$E\ 2\text{½-inch} = L/C$$

where

$$L = 500$$
$$C = 2.5 \text{ (from Table 8.1)}$$

Then

$$E\ 2\text{½-inch} = 500/2.5$$
$$= 200 \text{ feet}$$

The layout now appears as shown in Figure 8.30. Using the revised layout, solve for the friction by using Table 7.1:

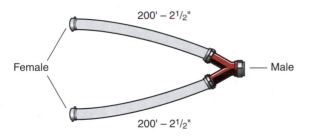

200' – 2½"

Female

200' – 2½"

Friction loss = ?

FIGURE 8.30 ◆

ANSWER

$$FL = CQ^2L$$

where

$$C = .5$$
$$Q = 300/100 = 3$$
$$L = 200/100 = 2$$

Then

$$FL = (.5)(3)(3)(2)$$
$$= 9 \text{ psi}$$

Note: The comparable answers will not always be exact; however, the answer determined by changing the hose into the 2½-inch equivalent will be sufficiently close for all practical purposes. The next question gives an example. ■

QUESTION Determine the friction loss in a 600-foot siamesed assembly when two of the lines are 3-inch lines with 2½- inch couplings and one is a 2½-inch line. The total flow in the assembly is 500 gpm (Figure 8.31).

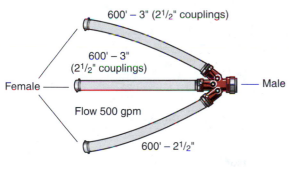

600' – 3" (2½" couplings)

600' – 3"
(2½" couplings)

Female

Flow 500 gpm

Male

600' – 2½"

Friction loss = ?

FIGURE 8.31 ◆

ANSWER   The friction loss is given as

$$FL = CQ^2L$$

where

$$C = .12 \text{ (from Table 7.1)}$$
$$Q = 500/100 = 5$$
$$L = 600/100 = 6$$

Then

$$FL = (.12)(5)(5)(6)$$
$$= 18 \text{ psi}$$

■

QUESTION Consider the same situation as in the foregoing question. Convert the lines into 2½-inch equivalents and solve for the friction loss.

ANSWER First, convert the 600 feet of 3-inch hose into an equivalent 2½-inch line:

$$E \ 2\tfrac{1}{2}\text{-inch} = L/C$$

where $L = 600$

$C = 2.5$ (from Table 8.1)

Then $E \ 2\tfrac{1}{2}\text{-inch} = 600/2.5$

$= 240$ feet

The layout appears as shown in Figure 8.32.

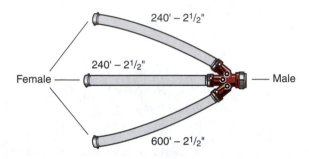

240' – 2¹/₂"

240' – 2¹/₂"

Female ———— ———— Male

600' – 2¹/₂"

FIGURE 8.32 ◆   Friction loss = ?

Next, determine the average length of hose in the siamese. To determine this, add the lengths of the three lines and divide by 3:

$$\text{Average length} = \frac{\text{Line 1 + Line 2 + Line 3}}{3}$$

where Line 1 = 240 feet

Line 2 = 240 feet

Line 3 = 600 feet

Then $\text{Average length} = \dfrac{240 + 240 + 600}{3}$

$= \dfrac{1080}{3}$

$= 360$ feet

The siamese assembly now appears as shown in Figure 8.33.

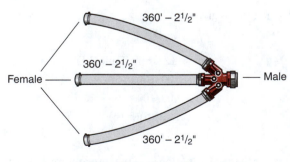

360' – 2¹/₂"

360' – 2¹/₂"

Female ———— ———— Male

360' – 2¹/₂"

FIGURE 8.33 ◆   Friction loss = ?

Now, solve for the friction loss in the assembly:

$$FL = CQ^2L$$

where
$$C = .22 \text{ (from Table 7.1)}$$
$$Q = 500/100 = 5$$
$$L = 360/100 = 3.6$$
Then
$$FL = (.22)(5)(5)(3.6)$$
$$= 19.8 \text{ psi}$$

A difference, but not a significant difference. ■

◆ SOLVING SIAMESED PROBLEMS INVOLVING DIFFERENT SIZES AND LENGTHS

The information contained in Chapters 5 through 7 plus that given in this chapter is now used for solving problems in this area. All the layouts presented in this section first have to be converted into 2½-inch hose before the problem can be solved.

QUESTION Two lines have been taken off Engine 1 and siamesed into a single 1½-inch line. One of the lines is 400 feet of 2½-inch hose and one is 600 feet of 2-inch hose. Water in the 1½-inch line is flowing at 80 gpm. Solve for the friction loss in the siamese assembly (Figure 8.34).

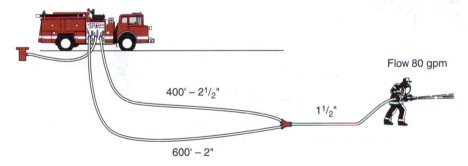

Flow 80 gpm

400' – 2½"

1½"

600' – 2"

Friction loss in siamese assembly = ?

FIGURE 8.34 ◆

ANSWER First, convert the two-inch line into a 2½-inch equivalent:

$$E\ 2\tfrac{1}{2}\text{-inch} = L/C$$

where
$$L = 600$$
$$C = .25 \text{ (Table 8.1)}$$
Then
$$E\ 2\tfrac{1}{2}\text{-inch} = 600/.25$$
$$= 2400 \text{ feet}$$

The layout now appears as shown in Figure 8.35.

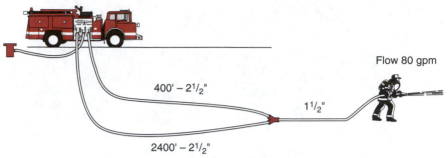

Flow 80 gpm

400' – 2½"

1½"

2400' – 2½"

Friction loss in siamese assembly = ?

FIGURE 8.35 ◆

Now add the two lines and divide by 2 to determine the average length of hose in the layout:

$$\text{Average length} = \frac{400 + 2400}{2}$$

$$= \frac{2800}{2}$$

$$= 1400 \text{ feet}$$

The layout now appears as shown in Figure 8.36. The friction loss in the layout can now be determined by using Table 7.1:

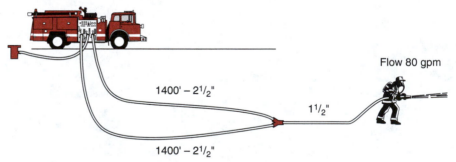

Flow 80 gpm

1400' – 2½"

1½"

1400' – 2½"

Friction loss in siamese assembly = ?

FIGURE 8.36 ◆

$$FL = CQ^2L$$

where

$C = .5$

$Q = 80/100 = .8$

$L = 1400/100 = 14$

Then

$FL = (.5)(.8)(.8)(14)$

$= 4.48 \text{ psi}$ ∎

QUESTION Three lines have been taken off Engine 12 and fed into the inlets of an elevated platform. One of the lines is 500 feet of 5-inch hose; another is 400 feet of 4-inch hose; and the third is 550 feet of 3½-inch hose. The platform is operating at a height of 50 feet and is discharging 1000 gpm at 100 psi through a fixed-gallonage nozzle. The friction loss in the fixed piping is 20 psi. What is the required pump discharge pressure for Engine 12 (Figure 8.37)?

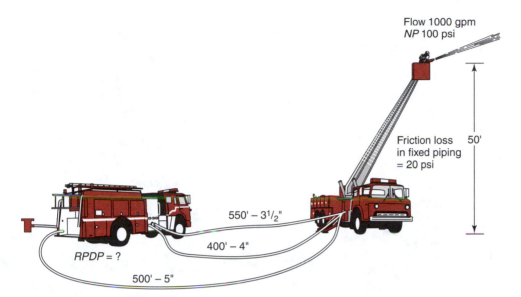

FIGURE 8.37 ◆

ANSWER The solution to this problem is broken down into three parts. Part one requires the determination of the friction loss in the hose. Part two solves for the required back pressure. Part three determines the required pump discharge pressure for Engine 12.

Part one. Change the 5-inch hose into an equivalent length of 2½-inch hose:

$$E\ 2\tfrac{1}{2}\text{-inch} = L/C$$

where
$$L = 500 \text{ feet}$$
$$C = 25 \ (\text{from Table 8.1})$$

Then
$$E\ 2\tfrac{1}{2}\text{-inch} = 500/25$$
$$= 20 \text{ feet}$$

Change the 4-inch hose into an equivalent length of 2½-inch hose:

$$E\ 2\tfrac{1}{2}\text{-inch} = L/C$$

where
$$L = 400 \text{ feet}$$
$$C = 10 \ (\text{from Table 8.1})$$

Then
$$E\ 2\tfrac{1}{2}\text{-inch} = 400/10$$
$$= 40 \text{ feet}$$

Change the 3½-inch hose into an equivalent length of 2½-inch hose:

$$E\ 2\tfrac{1}{2}\text{-inch} = L/Q$$

where
$$L = 550 \text{ feet}$$
$$Q = 5.88 \ (\text{from Table 8.1})$$

Then
$$E\ 2\tfrac{1}{2}\text{-inch} = 550/5.88$$
$$= 93.54 \text{ or } 94 \text{ feet}$$

The layout now appears as shown in Figure 8.38.

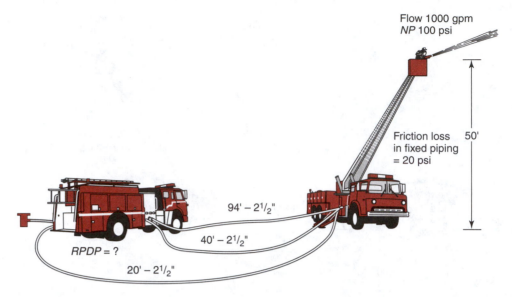

Flow 1000 gpm
NP 100 psi

Friction loss
in fixed piping
= 20 psi

50'

94' – 2½"

40' – 2½"

RPDP = ?

20' – 2½"

FIGURE 8.38 ◆

Next, determine the average length of the three lines. This can be found by adding the three lengths and dividing by three:

$$\text{Average length} = \frac{20 + 40 + 94}{3}$$

$$= \frac{154}{3}$$

$$= 51.34 \text{ or } 51 \text{ feet}$$

The layout now appears as shown in Figure 8.39.

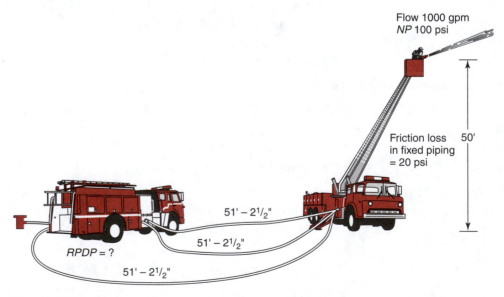

Flow 1000 gpm
NP 100 psi

Friction loss
in fixed piping
= 20 psi

50'

51' – 2½"

51' – 2½"

RPDP = ?

51' – 2½"

FIGURE 8.39 ◆

Now solve for the friction loss in the layout.

$$FL = CQ^2L$$

where

$$C = .22 \text{ (from Table 7.1)}$$
$$Q = 1000/100 = 10$$
$$L = 51/100 = .51$$

Then

$$FL = (.22)(10)(10)(.51)$$
$$= 11.22 \text{ psi}$$

Part two. The platform is working at a height of 50 feet. Therefore, the back pressure is

$$BP = (.5)(50)$$
$$= 25 \text{ psi}$$

Part three. The required pump discharge pressure can be determined by using the formula

$$RPDP = NP + FL + BP + AFL + \text{piping loss}$$

where

$$NP = 100 \text{ psi}$$
$$FL = 11.22 \text{ psi}$$
$$BP = 25 \text{ psi}$$
$$AFL = 25 \text{ psi (for master stream)}$$
$$\text{Piping loss} = \text{loss in fixed piping on apparatus (20 psi)}$$

Then

$$RPDP = 100 + 11.2 + 25 + 25 + 20$$
$$= 181.22 \text{ psi}$$ ■

QUESTION Two lines have been laid from Engine 7 to a portable manifold. One of the lines is 400 feet of 3-inch hose with 2½-inch couplings. The other is 500 feet of 3½-inch hose. Three lines have been taken off the manifold. One is a single line of 2½-inch hose 500 feet in length. It is equipped with a selectable-gallonage nozzle with a flow of 250 gpm at 100 psi nozzle pressure. The other two lines have been siamesed into a portable monitor equipped with a 1⅜-inch smooth-bore tip at 80 psi nozzle pressure. One of the lines is a 2½-inch line 450 feet in length. The other is 400 feet of 3-inch line with 3-inch couplings. What is the required pump discharge pressure for Engine 7?

ANSWER This is a problem where the portable manifold assumes the role of an attack pumper. The assumed layout appears as shown in Figure 8.40. The problem is broken down into four parts. Part one requires solving for the pump discharge pressure at the manifold for the portable monitor. Part two requires solving for the pump discharge pressure at the manifold for the single 2½-inch line. Part three requires solving for the friction loss in the siamese layout taken off Engine 7. Part four requires solving for the required pump discharge pressure for Engine 7.

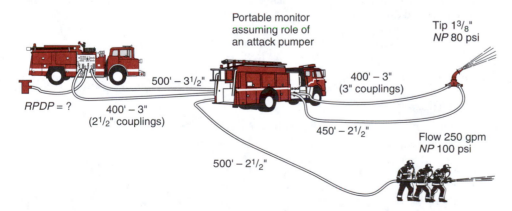

FIGURE 8.40 ◆ Portable monitor assuming the role of an attack pumper.

Part one. This part requires solving for the friction loss in the siamese assembly taken off the manifold and the pressure required at the manifold for the portable monitor. First, solve for the discharge from the 1⅜-inch tip:

$$\text{Discharge} = 29.7D^2\sqrt{P}$$

where

$$D = 1\tfrac{3}{8} \text{ or } 1.375 \text{ inches}$$
$$P = 80 \text{ psi}$$

Then

$$\text{Discharge} = (29.7)(1.375)^2(\sqrt{80})$$
$$= (29.7)(1.375)(1.375)(8.94)$$
$$= 502 \text{ gpm}$$

Second, change the siamese lines into 2½-inch equivalents. Since one of the lines is a 2½-inch line, it is only necessary to determine the equivalent length for the 3-inch line. The equivalent line can be solved by using

$$E \; 2\tfrac{1}{2}\text{-inch} = L/C$$

where

$$L = 400 \text{ feet}$$
$$C = 2.95 \text{ (from Table 8.1)}$$

Then

$$E \; 2\tfrac{1}{2}\text{-inch} = 400/2.95$$
$$= 135.6 \text{ or } 136 \text{ feet}$$

The layout now appears as shown in Figure 8.41.

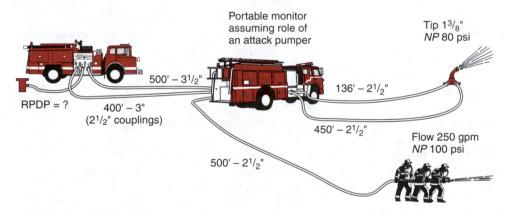

FIGURE 8.41 ◆ Portable monitor assuming the role of an attack pumper.

Next, average the length of the lines in the siamese assembly. This is done by adding the lengths of the two lines and dividing by two:

$$\text{Average length} = \frac{450 + 136}{2}$$
$$= \frac{586}{2}$$
$$= 293 \text{ feet}$$

The layout now appears as shown in Figure 8.42. Now, solve for the friction loss in the siamese:

$$FL = CQ^2L$$

where

$$C = .5 \text{ (from Table 7.1)}$$
$$Q = 502/100 = 5.02$$
$$L = 293/100 = 2.93$$

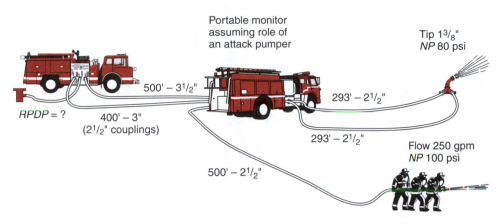

Portable monitor
assuming role of
an attack pumper

Tip 1³/₈"
NP 80 psi

500' – 3¹/₂"

293' – 2¹/₂"

RPDP = ?

400' – 3"
(2¹/₂" couplings)

293' – 2¹/₂"

Flow 250 gpm
NP 100 psi

500' – 2¹/₂"

FIGURE 8.42 ◆

Then $$FL = (.5)(5.02)(5.02)(2.93)$$
$$= 36.92 \text{ psi}$$

Next, solve for the required pump discharge pressure for the portable monitor, using the formula

$$RPDP = NP + FL + AFL$$

where $$NP = 80 \text{ psi}$$
$$FL = 36.92 \text{ psi}$$
$$AFL = 25 \text{ psi (master stream)}$$

Then $$RPDP = 80 + 36.92 + 25$$
$$= 141.92 \text{ psi}$$

Part two. This requires determining the pump discharge pressure for the single 2½-inch line. First, determine the friction loss in the line:

$$FL = CQ^2L$$

where $$C = 2 \text{ (Table 6.1)}$$
$$Q = 250/100 = 2.5$$
$$L = 500/100 = 5$$

Then $$FL = (2)(2.5)(2.5)(5)$$
$$= 62.5 \text{ psi}$$

Next, solve for the required pump discharge pressure, using the formula

$$RPDP = NP + FL$$

where $$NP = 100 \text{ psi}$$
$$FL = 62.5 \text{ psi}$$

Then $$RPDP = 100 + 62.5$$
$$= 162.5 \text{ psi}$$

Part three. This part requires determining the friction loss in the siamese assembly taken off Engine 7. It is first necessary to change the lines in the layout into 2½-inch equivalents. First change the 3-inch line:

$$E \text{ 2½-inch} = L/C$$

where $\qquad\qquad\qquad\qquad\qquad\qquad$ $L = 400$ feet

$\qquad\qquad\qquad\qquad\qquad\qquad\qquad$ $C = 2.5$ (Table 8.1)

Then $\qquad\qquad\qquad\qquad\qquad$ $E\ 2\frac{1}{2}\text{-inch} = 400/2.5$

$\qquad\qquad\qquad\qquad\qquad\qquad\qquad$ $= 160$ feet

Now, change the 3½-inch line:

$$E\ 2\tfrac{1}{2}\text{-inch} = L/C$$

where $\qquad\qquad\qquad\qquad\qquad\qquad$ $L = 500$

$\qquad\qquad\qquad\qquad\qquad\qquad\qquad$ $C = 5.88$ (Table 8.1)

Then $\qquad\qquad\qquad\qquad\qquad$ $E\ 2\frac{1}{2}\text{-inch} = 500/5.88$

$\qquad\qquad\qquad\qquad\qquad\qquad\qquad$ $= 85.03$ or 85 feet

The layout now appears as shown in Figure 8.43.

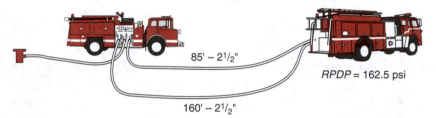

FIGURE 8.43 ◆

The next step is to average the lengths in the siamese assembly. This is accomplished by adding the two revised lengths and dividing by two:

$$\text{Average length} = \frac{160 + 85}{2}$$

$$= \frac{245}{2}$$

$$= 122.5 \text{ or } 123 \text{ feet}$$

The siamese assembly now appears as shown in Figure 8.44.

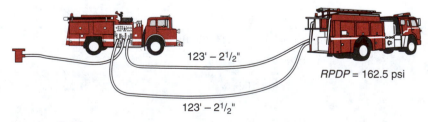

FIGURE 8.44 ◆

The assembly is now in a form to solve for the friction loss:

$$FL = CQ^2 L$$

where $\qquad\qquad\qquad\qquad\qquad$ $C = .5$ (from Table 7.1)

$$Q = \text{flow from monitor} + \text{flow from single}$$
$$2\tfrac{1}{2}\text{-inch line divided by } 100$$
$$= \frac{502 + 250}{100}$$
$$= 752/100 = 7.52$$
$$L = 123/100 = 1.23$$

Then
$$FL = (.5)(7.52)(7.52)(1.23)$$
$$= 34.78 \text{ psi}$$

Part four. Now the problem can be solved for the required pump discharge pressure for Engine 7. The pressure requirement that needs to be inserted into the formula is the greatest between that required for the portable monitor and that required for the single 2½-inch line. This pressure requirement has to be added to the friction loss in the siamese assembly from Engine 7 plus the appliance friction loss in the portable manifold. In this problem, the single 2½-inch line requires the greater pressure. The formula to use is

$$RPDP = \text{pressure requirement for } 2\tfrac{1}{2}\text{-inch line} + FL + AFL$$

where
$$\text{Pressure requirement} = 162.5 \text{ psi}$$
$$FL = 34.78$$
$$AFL = 10 \text{ psi}$$

Then
$$RPDP = 162.5 + 34.78 + 10$$
$$= 207.28 \text{ psi} \quad ■$$

QUESTION   This problem is not a practical layout, but is set up for demonstrative purposes. Three lines have been taken off Engine 5 and extended down a 12 percent grade, where they are siamesed into a single 2-inch line 350 feet in length. Two firefighters are using the 2-inch line, through which is flowing 200 gpm of water at 100 psi. The three lines taken off Engine 5 are 400 feet of 2-inch hose, 350 feet of 1½-inch hose, and 300 feet of 1¾-inch hose. What is the required pump discharge pressure for Engine 5 (Figure 8.45)?

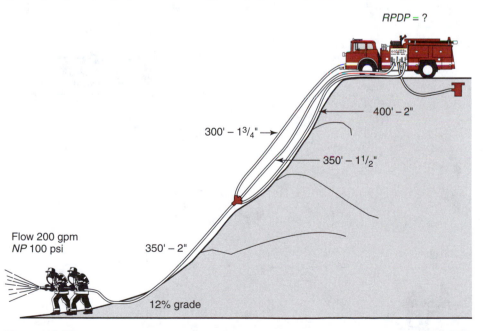

RPDP = ?

400' – 2"

300' – 1³/₄"

350' – 1¹/₂"

Flow 200 gpm
NP 100 psi

350' – 2"

12% grade

FIGURE 8.45   ◆

ANSWER This problem is divided into three parts. Part one requires solving for the friction loss in the siamese assembly. Part two determines the friction loss in the single 2-inch line. Part three requires solving for the required pump discharge pressure for Engine 5.

Part one. This part requires that each of the lines in the siamese assembly be changed to 2½-inch equivalents. First, change the 400 feet of the 2-inch line:

$$E\ 2\text{½-inch} = L/C$$

where
$$L = 400 \text{ feet}$$
$$C = .25 \text{ (from Table 8.1)}$$

Then
$$E\ 2\text{½-inch} = 400/.25$$
$$= 1600 \text{ feet}$$

Next, solve for the 350 feet of 1½-inch hose:

$$E\ 2\text{½-inch} = L/C$$

where
$$L = 350$$
$$C = .08 \text{ (from Table 8.1)}$$

Then
$$E\ 2\text{½-inch} = 350/.08$$
$$= 4375 \text{ feet}$$

Next, change the 300 feet of 1¾-inch hose:

$$E\ 2\text{½-inch} = L/C$$

where
$$L = 300$$
$$C = .13 \text{ (from Table 8.1)}$$

Then
$$E\ 2\text{½-inch} = 300/.13$$
$$= 2307.69 \text{ or } 2308 \text{ feet}$$

The siamese assembly now appears as shown in Figure 8.46.

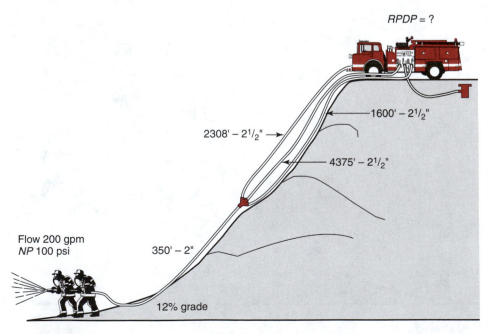

FIGURE 8.46 ◆

Now average the length of the changed lines by adding the three lines and dividing by three:

$$\text{Average } 2\tfrac{1}{2}\text{-inch} = \frac{\text{Line 1 + Line 2 + Line 3}}{3}$$

$$= \frac{1600 + 4375 + 2308}{3}$$

$$= \frac{8283}{3}$$

$$= 2761 \text{ feet}$$

The siamese assembly now appears as shown in Figure 8.47.

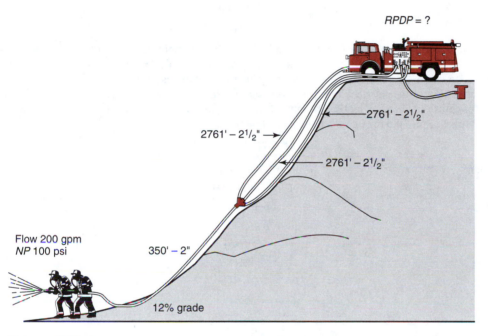

RPDP = ?

2761' – 2¹/₂"

2761' – 2¹/₂"

2761' – 2¹/₂"

Flow 200 gpm
NP 100 psi

350' – 2"

12% grade

FIGURE 8.47 ◆

Now solve for the friction loss in the siamese:

$$FL = CQ^2L$$

where

$$C = .22 \text{ (from Table 7.1)}$$

$$Q = 200/100 = 2$$

$$L = 2761/100 = 27.61$$

Then

$$FL = (.22)(2)(2)(27.61)$$

$$= 24.3 \text{ psi}$$

Part two. This part requires solving for the friction loss in the single 2-inch line used by the firefighters:

$$FL = CQ^2L$$

where

$$C = 8 \text{ (from Table 6.1)}$$

$$Q = 200/100 = 2$$

$$L = 350/100 = 3.5$$

Then
$$FL = (8)(2)(2)(3.5)$$
$$= 112 \text{ psi}$$

Part three. This part involves determining the required pump discharge pressure for Engine 5. The formula to use is

$$RPDP = NP + TFL - FP$$

No appliance friction loss is involved since the flow is less than 350 gpm.

The total friction loss is that in the single 2-inch line plus that in the siamese assembly:

$$TFL = FL \text{ in 2-inch line}$$
$$+ FL \text{ in siamese assembly}$$
$$= 112 + 24.3$$
$$= 136.3 \text{ psi}$$

Determining the forward pressure is a little unusual. It might be thought that the working line is being used at a point corresponding to the longest line laid in the siamese assembly that is laid downhill. However, that is not the case. The working line is actually being used at a point equal to that of the shortest line. Therefore, the length of the shortest line in the siamese assembly plus the line used by the firefighters is 300 + 350 = 650 feet. In actual practice, it is less than this because the firefighters need some line to maneuver. Use

$$H = GL$$

where
$$G = 12 \text{ percent}$$
$$L = \text{length}/100$$
$$= 650/100$$
$$= 6.5$$

Then
$$H = (12)(6.5)$$
$$= 78 \text{ feet}$$

Now,
$$FP = .5H$$
$$= (.5)(78)$$
$$= 39 \text{ psi}$$

Now, solve for the required pump discharge pressure:

$$RPDP = NP + FL - FP$$

where
$$NP = 100 \text{ psi}$$
$$FL = 136.3 \text{ psi}$$
$$FP = 39 \text{ psi}$$

Then
$$RPDP = 100 + 136.3 - 39$$
$$= 197.3 \text{ psi}$$

QUESTION Pumper A is connected to a hydrant and is supplying Pumper B through 800 feet of 5-inch hose. Pumper B is supplying water to a detachable ladder pipe, which is raised to a height of 60 feet and through which water is flowing at 500 gpm at a nozzle pressure of 100 psi. A 100-foot length of 3-inch line with 2½-inch couplings extends from the ladder pipe to the ground, where it is supplied by three lines from Pumper B. Two of the lines are 3½-inch lines each 400 feet in length. The third is 500 feet of 3-inch line with 3-inch couplings. What is the required pump discharge pressure on Pumper A to provide Pumper B with a 20-psi suction pressure (Figure 8.48)?

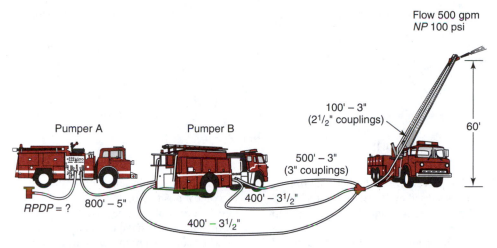

Flow 500 gpm
NP 100 psi

Pumper A

Pumper B

100' – 3"
(2¹/₂" couplings)

500' – 3"
(3" couplings)

60'

RPDP = ?

800' – 5"

400' – 3¹/₂"

400' – 3¹/₂"

FIGURE 8.48 ◆

ANSWER This problem has been inserted to emphasize the need to analyze a problem prior to trying to solve it. Once the question is analyzed, it becomes apparent that all the information necessary to solve the problem as presented is available without the necessity of bothering with the problems with which Pumper B is concerned. The pump discharge pressure at Pumper A is that necessary to supply the amount of water required to Pumper B. All that is necessary to solve the problem as presented is to determine the friction loss in the 800 feet of 5-inch hose and add the 20 psi required as a suction pressure at Pumper B. Thus,

$$FL = CQ^2L$$

where

$$C = .08 \text{ (from Table 6.1)}$$
$$Q = 500/100 = 5$$
$$L = 800/100 = 8$$

Then

$$FL = (.08)(5)(5)(8)$$
$$= 16 \text{ psi}$$

Now solve for the required pump discharge pressure at Pumper A. Use the formula

$$RPDP = FL + 20$$
$$= 16 + 20$$
$$= 36 \text{ psi}$$

With a strong hydrant, the results could be obtained without the use of Pumper A. ■

■ ■

Summary of Chapter Formulas

To change hose other than 2½-inch hose into 2½-inch hose equivalent, use

$$E\ 2\tfrac{1}{2}\text{-inch} = L/C$$

where $E\ 2\tfrac{1}{2}$-inch = equivalent length of 2½-inch hose for the line being considered

L = length of line being considered

C = coefficient from Table 8.1

An additional formula for the same purpose is

$$E\ 2\frac{1}{2}\text{-inch} = (L/100) \times C2$$

where L = length of hose to be changed

$C2$ = equivalent length from Table 8.2

- -

Review Questions

1. What percentage of the national fire loss is caused by 10 percent of the fires?
2. What can cause complacency in a pump operator?
3. Define an attack pumper as used in this book.
4. Name three common methods of supplying water to an attack pumper.
5. What appliances may be used in the role of an attack pumper?
6. What is the general procedure for solving problems when several single lines are taken off a pumper?
7. What is meant by feathering an outlet?
8. When does a feathering operation become a positive operation?
9. What formula can be used to ensure that all factors have been considered when solving a pump discharge pressure problem?
10. Under what condition does a pumper that has several lines taken off it assume the role of a supply pumper?
11. What is the formula for changing hose other than 2½-inch hose into the 2½-inch-hose equivalent when using Table 8.1?
12. What is a second formula that can be used for this purpose when using Table 8.2?

- -

Test Eight

It is suggested that drawings be made when solving these problems.

1. Three lines have been taken off Engine 2. One is a single 2½-inch line 400 feet in length. It is equipped with a 1¼-inch tip at 50 psi. The second is 200 feet of 2-inch hose through which water is flowing at 175 gpm at 100 psi. The third is 250 feet of 1½-inch hose through which water is flowing at 95 gpm at 100 psi. What is the required pump discharge pressure for Engine 2?

2. Engine 6 is supplying a portable manifold through 400 feet of 3-inch hose with 2½-inch couplings. Three lines have been taken off the manifold. One is 100 feet of 1½-inch hose through which water is flowing at 60 gpm at 100 psi nozzle pressure. The other two lines are 150 feet of 1½-inch hose. One is equipped with a ½-inch tip at 50 psi nozzle pressure. Water is flowing through the other at 110 gpm at 100 psi nozzle pressure. What is the required pump discharge pressure for Engine 6?

3. A 400-foot length of 2½-inch hose is laid from Engine 2 and wyed into two lines. One of the lines is 150 feet of 2-inch hose laid up a 12 percent grade from the wye. Water flows through it at 190 gpm at 75 psi nozzle pressure. The second line is laid downhill from the wye to a point 20 feet below the wye. Water flows through it at 95 gpm at 75 psi nozzle pressure through 200 feet of 1½-inch hose. What is the required pump discharge pressure at Engine 2?

4. Engine 7 has laid 400 feet of 3½-inch hose into the standpipe of a twelve-story building. A single 2½-inch line 150 feet in length and equipped with a 1-inch tip operating at 50 psi nozzle pressure has been

taken off the outlet on the sixth floor. The crew is working on a fire on that floor. A 150-foot line of 1½-inch hose has been taken off the outlet on the seventh floor and water is flowing through it at 125 gpm at 100 psi nozzle pressure to prevent the fire from extending up the interior stairwell. What is the required pump discharge pressure for Engine 7?

5. A 700-foot length of 3½-inch hose has been laid down a 12 percent grade into the standpipe of a ten-story building. A line has been taken off on the eighth floor and is working on a fire there. The line is a 150-foot length of 2½-inch hose through which water is flowing at 210 gpm at 100 psi nozzle pressure. What is the required pump discharge pressure for the pumper supplying the 3½-inch line?

6. The crew from Engine 4 has laid 500 feet of 5-inch hose from the pumper into a portable manifold. Three lines have been taken off the manifold. One is a single 2½-inch 200 feet in length through which water is flowing at 225 gpm at 100 psi nozzle pressure. The other two lines have been laid into a portable monitor equipped with a 1½-inch smooth-bore nozzle with a nozzle pressure of 80 psi. The two lines are each 350 feet of 2½-inch hose. What is the required discharge pressure for Engine 4?

7. What is the equivalent length of 2½-inch hose for 800 feet of 4-inch hose?

8. What is the equivalent length of 2½-inch hose for 300 feet of 1¾-inch hose?

9. Change 400 feet of 3-inch hose with 2½-inch couplings into an equivalent length of 2½-inch hose.

10. Three lines have been taken off Pumper A and siamesed into a single 3-inch line. One of the lines is 300 feet of 3-inch hose with 3-inch couplings. Another is

350 feet of 3-inch hose with 2½-inch couplings, and the third is 400 feet of 3½-inch hose. If this layout is changed into a layout of equivalent 2½-inch hose, what is the length of each of the lines in the 2½-inch siamese?

11. The three lines in a siamese assembly are 200 feet of 2-inch hose, 250 feet of 1¾-inch hose, and 150 feet of 1½-inch hose. What is the length of each of the lines in a 2½-inch siamese that is equivalent to this layout?

12. What is the length of the lines in a 2½-inch siamese if the friction loss is equivalent to that in a siamese assembly that contains one 450-foot length of 3½-inch hose, one 400-foot length of 4-inch hose, and one 500-foot length of 5-inch hose?

13. Three lines have been taken off Pumper B and laid into the connections on an elevated platform that is raised to a height of 80 feet and discharging 800 gpm at 100 psi nozzle pressure. One of the lines is 400 feet of 4-inch hose, one is 350 feet of 3½-inch hose, and one is 450 feet of 3-inch hose with 3-inch couplings. What is the required pump discharge pressure for Pumper B?

14. Two 400-foot lengths of 4-inch hose have been taken from Engine 9 and laid into a portable manifold. Four lines have been taken off the manifold. Two of them are 2½-inch lines each 300 feet in length that are used to supply a portable monitor through which water is flowing at 500 gpm at 100 psi. One of the other two lines is a 3-inch line with 2½-inch couplings. It is 350 feet in length. The fourth line is 400 feet of 3½-inch hose. These two lines are laid into a deck gun that is 9 feet off the ground with a flow of 600 gpm at 100 psi. What is the required pump discharge pressure for Engine 9?

CHAPTER 9

Pumping Capacity and Drafting Operations

Objectives

Upon completing this chapter, the reader should:

- Understand the principles of pump capacity.
- Understand what happens when a net pump pressure of 150 psi is exceeded.
- Understand the work capability of a pump.
- Be able to determine the approximate amount of water that a pump is capable of delivering when the discharge pressure exceeds 150 psi net pump pressure.
- Understand what happens and how to recognize when a pump is cavitating.
- Know the difference between absolute and relative pressure.
- Be able to determine the pressure reduction within a pump by reading the number of inches of mercury on the compound gage.
- Be able to explain the principle of lifting water when drafting.
- Be able to determine approximately how high water can be lifted by reading the number of inches of mercury on the compound gage.
- Be able to determine the number of inches of mercury required on the compound gage in order to lift water a given distance.
- Be able to explain the effects of altitude and weather on drafting operations.
- Be able to determine the net pump pressure when taking water from a hydrant or when conducting drafting operations.
- Be able to explain the steps required when conducting drafting operations.
- Be able to explain how to set up drafting operations from a broken connection.

◆ PUMP CAPACITY

Exceeding the pump capacity does not normally present a problem in the day-to-day operations of most fire departments. In fact, in some communities a pump operator might never encounter a situation where the limitation of his or her pump may be tested. When these situations do exist, however, a thorough knowledge of pump limitations is required. The limitation of a pump is most likely to be challenged when a master stream has been placed into operation.

Of all the types of fire streams used, the chance of an ineffective stream being developed is most likely to occur with a master stream. Insufficient nozzle pressure has traditionally been given as the cause of these ineffective streams; however, in many cases they are caused by the pump operator exceeding the capacity limitation of the pump and not recognizing it.

Most pumpers in general use are rated at a given volume at 150 psi net pump pressure; at 70 percent of the rated capacity at 200 psi net pump pressure; and at 50 percent of the rated capacity at 250 psi net pump pressure. A complete discussion of **net pump pressure** is included in the section on drafting operations. At this point is should be understood to mean the pressure actually produced by the pump. It is important that pump operators and fire officials have a good foundation on pump ratings and how they are related to the net pump pressure.

It should be apparent from a quick glance at the ratings that less water can be discharged by a pump whenever the net pump pressure exceeds 150 psi than when the net pump pressure is kept at 150 psi or less. However, pump operators can be trapped into trying to discharge additional water by increasing the discharge pressure when, in fact, an increase in pressure will have the opposite effect. Following is an example of how a pump operator can be trapped into exceeding the capacity of his or her pump.

A pump operator assigned to a 1250-gpm pumper at a fire has been given the responsibility of supplying a 2-inch smooth-bore tip. The pump operator calculates that a certain discharge pressure is required in order to obtain the desired nozzle pressure. A 2-inch tip discharges approximately 1000 gpm at a nozzle pressure of 80 psi. If the pump operator determines that the discharge pressure required to produce an effective stream is one where the net pump pressure is 150 psi, then when this pressure is set on the discharge gate, the stream produced from the 2-inch tip will be too weak to be considered effective. This is probably because the friction loss in the hose is too great.

When the water does not reach the fire, the natural reaction of the person operating the nozzle is to relay to the pump operator that additional pressure is required. The pump operator tries to comply by increasing the discharge pressure, but does not consider the limitation of the pump. The pump operator's action pushes the net pump pressure beyond 150 psi and results in a reduction of the water supply to the tip, causing the stream to become even less effective. The firefighter operating the nozzle sees that the stream is losing some of its power rather than increasing the reach as expected. The nozzle operator thinks that the pump operator must have misunderstood what he or she wanted; therefore, he or she repeats the request for more pressure. When the pump operator tries to comply, the situation worsens.

If the pump operator had recognized at the first request that the pump was operating at its limit, the operator would not have complicated the problem by trying to comply. Instead he or she would have taken different steps to try to correct the situation. Perhaps a reduction in the tip size or an additional line laid from the pumper into the appliance might have resulted in an effective stream. If these measures had been ineffective, an additional pumper could have been requested to assist in supplying the appliance. The result would have been more water on the fire at the proper nozzle pressure. However, because the pump operator did not recognize that the pump limitation had been reached, conditions deteriorated. Knowing when the pump limitation has been reached requires a thorough understanding of the basic concept of pump capability.

PUMP CAPABILITY

A pump is like any other machine—it is capable of doing a certain amount of work and no more. For example, 1 horsepower is the power required to lift 550 pounds 1 foot in 1 second. A 4-horsepower engine can then be said to be capable of 2200 foot-pounds of work in 1 second. A 4-horsepower engine can lift 1100 pounds 2 feet in 1 second or 4400 pounds ½ foot in 1 second, as the capacity in either instance is 2200 foot-pounds per second.

A pump, on the other hand, should be considered capable of producing a given number of pound-gallons per minute. The pound-gallons capability of a pump can be determined by multiplying the rated capacity of the pump by the rated pressure. As an example, a pump rated at 1000 gpm at 150 psi net pump pressure has a work capability of 150,000 pound-gallons (1000 × 150) in 1 minute. A 1250-gpm pump has a work capability of 187,500 pound-gallons (1250 × 150) in 1 minute.

This work capability of a pump may be somewhat misleading, for a pump is not capable of delivering more water than its rated capacity; however, it does deliver less water whenever the initial rated net pump pressure is exceeded. As an example, a pump rated at 1000 gpm at 150 psi net pump pressure is not capable of delivering 1500 gpm; however, it will deliver less than the 1000 gpm whenever the net pump pressure of 150 psi is exceeded. The approximate amount that can be delivered can be determined by using the following formula:

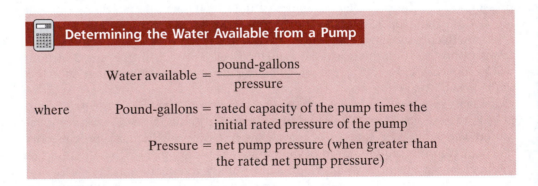

Determining the Water Available from a Pump

$$\text{Water available} = \frac{\text{pound-gallons}}{\text{pressure}}$$

where

Pound-gallons = rated capacity of the pump times the initial rated pressure of the pump

Pressure = net pump pressure (when greater than the rated net pump pressure)

The application of this theory can best be illustrated by an example.

QUESTION Approximately how much water can be supplied at 200 psi net pump pressure by a pump rated at 1000 gpm at 150 psi net pump pressure?

ANSWER

$$\text{Water available} = \frac{\text{pound-gallons}}{\text{pressure}}$$

where

Pound-gallons = 150,000

Pressure = 200 psi

Then

$$\text{Water available} = \frac{150,000}{200}$$

$$= 750 \text{ gpm}$$

■

QUESTION Approximately how much water can be supplied at 250 psi net pump pressure by a pump rated at 1000 gpm at 150 psi net pump pressure?

ANSWER

$$\text{Water available} = \frac{\text{pound-gallons}}{\text{pressure}}$$

where

$$\text{Pound-gallons} = 150{,}000$$
$$\text{Pressure} = 250 \text{ psi}$$

Then

$$\text{Water available} = \frac{150{,}000}{250}$$

$$= 600 \text{ gpm} \quad\blacksquare$$

It should be noted that according to these two problems the number of gallons available is greater than can be expected according to pump ratings. The formulas indicate that 750 gpm can be supplied at 200 psi and 600 gpm at 250 psi, whereas pump ratings are 700 gpm and 500 gpm, respectively. The primary reason for the difference is that the formula is based on the theory of 100 percent efficiency, whereas the actual pump ratings consider pump slippage and other operational factors that result in less than 100 percent efficiency. The chief reason for introducing the capacity formula is to provide a foundation for the understanding of pump limitations.

Pump operators and fire officials should be aware of the critical nature of pump limitations. The seriousness is far greater when working from draft than from a hydrant. When pumping from draft, the pump must work to lift the water from the source into the pump as well as to discharge the water. When water is taken from a hydrant, on the other hand, the pump is relieved of some of the work (see the section Net Pump Pressure later in this chapter).

It should be noted that whenever a pump is drafting, the net pump pressure is greater than the discharge pressure. This means that the rated capacity of the pump is not available at a discharge pressure of 150 psi. For example, the pressure required at the base of a ladder pipe that is raised to 60 feet and supplied by a 100-foot section of 3-inch line is approximately 45 psi. Therefore, it is theoretically impossible for a 1000-gpm pumper to supply a 2-inch tip on one of these appliances when the supplying pumper is taking water from draft. In fact, it would be difficult to supply a 1¾-inch tip under these conditions. It is therefore wise to limit the tip size to 1½ inches whenever the pump used to supply a ladder pipe is rated at 1000 gpm and water is being taken from draft. Of course, if a larger tip is required, it is best to supply the appliance through the use of two pumpers.

Operating from a positive water source such as a hydrant gives the pump operator more leeway; however, caution must still be taken to ensure that the limitation of the pump is not exceeded. For example, a 1250-gpm pump cannot supply a 1000-gpm fixed-gallonage fog nozzle if the required discharge pressure is 200 psi and the incoming pressure from the hydrant is bordering on zero at that point. It would take a residual pressure of 30 to 40 psi for the pump operator to feel safe.

Pump operators who are assigned to 1500-gpm pumpers and consistently work from a strong water supply system can generally meet the requirements of 2-inch tips and 1000-gpm fixed-gallonage fog nozzles. However, even these operators may be challenged if small supply lines to the appliance and excessively long lays are used. No pump operator can be completely sure that he or she will be capable of meeting the demands placed on the pump under all circumstances. Every operator must be aware of the pump's limitations and know how to cope with them.

CAVITATION

When a pump operator attempts to discharge more water than the pump is capable of producing, it is said that he or she is running away from the water. This means that an attempt is being made to discharge more water than is entering the pump. This results in a condition referred to as **cavitation**. Cavitation can occur when operating from a positive pressure source, but is more likely to happen when drafting.

Attempting to discharge more water than is entering the pump results in cavities being created in the water as it enters the eye of the impeller. Cavitation can cause serious damage to a pump.

There are several indicators to the pump operator that cavitation is taking place or about to take place. Some people suggest that cavitation sounds like a bunch of rocks rattling around inside the pump. A better indication than listening for this sound is to pay attention to the discharge pressure gage. If the number of revolutions per minute (rpm's) of the pump is increased with no increase of discharge pressure, cavitation is taking place. Another indication is the intake suction hose starting to go soft when one is working from a positive-pressure source.

◆ DRAFTING OPERATIONS

Drafting operations involve taking water from a source other than a pressure hydrant or from a source that does not deliver water to the intake of the pump under pressure. Ponds, swimming pools, rivers, the ocean, and other such bodies of water are drafting sources. To successfully obtain water from these sources, a thorough knowledge of drafting principles and procedures is required.

ABSOLUTE AND RELATIVE PRESSURE

Pumpers are equipped with a minimum of two gages, one connected to the discharge side of the pump and one connected to the inlet side of the pump. The gage connected to the discharge side of the pump is generally referred to as a pressure gage and the one connected to the inlet side is generally referred to as a **compound gage**. The pressure gage registers pressure above atmospheric pressure, whereas the compound gage registers pressure both above and below atmospheric pressure.

Absolute zero pressure refers to a complete absence of pressure, or a perfect vacuum. **Absolute pressure** is pressure above absolute zero and is denoted psia.

Atmospheric pressure is measured in terms of absolute pressure. Normal atmospheric pressure at sea level is 14.7 psi.

Gage pressure is pressure above atmospheric pressure and is denoted psig. At sea level, zero gage pressure is an absolute pressure of 14.7 psi.

The zero reading on both the compound gage and the pressure gage on a pumper is measured in terms of gage pressure—an absolute pressure equal to the atmospheric pressure. The pressure readings on both gages are in psi or in pressure relative to atmospheric pressure. At sea level, for example, a gage pressure of 50 psi is equal to an absolute pressure of 64.7 psi (50 + 14.7)(Table 9.1).

TABLE 9.1 ◆ Comparison of Absolute and Relative Pressures	
39.7	25
29.7	15
19.7	5
14.7	0
10	10″
5	20″
0	30″
Absolute Pressure	Relative Pressure

The readings below zero on the compound gage are measured in **inches of mercury** (Hg). One inch of mercury is equal to an absolute pressure of .49 psi. A pressure of .49 psi at the bottom of a 1-square-inch container will support a column of water approximately 1.13 feet high.

An atmospheric pressure of 14.7 psi is equal to 29.92 inches of mercury (commonly rounded to 30 inches). A pressure of 14.7 psi at the base of a 1-square-inch container will support a column of water approximately 33.9 feet high.

Pressure above atmospheric pressure is generally referred to as *positive pressure* and pressure below atmospheric pressure is generally referred to as *negative pressure*. Another way of expressing these pressures is to refer to them as **relative pressures**, that is, pressures relative to atmospheric pressure. A comparison between absolute pressure and relative pressures at sea level is shown in Table 9.1.

The compound gage on a pumper registers pressure similar to that shown as relative pressure in Table 9.1. It can be seen from Table 9.1 that a reading of 10 inches of mercury on the compound gage approximates a pressure reduction within the pump of 5 psi and a reading of 20 inches of mercury approximates a reduction of 10 psi. The actual reduction can be determined by establishing a proportion.

QUESTION What pressure reduction is equal to a reading of 10 inches of Hg on the compound gage?

ANSWER

$$\frac{10}{30} = \frac{x}{14.7}$$

Cross-multiply to get

$$\frac{10}{30} \diagdown \frac{x}{14.7}$$

$$30x = (10)(14.7)$$
$$30x = 147$$
$$x = 4.9 \text{ psi}$$

Note: This problem was left in its original condition for demonstrative purposes. The fraction on the left-hand side (10/30) was not reduced to its lowest term. ■

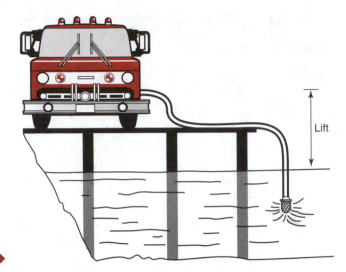

FIGURE 9.1 ◆

LIFT

Lift is the vertical distance from the surface of the water to the center of the pump when the pump is drafting (Figure 9.1). **Maximum lift** is the maximum height a pumper can draft water. Maximum lift is affected by the design and condition of the pump, the adequacy and the condition of the pumping engine, the size and the condition of the suction hose and the strainer, the atmospheric pressure, and the temperature of the water.

Theoretically, if a pump at sea level could produce a perfect **vacuum**, it could lift water to a height of 33.9 feet; however, no pump on a fire apparatus is capable of producing a perfect vacuum. Even if a pump could produce a perfect vacuum, the friction loss in the suction hose would restrict the height of the lift. If a pump is in excellent condition and drafting at sea level, it should be capable of obtaining a lift of 25 feet. However, under practical conditions, it is best to assume that the maximum lift is restricted to approximately 23 feet.

THE PRINCIPLE OF LIFTING WATER

Water will move from one location to another whenever a difference of pressure exists between the locations and there is a clear path of travel between them.

In the normal procedure for drafting operations, the pump operator engages a positive-displacement priming pump and then opens a line between the main **centrifugal pump** and the **priming pump**. What happens is best illustrated in Figure 9.2, where the following steps can be observed:

1. The pump operator engages **positive-displacement pump** *A*. Pump *A* is referred to as the priming pump. Positive-displacement pumps are capable of pumping air as well as water, and thus they are capable of creating a **partial vacuum** or a reduction of pressure.
2. The pump operator opens valve *C* between the priming pump *A* and the main pump *B*. On some pumps, a valve opens automatically when the priming pump is engaged.
3. The priming pump turns and removes air from the main pump *B* and discharges it through hose *D*.

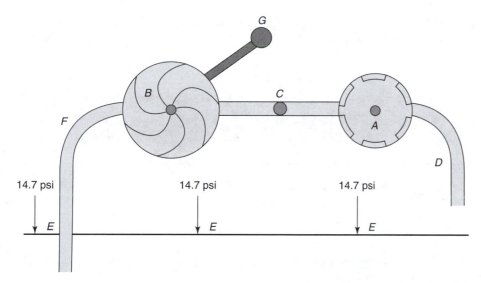

FIGURE 9.2 ◆ The principle of lifting water. (*A*) Positive-displacement priming pump; (*B*) main centrifugal pump; (*C*) valve between priming pump and main pump; (*D*) discharge outlet from priming pump; (*E*) atmospheric pressure; (*F*) suction hose from main pump to body of water (the end of the suction hose should be at least 2 feet below the water surface).

4. The result of the removal of air from pump *B* is a reduction of pressure within pump *B*, which is registered in inches of mercury on compound gage *G*. For example, suppose that the compound gage reads 18 inches of mercury. This is a pressure reduction of

$$\frac{18}{30} = \frac{x}{14.7}$$

First, reduce the fraction 18/30 to its lowest terms by dividing both the numerator and denominator by 6. This results in the fraction 3/5. Then cross-multiply:

$$\frac{3}{5} \diagdown\!\!\!\!\diagup \frac{x}{14.7}$$

$$5x = (3)(14.7)$$
$$5x = 44.1$$
$$x = 8.82 \text{ psi}$$

The remaining pressure within the pump is then 14.7 − 8.8, or approximately 5.9 psi.

5. The pressure differential between the 14.7 psi atmospheric pressure at the surface of the water and the 5.9 psi pressure within the main pump *B* causes the water to move up the suction hose (*F*) and into pump *B*.
6. The water continues through valve *C* into priming pump *A* and is discharged through hose *D*.
7. When a solid stream of water is discharged from hose *D*, priming pump *A* is disengaged and valve *C* is closed.
8. Main pump *B* is engaged, which makes water available at the discharge gates.

HOW FAR WILL WATER RISE?

Theoretically, water will rise 2.3 feet for each pound of pressure reduction. The amount of pressure reduction can be determined by multiplying the vacuum reading in inches of mercury by .49. Water will rise 1.13 feet for each inch-of-mercury reduction (.49 × 2.3), so the height to which water will theoretically rise can be determined by using the following formula:

Determining the Theoretical Height to Which Water will Rise When Drafting

$$h = 1.13\,Hg$$

where

$$h = \text{height in feet}$$

$$Hg = \text{inches of mercury}$$

QUESTION The compound gage on a pumper reads 18 inches of mercury. To what height will water theoretically rise?

ANSWER $h = 1.13Hg$

where $Hg = 18$

Then $h = (1.13)(18)$

$= 20.34 \text{ feet}$ ■

The figures obtained by using this formula are the theoretical heights to which water will rise. In actual practice, such heights are not obtained due to minute **air leaks**, friction loss in the hose, and so forth. A good rule of thumb is to estimate a rise of approximately 1 foot per inch of mercury.

Theoretically, it might be desirable to know what reading is required in order to lift water to a given height. This information can be obtained by using the following formula:

Determining the Pressure Reduction Required to Lift Water a Given Height

$$\text{Required } Hg = .885h$$

where

$$Hg = \text{inches of mercury}$$

$$h = \text{height}$$

QUESTION What reading is required on the compound gage in order to lift water 20 feet?

ANSWER $Hg = .885h$

where $h = 20$

Then $Hg = (.885)(20)$

$= 17.7$ ■

EFFECTS OF ALTITUDE ON DRAFTING OPERATIONS

There is a reduction of approximately ½ pound of atmospheric pressure for each 1000 feet of elevation. The reduction in atmospheric pressure directly affects the height to which water can be lifted. The approximate reduction in maximum lift due to elevation can be determined by establishing a proportion formula.

QUESTION   A pumper at sea level can lift water to a maximum height of 22 feet. What is the maximum height-lifting capability of the pump at an elevation of 4000 feet, where the atmospheric pressure is 12.7 psi?

ANSWER

$$\frac{12.7}{14.7} = \frac{x}{22}$$

Cross-multiply to get

$$\frac{12.7}{14.7} \diagdown \frac{x}{22}$$

$$14.7x = (12.7)(22)$$

$$14.7x = 279.4$$

$$x = 19 \text{ feet}$$

■

Nineteen feet is the maximum theoretical height to which the pumper can lift water at the 4000-foot elevation, assuming that the engine of the pumper is adequate at all elevations. However, the power generated by an internal combustion engine will decrease approximately 3½ percent for each 1000 feet of elevation. Therefore, an engine that was just adequate at sea level would be about 14 percent deficient at an altitude of 4000 feet.

EFFECTS OF WEATHER CHANGES

Increases or decreases in the barometric pressure due to movement of air masses will have the same effect on the drafting ability of a pump as will a change in elevation. Barometric pressure increases during good weather and decreases during poor weather. A pumper operating on a clear day may be capable of lifting water 1 foot higher than on a rainy day.

QUANTITY OF LIFT

The amount of water that a pumper can be expected to lift depends upon its rated capacity, the size of the suction hose, and the lift distance. The minimum discharge that should be expected of a 1500-gpm pumper in good condition operating at draft with a 20-foot lift and using a 6-inch suction hose is approximately 950 gpm. Full capacity with a 10-foot lift can be achieved with a 6-inch suction hose. Full capacity of 1250-gpm and larger pumpers cannot be achieved when any single suction hose less than 6 inches in size is used; however, in practical operations, maximum capacity can be approached or even achieved at low lifts through the use of dual 4-inch suctions. The dual 4-inch suctions have an equivalent approximate size of a 5.6-inch suction hose.

NET PUMP PRESSURE

Net pump pressure is the amount of pressure actually being produced by the pump. When a pump is connected to a hydrant, the net pump pressure is the difference between the discharge pressure and the intake pressure. For example, if a pumper is producing a satisfactory hose stream from a 1½-inch tip while the discharge pressure is 190 psi and the intake pressure is 40 psi, then the tip is being supplied while the pumper is pumping at a net pump pressure of 150 psi. The discharge from a 1½-inch tip at 80 psi nozzle pressure is approximately 600 gpm. If the pumper is rated at 1000 gpm, then an additional 400 gpm is available for other uses; this is so long as the net pump pressure does not exceed 150 psi, even if the discharge pressure is 190 psi. Of course, it is necessary that the water source be capable of supplying the required flow.

The net pump pressure, then, is the result of the actual work being done by the pump. The pump does just as much work while drafting as it does while forcing water to a height equivalent to the lift. It requires the same amount of effort to lift 500 gpm of water 10 feet while drafting as it does to force 500 gpm to a height of 10 feet. A pumper operating from draft with a 10-foot lift that takes the water and forces it up an additional 30 feet would do the same work as it would if it were forcing water to a height of 40 feet.

When pumping from draft, a pump must produce sufficient pressure to do the following:

1. Provide the desired discharge pressure.
2. Lift the water from the surface of the water to the center of the pump.
3. Overcome the friction loss in the suction hose.

When determining the actual discharge pressure, the gage pressure must be correct for any gage error. The allowances for lift and friction loss in the suction hose are usually given in feet and must be converted to pounds per square inch in order to determine the net pump pressure. The overall result is that there is a difference in the capability of a pumper operating from a hydrant and the same pumper working from draft. As an example, a 1000-gpm pumper operating from a hydrant can supply a 1½-inch tip and have the residual capability of supplying an additional 1¼-inch tip. However, the same pump operating from draft may be able to supply the 1½-inch tip, but would probably not have the residual capability of supplying an additional 2½-inch line. However, a couple of small lines could probably be taken off the pump, but this might push the pump to its capability limitation.

◆ DRAFTING PROCEDURES

A company officer preparing to establish drafting operations should select a spot carefully for placing the pumper. The spot should permit the pumper to be set on level ground and should be able to support the weight of the apparatus under conditions of considerable vibration. If several similar spots are available, it is best to choose the one requiring the lowest lift. Whereas the full capacity of the pumper is available with a lift of 10 feet or less, only approximately 60 percent of the capacity is available with a lift of 20 feet.

The pumper should be placed as close as possible to the water's edge, with the pumping panel preferably located away from the water's edge. The entire crew should assist in setting up for drafting if hard suction hose is used, as the manipulation of the suction hose is a heavy and cumbersome task. Using corrugated hard suction hose or flexible suction hose with light couplings makes the task easier, but it still requires extra help.

In most cases it will take at least two 10-foot lengths of suction hose to reach the water. When selecting the number of lengths to use, allowance should be made for tide movements or other changes in the water level. It is important to remember that if the source of water is limited (such as a swimming pool), the height of the lift will continue to increase, while the depth of the strainer beneath the water will continuously be reduced.

If possible, plans should be made to keep the **suction strainer** at least 2 feet below the surface of the water and at least 1 foot off the bottom at all times. A distance of less than 2 feet will allow whirlpools to form, which will result in the pumper losing its prime.

If early evaluation indicates that it will be difficult to maintain a distance of 2 feet below the surface of the water, then a door or other large, flat surface that will float on the water should be used to break the whirlpool effect. The object selected should, if possible, be placed around the suction hose before the hose is lowered into the water.

The suction hose should be checked for gaskets prior to coupling the sections together. A small leak in a coupling could result in the inability to properly prime the pump. The suction hose should be coupled together, the suction strainer placed on the end, and the tie rope attached to the suction hose before connecting the suction hose to the pump inlet and lowering the hose into the water. Suction hose couplings must be airtight. A rubber mallet can be used to tighten the couplings.

The pump operator should check all **pump drains**, discharge gates, valves, **bleeder valve**, and any other openings into the pump to ensure that they are closed. Caps on unused suction inlets should be tightened with a rubber mallet. Any leakage on either the suction or the discharge side of the pump, no matter how small, might result in failure to prime the pump.

After checking the pump to ensure that all openings are closed and that all couplings and caps are tightened, the priming pump should be engaged and the throttle advanced to the recommended rpm. If the pumper is equipped with a **priming valve**, the valve should be engaged. This will open a line between the priming pump and the main pump. All pumpers do not have a separate priming valve; with some pumpers, the function takes place automatically when the priming pump is engaged. Some pumpers require that the main pump be engaged during priming operations and others do not.

While waiting for water, the pump operator should remain alert for air leaks. A high whistling sound is often an indication of such air leaks.

Water should be received within 30 seconds on pumpers with a capacity of less than 1500 gpm. It may take as long as 45 seconds with 1500-gpm and larger pumpers. If water is not received within the allocated time, the priming pump should be disengaged and the source of trouble sought. The problem is usually one or more of the following: (1) the priming pump is dry and needs lubricating; (2) there is an air leak; (3) the strainer is obstructed; (4) the lift is too high.

If the priming pump is running dry, a quick check of the oil reservoir for the pump will disclose the trouble.

The air leak could be at a number of locations. However, if the pump operator made a complete inspection of all discharge gates, valves, and so on, the leak is probably at a coupling on the suction hose. The most likely place for a leak is the first coupling below the suction inlet.

If the lift is too high or the strainer is obstructed, there may be an extra-high reading on the vacuum gage.

The first indication that water has entered the main pump will be a discharge of water from the priming pump. This will generally be a mixture of water and air. The water should be allowed to flow steadily for several seconds before the priming valve is closed. The main pump should then be engaged and the priming pump disengaged, the process depending upon the configuration of the pumping arrangement.

A discharge gate should then be opened slowly. At the same time, the throttle should be advanced to take up the load. The discharge gate or gates should be opened fully and the throttle advanced until a satisfactory stream is flowing from the working lines. The pressure increase on the discharge gage should be watched until the desired pressure is reached. The suction gage should continue to show a vacuum.

The discharge pressure should continue to increase as the throttle is advanced. If the throttle is advanced without a corresponding increase in pressure, one of two things has happened:

1. The most efficient point of the pump has been passed. The throttle should be backed down to the point where the discharge pressure commences to drop. Operation should be at this pressure. It is useless to increase the throttle beyond this point.
2. The pump has not been completely primed. The discharge pressure should be backed down to a low pressure and the transfer valve shifted from volume to pressure and back to volume. This may clear the **air lock**. If it does not, the pump should be reprimed. Once the pump is operating satisfactorily, the relief valve should be set.

◆ DRAFTING FROM A BROKEN CONNECTION

A **broken connection** is one in which water is taken from a hydrant or other positive pressure source, discharged into an open container, and then taken from this container by a drafting operation (Figure 9.3).

This type of operation may become necessary in the case of a break in a flammable liquid line, where it is necessary to pump water into the break to force the liquid back into the piping. Such an operation generally continues until the flammable liquid lines have been shut down and the liquid drained off.

Some flammable liquids, such as certain types of liquefied petroleum gases, have extremely high vapor pressures. It is possible that the vapor pressure will force the liquid into the hose lines and subsequently back into the water mains. This is not just a theory, it has actually occurred in at least one large city. This can be prevented by the use of broken connections.

Several possible makeshift reservoirs can be used for this type of operation. A hole can be dug in the ground and a salvage cover can be used as a lining; several ladders can be used to form a basin with salvage covers used for linings; or barrels, if

FIGURE 9.3 ◆ Drafting from a broken connection.

available, can be used. The makeshift reservoir can be kept full by the use of open-butt lines from hydrants.

Drafting procedures with this type of operation are the same as routine drafting operations. However, care must be taken to ensure that the makeshift reservoir is kept full, so that the pump does not run away from the water. Care must also be taken that the line used to fill the makeshift reservoir is not directed at the strainer of the suction hose. This causes air to be introduced into the pump, and the pump will lose its prime.

Summary of Chapter Formulas

Determining the Water Available from a Pump

$$\text{Water available} = \frac{\text{pound-gallons}}{\text{pressure}}$$

where   Pound-gallons = rated capacity of the pump times the initial rated pressure of the pump

Pressure = net pump pressure (when greater than rated net pump pressure)

Theoretical Height to Which Water Will Rise

$$h = 1.13 Hg$$

where h = height in feet

Hg = inches of mercury

Pressure Reduction Required to Lift Water a Given Amount

$$Hg = .885h$$

where Hg = inches of mercury

h = height

Review Questions

1. In what circumstance are the limits of a pump most likely to be challenged?
2. What has traditionally been given as the cause of ineffective master streams?
3. What are the ratings of most pumpers in general use?
4. What happens when a net pump pressure of 150 psi is exceeded?
5. How is the capability of a pump determined?
6. What formula can be used to determine the theoretical discharge capacity of a pump whenever the net pump pressure of 150 psi is exceeded?
7. Why is there a difference between the theoretical amount of water that can be delivered by a pump at a given net pump pressure and the rated capacity of the pump at that net pump pressure?
8. Is it most likely that the pump capacity will be exceeded from draft or from a hydrant?
9. Is it safe to assume that a 1250-gpm pump is capable of supplying a 1000-gpm fixed-gallonage fog nozzle?
10. Under what conditions might the capability of a 1500-gpm pumper be challenged?
11. What are drafting operations?
12. What are some of the sources of water for drafting operations?
13. What is the difference between the gage on the discharge side of a pumper and the gage on the intake side of a pumper?
14. What is meant by absolute zero pressure?
15. What is absolute pressure?
16. How is atmospheric pressure measured?

17. How is the pressure on the compound gage measured?

18. What is the equivalent of 1 inch of mercury in absolute pressure?

19. How many inches of mercury does atmospheric pressure equal at sea level?

20. What is meant by relative pressure?

21. What is the definition of lift?

22. What is meant by maximum lift?

23. What factors affect the maximum lift?

24. Theoretically, how high could a pumper at sea level lift water if it could produce a perfect vacuum?

25. From a practical standpoint, what should be considered the maximum lift for a pumper?

26. What causes water to move up into the pump when drafting?

27. How is the condition in Question 26 brought about?

28. Theoretically, how far will water rise for each pound of pressure reduction?

29. What is the formula for finding the inches of mercury required to achieve a given lift?

30. What is the formula for finding the height water will rise when the inches of mercury is known?

31. What effect does altitude have on the ability of a pumper to lift water?

32. What effect do weather changes have on the ability of a pumper to lift water?

33. Can a pumper lift water higher on a clear day or a rainy day? Why?

34. What are the three factors that affect the amount of water a pumper can be expected to lift?

35. What size of suction hose is needed to achieve full lift capacity with a 1250-gpm pumper?

36. What is net pump pressure?

37. How is the net pump pressure determined when a pumper is working from a hydrant?

38. How is the net pump pressure determined when a pumper is working from draft?

39. What is the difference in the amount of work a pumper does when lifting 500 gpm of water 10 feet as compared with forcing 500 gpm of water to a height of 10 feet?

40. What type of spot should be selected for the pumper when preparing to draft?

41. On which side should the pump panel be placed in relationship to the water when setting up for drafting operations?

42. How many sections of 10-foot lengths of hard suction hose are required for drafting?

43. What should be remembered about tide movements, swimming pools, and such when preparing to draft?

44. What are the ideal minimum distance below the surface and the ideal minimum distance off the bottom that the suction strainer should be placed?

45. What procedure can be used if the minimum distance below the surface cannot be maintained?

46. Within what length of time should water be received into the main pump when the pump is being primed?

47. If water is not received within this time limit, what are the four most likely causes?

48. What might an extra-high reading on the vacuum gage mean when a pump is being primed?

49. If a pumper cannot be primed due to an air leak and the operator has satisfactorily checked all openings into the pump, where is the leak most likely to be?

50. What is the first indication for the pump operator that water has entered the main pump when drafting?

51. What procedure should be followed once water is received in the main pump?

52. What are the possible causes for the discharge pressure not to increase with a corresponding advancement of the throttle?

53. What should be done to correct these problems?

54. What is meant by drafting from a broken connection?

55. How do drafting procedures from a broken connection differ from standard drafting procedures?

56. What is meant by running away from the water?

57. Is cavitation more likely to occur when working from a hydrant or working from draft?

58. What is happening within the pump when a pump is cavitating?

59. What are some indications that a pump is cavitating or about to cavitate?

Test Nine

1. What pressure reduction is equal to a reading of 14 inches of mercury on the compound gage?
2. The compound gage on a pumper reads 16 inches of mercury. To what height will water theoretically rise due to this reduction?
3. What reading is required on the compound gage in order to theoretically lift water 18 feet?
4. A pumper at sea level can lift water to a maximum height of 23 feet. What is the maximum height-lifting capability of the pump at an elevation of 5000 feet, where the atmospheric pressure is 12.2 psi (discounting power reduction of the engine)?
5. A pumper is attached to a hydrant and is discharging water to one 1¼-inch tip. The discharge pressure is 155 psi, while the residual pressure on the hydrant is 20 psi. What is the net pump pressure?

CHAPTER **10** **Relay Operations**

Objectives

Upon completing this chapter, the reader should:

- Understand the various factors that should be considered in relay operations.
- Understand the recommended relay standards.
- Know the difference between small-quantity and large-quantity relays.
- Be able to determine the maximum allowable distance between pumpers in a relay when lines are laid at ground level.
- Be able to determine the maximum allowable distance between pumpers in a relay when lines are laid either uphill or downhill.
- Be able to determine the number of pumpers required for a relay operation.
- Be able to determine where the attack pumper supplying the appliance should be placed at a large-quantity relay.
- Be able to determine whether a single-series or a dual-series relay is required in a relay operation.
- Be able to explain the setting up of a relay operation.

Relay operations involve the movement of water using two or more pumpers in series, where the water discharged from one pumper passes into the intake side of another. Such operations are necessary whenever the conditions encountered require more than a single pumper to produce an effective stream. These conditions generally arise because of excessive friction loss in long hose layouts or back pressure due to elevation.

Relaying of water is part of the normal operations of some stationary fire protection systems; for example, in high-rise buildings it is necessary to place pumps at intermediate floor levels in order to move water from the ground to upper floors. Relay operations are not part of the normal day-to-day operations of most fire departments because an adequate hydrant distribution system is generally provided in built-up areas, and tankers, dry hydrants, or static water is available to departments providing protection in rural areas. However, any fire department may, at any moment, find it necessary to set up such operations. Although the need for relays is normally restricted

258

either to fires involving brush or at rural fire operations, the need may arise in the heart of any metropolitan area should there be damage to the water system caused by earthquakes or other disasters. Relay operations may also become necessary in high-density metropolitan areas because of incidents on freeways, where the emergency can be far from the nearest water supply. It is therefore imperative that all fire officers and pump operators be thoroughly familiar with the theory of relay principles and be able to properly set up such operations on the fire ground when the need arises.

Whereas relay operations may be an occasional affair in cities having a well-designed positive-pressure water system, they may be part of the routine day-to-day operations for those departments protecting rural areas, where static water and drafting operations are the norm. More and more departments protecting these areas are designing and installing static dry hydrants for use in firefighting operations. It is not unusual in these areas to see relay operations set up from a drafting hydrant to the fire or from the drafting hydrant to the fill-up point for tankers. However, the use of LDH has had a big part in reducing the need to set up relay operations.

As important as relay operations may be to the success of some unusual situations, few departments schedule regular drills to provide the training essential for these operations. Relay drills should be part of the training program for all departments. If mutual aid companies may become involved in these operations, combined drills should be conducted periodically.

◆ PUMPERS USED IN RELAY OPERATIONS

The pumpers used in relay operations are normally those staffed and equipped for day-to-day operations within a community. When assigned to relay operations, a pumper may be designated as one of three types.

One or more pumpers are designated as supply pumpers. Supply pumpers are those that take water from the supply source and pump it to the first pumper in the relay. The first pumper may be a **relay pumper** or an attack pumper.

The attack pumper or pumpers are those used to supply water to those firefighters staffing handlines or to lines supplying master-stream appliances. Normally only one attack pumper is used in a relay operation, but at times two or more may be required.

A pumper in line between the supply pumper and the attack pumper is designated as a relay pumper. There may be one or more relay pumpers in the operation, depending on the amount of water required at the fire, the distance from the water source to the fire, and the topography.

◆ FACTORS TO BE CONSIDERED IN RELAY OPERATIONS

It is doubtful that any two relay operations will be exactly the same, so it is not practical to establish rigid rules governing all such operations. It is best to consider some of the basic principles together with some general requirements and limitations that can be tailored to fit various situations that could occur.

Time is a critical factor in most fire situations. Setting up relay operations is generally a slow process. Prior to committing his or her forces, the officer in charge of a fire operation should be sure that a relay is essential. Perhaps another source of water is available or the total flow required is so minor that a relay is not necessary even when the distance from the water source to the fire appears extreme. However, once a decision is made to relay water, then the operation should proceed without delay.

The time required to set up operations depends largely on the amount of water required, the distance between the source of water and the intended point of use, the pumping capacity of the available apparatus, the size of hose available, and the topography.

THE AMOUNT OF WATER TO BE RELAYED

The amount of water required at the fire is one of the primary factors in the determination of the pumper spacing and the size of hose to be used in the relay. The officer in charge of relay operations should consider carefully the amount of water required at the fire. Once hose has been laid and pumpers placed, the maximum amount of flow cannot be increased without laying additional lines between pumpers or setting up a second, parallel relay. It is probably better to overestimate the amount of water required than to underestimate it.

Very small quantity flows do not normally present a serious relay problem. In fact, relays may not be required even when there is an exceptionally long distance between the source of water and the point of use, except perhaps where elevation is involved which results in back pressure problems. For example, the discharge from a ⅝-inch tip at 50 psi nozzle pressure is approximately 80 gpm. This flow results in a friction loss of about 1.2 psi per 100 feet of a single 2½-inch hose. Water can be pumped about 11,000 feet or approximately 2 miles with this flow and still provide a nozzle pressure of 50 psi without exceeding a pump pressure of 185 psi. The total flow is about 160 gpm if a wyed assembly consisting of two 100-foot lengths of 1½-inch hose with ⅝-inch tips is placed at the end of the 2½-inch hose. The total friction loss in the wyed assembly (including the nozzle pressure) is about 65 psi. This means that 120 psi is available for friction loss if the pump pressure is 185 psi. The friction loss per 100 feet of single 2½-inch hose with this flow is approximately 5 psi, which allows 2400 feet of 2½-inch hose to be attached to the assembly and stay within the 185-psi pump limitation.

The problem changes completely when it becomes necessary to transfer large quantities of water over a long distance. Friction loss can become excessive in small lines or in single lines stretched between pumpers. Large-quantity relays require larger hose, more pumpers, and more critical decisions. Large-quantity relays are discussed more thoroughly later in the chapter.

Because hose can be laid a long distance without relaying when small tips are placed into use, only the flows from 1-inch and larger tips are used for illustrating the principles of relaying in this chapter (Table 10.1).

THE DISTANCE BETWEEN THE SOURCE OF WATER AND THE POINT OF USE

The criticalness of the distance between the source of water and the point of use increases in relationship to the topography, the amount of water required, and the size of the hose used for relay operations. For example, no difficulty is encountered in supplying 80 gpm through a single 2½-inch line over a distance of 1800 feet if the line is laid at ground level. A relay operation is required, however, if the point of use is located 400 feet above the level of the pumper supplying the line. Given these conditions, if it then becomes necessary to use a 1-inch tip at the fire, an intermediate pumper has to be put into use. Of course, if the layout is 1700 feet of a single 3½-inch

TABLE 10.1 ◆ Flows and Friction Loss for Single Lines for Relay Operations

Tip Size (in.)	Flow (gpm)	Nozzle Pressure (psi)	Friction Loss per 100 Feet						
			2½"	3"*	3½"	4"	4½"	5"	6"
1	210	50	9	4	2				
1⅛	265	50	14	6	3				
1¼	325	50	21	9	4				
1½	600	50		28	12	7	4	3	2
1¾	800	80		51	22	13	7	5	3
2	1000	80			34	20	10	8	5
2¼	1350	80			62	36	18	15	9
2½	1650	80				54	27	22	14
3	2400	80					58	46	29

Flows are for smooth-bore tips or fog nozzle equivalents (100 psi NP). The flows shown are not the actual amounts of water discharged at the nozzle pressure indicated; however, the flows are sufficiently accurate for use in determining relay requirements. The friction losses have been rounded off for ease of computation.

* 2½-inch couplings.

hose reduced to 100 feet of a single 2½-inch hose to which the 1-inch tip is attached, then an adequate stream can be provided because the friction loss in the 3½-inch line is only about 1½ psi per 100 feet.

The problem becomes much more critical if a master stream is used at the fire. If a 1¾-inch tip is used, the flow required is approximately 800 gpm. With this flow, the friction loss in a single 3½-inch line is about 22 psi per 100 feet. Even if two siamesed 3½-inch lines are used, the total pressure requirement is over 200 psi.

THE CAPACITY OF AVAILABLE PUMPERS

The pumping capacity of the available apparatus is a primary factor influencing the amount of water to be relayed. The amount of water is limited to the pumping capacity of the smallest pumper in the relay. When considering relay operations, it should be remembered that pumpers are rated at full capacity at 150 psi net pressure and at 70 percent of rated capacity at 200 psi net pump pressure. The NFPA standards recommend that an annual test of relaying hose of sizes 3½ to 5 inches should be conducted at 200 psi and a maximum operating pressure when a pumper is used in a relay of 185 psi. When a relay pumper is discharging water at a pressure of 185 psi with an intake pressure of 20 psi, the net pump pressure is 165 psi (185 − 20). Because of this, it is wise to consider the maximum flow of a relay pumper to be 75 percent of its rated capacity. The maximum dependable flows from pumpers of various sizes are shown in Table 10.2.

Thus, when the smallest pumper in the relay is rated at 500 gpm, the maximum flow that can be depended upon is 375 gpm; when the smallest pumper is rated at 750 gpm, the maximum dependable flow is 560 gpm; and when the smallest pumper is rated at 1000 gpm, the maximum dependable flow is 750 gpm.

The maximum dependable flows shown in Table 10.2 can be increased somewhat for the supply pumper if the pumper is attached to a hydrant with a strong

TABLE 10.2 ◆ Maximum Dependable Flows from Pumpers in a Relay

Rated Capacity of Pumper (gpm)	Maximum Dependable Flow (gpm)	Largest Tip to Be Supplied (or Fog Nozzle Equivalent)
500	375	1¼″
750	560	500-gpm fog nozzle
1000	750	1½″
1250	935	1¾″
1500	1125	2″ or 1000-gpm fog nozzle

This table is based upon a discharge pressure of 185 psi and an inlet pressure of 20 psi for the relay pumpers. This configuration will provide a net pump pressure of 165. The maximum flows are based upon a 75 percent portion of the rated capacity of the pumper.

residual pressure. For example, if the residual pressure is 35 psi or greater when the pumper is discharging water at 185 psi, the net pump pressure is 150 psi or less. This allows the pumper to provide its full rated capacity. However, the opposite is true if the supply pumper is working from draft. The pressure loss due to moving the water from the static source to the pumper has to be added to the 185-psi discharge pressure to determine the net pump pressure. It is probably near 200 psi, which further reduces the dependable flow from the pumper. For this reason, it is better to place the lowest capacity pumper at the hydrant if the supply is from a positive-pressure source, but used it as a relay pumper somewhere in the line if water is being taken from a static source.

THE SIZE OF HOSE AVAILABLE

Selection of the proper size of hose for use in a relay greatly influences the effectiveness of the operation. If possible, hose that will result in excessive friction loss should not be selected. Because of the importance of reducing friction loss to a minimum, it is recommended that relay operations be restricted to the use of 2½-inch hose or larger. Table 10.3 lists the maximum amount of water that hose lines of various sizes can carry without introducing excessive friction losses.

An evaluation of the carrying capacity of hoses indicates that a single 2½-inch hose should not be used in a relay if the total consumption of water at the fire exceeds 300 gpm. Therefore, the use of a single 2½-inch hose between pumpers in a relay limits the tip size in this situation: Either a 1¼-inch tip, which produces an effective stream while discharging 300 gpm, or a 300-gpm fog-nozzle can be used. Of course, a number of smaller tips can be used if the total flow does not exceed 300 gpm. A number of smaller tips on 1-inch and 1½-inch lines can be placed in operation without exceeding the 300-gpm limitation.

When flows exceed 300 gpm, it becomes necessary to consider the use of lines other than a single 2½-inch line. A combination of satisfactory layouts is available, depending upon the total flow required. Table 10.4 lists some possible combinations.

TABLE 10.3 ◆ Maximum Efficient Carrying Capacity of Hose Lines

Hose Size (in.)	Maximum Efficient Carrying Capacity (gpm)	Friction Loss per 100 Feet
2½	300	18
3 (2½″ couplings)	500	20
3½	750	19
4	1000	20
5	1200	12
6	2000	20

Without a doubt, it is much more satisfactory to lay 5-inch hose than one of the combinations listed. Five-inch hose can normally be laid faster and farther when providing the same flow than can be achieved with multiple lines of hose. However, because all fire departments do not routinely carry 5-inch hose, Table 10.4 is provided to suggest layouts that can be achieved from the hose carried. Of course, if the department has a hose tender available that carries 5-inch hose, this should be given first consideration.

TABLE 10.4 ◆ Possible Hose Layouts for Large-Quantity Relay Operations

Tip Size (in.)	Discharge (gpm)	Possible Layouts	Total Carrying Capacity (gpm)
1½	600	Two 2½″	600
		One 2½″ and one 3″	800
		Two 3″	1000
		One 3½″	750
		One 4″	1000
1¾	800	One 2½″ and one 3″	800
		Three 2½″	900
		Two 3″	1000
		Two 3½″	1500
		One 4″	1000
2	1000	Two 3″	1000
		One 3″ and one 3½″	1250
		One 4″	1000
		Four 2½″	1200
		Two 3½″	1500

Note: For those companies carrying 5″ or 6″ hose, a better choice than the layouts given in this table is a single lay of either 5″ or 6″ hose, which will provide a greater flow with less friction loss. This should be a first choice in most cases.

TOPOGRAPHY

It may be possible to lay several thousand feet of hose without setting up a relay if the flow is small and lines are laid at ground level. However, when operating at fires in the hills or mountains, requirements change rapidly. Difference of elevation may call for a pumper every 300 or 400 feet, whereas the same layout on level ground could be completed without resorting to a relay.

Back pressure or forward pressure will almost always need to be considered when long lines are used at brush or other fires in mountainous areas. In fact, where the ultimate consumption of water is less than 200 gpm, the consideration given to back pressure in establishing the positioning of pumpers in a relay will be more important than that given to friction loss in the hose.

When making plans for relay operations in mountainous regions, it is not easy to determine the actual back pressure that will be encountered. All estimates are judgments based on the experience and knowledge of the officer responsible for setting up the relay.

It should be remembered that relay operations will generally be made along dedicated streets or highways, and that a definite effort has been made to hold the grade on these travelways to a maximum of 12½ percent. Without other known factors, or where it is obvious that the grade is different, the officer responsible for establishing the relay is usually safe in assuming a grade of 10 percent, which is a rise of 10 feet in 100 feet, and allowing a back pressure of 5 psi for each 100 feet of hose laid out. This amounts to an allowance of 50 psi for back pressure for each 1000 feet of hose laid.

The method presented later in the chapter for determining the distance between pumpers in a relay is based upon terrain conditions that permit the spacing indicated by the formula; however, spacing indicated by the formula is not usually possible in mountainous areas. Occasionally, lines are laid along narrow roads that must be kept open for the use of other apparatus. Many times relay pumpers must be placed at passing points, such points being short of the ideal spacing; consequently, more pumpers will be required than formulas indicate. If the officer in charge has time prior to committing forces, he or she should survey the area and determine the logical locations for placement of pumpers. Without such a survey, a pumper laying a line might exceed the maximum distance allowed between pumpers before the company officer locates a place where the pumper can be spotted.

Of course, terrain does not always work against the officer in charge. Sometimes it becomes an ally. If lines are laid downhill, the forward pressure assists the operation. Under certain conditions, in fact, the need for a relay operation may be completely eliminated by the extra advantage gained by the forward pressure. The same method of estimating the back pressure should be used for estimating the forward pressure.

RELAY STANDARDS

Since there is little possibility that two relay operations will be set up in exactly the same way, firm rules of operation cannot be established; however, there are certain standards and guidelines that should be considered in setting up relay operations:

1. Where possible, all pumpers in the relay, except the one supplying water to the fire (attack pumper), should pump at or near the same pressure.
2. If possible, pumpers in relay operations should restrict their discharge to a maximum of 185 psi. Smaller hose is normally tested annually at a pressure of 250 psi, whereas 3½-inch to

5-inch hose is tested to 200 psi. A sudden closing of a line or a discharge gate could increase the pressure above 200 psi and possibly result in a ruptured hose line before the pump operator has time to act. A ruptured hose line is possible even if relief valves are properly set.

3. Water should be provided to the next pumper in a relay at a minimum of 20 psi.
4. **Four-way valves**, if carried, should be used on all pumpers in the relay. The use of a four-way valve permits a pumper to be replaced or moved, or a change to be made from a hydrant stream to a pumper stream, without shutting off the flow of water.
5. The intake inlet that will restrict friction loss to a minimum should be used. It is possible that the friction loss in one inlet intake could be as much as 10 or 15 psi greater than that in other inlets due to elbows and other restrictions in the plumbing. The large intake on the side entering directly into the pump is generally the better of the inlets.

◆ **SMALL-QUANTITY RELAYS**

The basic factors to be considered in relay operations apply to both small- and large-quantity relays. The chief difference between the two types is the size and number of hose lines that are used between pumpers.

    Small-quantity relays can be thought of as those in which a single 2½-inch line can be used between relaying pumpers. Small-quantity relays, then, are basically those in which the ultimate flow is 300 gpm or less even if lines larger than 2½-inch are in use.

    The formulas and examples used in this section are for small-quantity relays. They are used mainly to illustrate the principles involved.

DETERMINING THE MAXIMUM DISTANCE BETWEEN PUMPERS

The maximum allowable distance between pumpers in a relay depends upon the amount of water flowing and the topography. With the pumper discharge pressure restricted to 185 psi and the inlet pressure established as a minimum of 20 psi, only 165 psi of pressure is available for use against friction loss in the hose and back pressure. Of course, when lines are laid downhill, the forward pressure will assist the pump, resulting in additional pressure available against friction loss in the hose. When lines are laid at ground level, the entire 165 psi of pressure is available against friction loss. These conditions can be expressed in the following formula:

Determining the Maximum Distance Between Pumpers in a Relay

$$\text{Maximum distance between pumpers} = \frac{165}{FL + BP - FP} \times 100$$

where
        FL = friction loss per 100 feet of hose layout
        BP = back pressure per 100 feet of hose layout
        FP = forward pressure per 100 feet of hose layout

Lines Laid at Ground Level

    The entire 165 psi of pressure is available for use against friction loss in the hose when lines are laid at ground level. The formula can then be expressed as follows:

Determining the Maximum Distance Between Pumpers in a Relay When Lines Are Laid at Ground Level

$$\text{Maximum distance between pumpers} = \frac{165}{FL} \times 100$$

where FL = friction loss per 100 feet of hose layout

QUESTION The attack pumper at the end of a relay is supplying a 1⅛-inch tip through a single 2½-inch line. Nozzle pressure is 50 psi. What is the maximum distance that the pumper in the relay should be placed from the attack pumper supplying the 1⅛-inch tip if single 2½-inch hose is stretched between the two pumpers (Figure 10.1)?

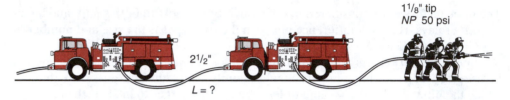

FIGURE 10.1 ◆ Determining the maximum distance between pumpers when lines are laid at ground level during relay operations.

ANSWER The friction loss in 100 feet of 2½-inch hose when a 1⅛-inch tip is used is 14 psi. Then

$$\text{Maximum distance between pumpers} = \frac{165}{14} \times 100$$
$$= 11.79 \times 100$$
$$= 1179 \text{ feet}$$

In practice, since sections of 2½-inch hose are 50 feet in length, the maximum distance needs to be reduced from 1179 feet to 1150 feet. ■

QUESTION How far apart can the pumpers be placed if it is decided that two 2½-inch lines are to be used between pumpers (Figure 10.2)?

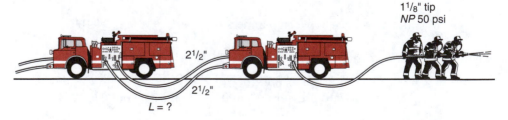

FIGURE 10.2 ◆ Determining the maximum distance between pumpers when lines are laid at ground level during relay operations.

ANSWER The friction loss for two 2½-inch siamesed lines is 3.5 psi. This means that, with the same flow in the two lines as in the single line, the two lines can be stretched four times the length of the single line (14/3.5 = 4). Then

$$\text{Maximum distance between pumpers } = (4)(1179)$$
$$= 4716$$

or, again considering that the hose comes in 50-foot sections, 4700 feet is the maximum distance. ■

Lines Laid Uphill

Back pressure must be considered in addition to the friction loss in the hose when the relay operation is laid uphill. The formula can be expressed as follows:

Determining the Maximum Distance Between Pumpers in a Relay When Lines Are Laid Uphill

$$\text{Maximum distance between pumpers} = \frac{165}{FL + BP} \times 100$$

where
FL = friction loss per 100 feet of hose layout
BP = back pressure per 100 feet of hose layout

QUESTION A number of small lines are being used off the attack pumper at the end of a relay. The total flow from the lines is 300 gpm. If a single 2½-inch hose is being used between pumpers and the layout is up a 10 percent grade, what is the maximum distance between the last pumper in the relay and the attack pumper (Figure 10.3)?

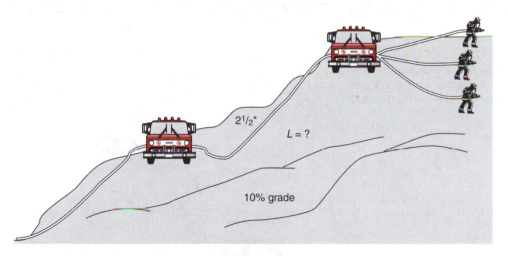

FIGURE 10.3 ◆ Determining the maximum distance between pumpers when lines are laid uphill during relay operations.

ANSWER The friction loss in 100 feet of the 2½-inch hose can be determined by using the formula

$$FL = CQ^2L$$

where
$$C = 2$$
$$Q = 300/100 = 3$$
$$L = 100/100 = 1$$

Then
$$FL = (2)(3)(3)(1)$$
$$= 18 \text{ psi}$$

The back pressure in 100 feet of hose is 5 psi. Then

$$\text{Maximum distance between pumpers} = \frac{165}{18 + 5} \times 100$$
$$= \frac{165}{23} \times 100$$
$$= 718 \text{ feet}$$

In practice this becomes a distance of 700 feet. ∎

Lines Laid Downhill

Forward pressure must be considered in addition to the friction loss in the hose when the relay operation is downhill. The forward pressure aids the pump. The formula can be expressed as follows:

Determining the Maximum Distance Between Pumpers in a Relay When Lines Are Laid Downhill

$$\text{Maximum distance between pumpers} = \frac{165}{FL - FP} \times 100$$

where

FL = friction loss per 100 feet of hose layout

FP = forward pressure per 100 feet of hose layout

QUESTION A fixed-gallonage fog nozzle set at 300 gpm is being used at the end of a relay. The nozzle pressure is 100 psi. The relay is down a 10 percent grade. What is the maximum distance between the attack pumper and the relay pumper if a single 2½-inch line is being used between pumpers (Figure 10.4)?

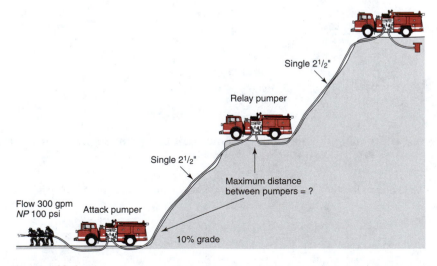

Single 2¹/₂"

Relay pumper

Single 2¹/₂"

Maximum distance between pumpers = ?

Flow 300 gpm
NP 100 psi

Attack pumper

10% grade

FIGURE 10.4 ◆

ANSWER The friction loss in 100 feet of the hose can be determined by using the formula

$$FL = CQ^2L$$

where
$$C = 2$$
$$Q = 300/100 = 3$$
$$L = 100/100 = 1$$

Then
$$FL = (2)(3)(3)(1)$$
$$= 18 \text{ psi}$$

The forward pressure is 5 psi per 100 feet or 5 psi. Then

$$\text{Maximum distance between pumpers} = \frac{165}{18 - 5} \times 100$$

$$= \frac{165}{13} \times 100$$

$$= 1270 \text{ feet}$$

Thus in practice the maximum distance is 1250 feet. ■

QUESTION How far apart could the pumpers be placed if it were decided to lay a single 3½-inch line between pumpers instead of a single 2½-inch line?

ANSWER The friction loss in 100 feet of the 3½-inch hose can be determined by using the formula

$$FL = CQ^2L$$

where
$$C = .34$$
$$Q = 300/100 = 3$$
$$L = 100/100 = 1$$

Then
$$FL = (.34)(3)(3)(1)$$
$$= 3.06 \text{ psi}$$

The forward pressure is 5 psi. Since the forward pressure in 100 feet of 3½-inch hose is greater than the friction loss in 100 feet of 3½-inch hose, the need for a relay operation is eliminated. ■

DETERMINING THE NUMBER OF PUMPERS REQUIRED

The objective of the relay is to provide the attack pumper with the water required at an inlet pressure of 20 psi or more. Although the theoretical ideal would be to space the pumpers in the relay all exactly the same distance apart, the practical situation seldom allows for this. The following procedure can be used to determine the number of pumpers required for the relay and the spacing distance between pumpers. The procedure results in equal spacing for all relay pumpers with the exception of the relay pumper that supplies the attack pumper. This spacing is normally less than that between the other pumpers in the relay. The overall result is that the attack pumper will receive the required amount of water at an inlet pressure of 20 psi or more.

The procedure requires a five-step process.

1. First, determine the maximum distance between pumpers by using the formulas previously introduced. This may require determining the back pressure or forward pressure per 100 feet of hose layout.
2. Divide the total distance of the layout by the maximum distance between pumpers. The result plus 1 equals the total number of pumpers required for the relay.
3. Multiply the maximum distance between pumpers by the total number of pumpers minus 2.

4. Subtract the result from the total length of the hose in the relay.

5. The result is the distance between the last relay pumper and the attack pumper.

QUESTION Water for a relay is being taken from a swimming pool. The distance from the pool to the fire is 3500 feet. A 1¼-inch tip at 50 psi nozzle pressure is being used on the fire. The relay is at ground level, with a single 2½-inch line being used throughout the relay. How many pumpers are required in the relay and how far apart should the pumpers be spaced?

ANSWER For the purpose of illustrating the principles involved, the steps will be taken one by one.

1. First, determine the maximum distance between pumpers by using the formulas previously introduced. The friction loss when a 1¼-inch tip is used at 50 psi nozzle pressure is 21 psi per 100 feet of 2½-inch hose. Then

$$\text{Maximum distance between pumpers} = \frac{165}{21} \times 100$$
$$= 786 \text{ feet}$$

which becomes 750 feet in practice.

2. Divide the total distance of the layout by the maximum distance between pumpers. The result plus 1 equals the total number of pumpers required for the relay:

$$\text{Total pumpers required} = \frac{3500}{750} + 1$$
$$= 4.67 + 1$$
$$= 5.67$$

Rounding this up gives six pumpers.

3. Multiply the maximum distance between pumpers by the total number of pumpers minus 2:
$$6 - 2 = 4$$
$$4 \times 750 = 3000 \text{ feet}$$

4. Subtract the result from the total length of the hose in the relay:
$$3500 - 3000 = 500 \text{ feet}$$

5. The result is the distance between the last relay pumper and the attack pumper. The relay will appear as shown in Figure 10.5. ■

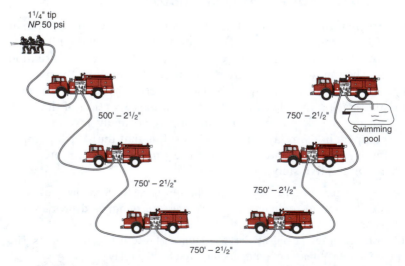

FIGURE 10.5 ◆

QUESTION The distance from the last hydrant to a fire is 3700 feet. The fire is located 210 feet above the hydrant. A 260-gpm fixed-gallonage nozzle at 100 psi nozzle pressure is to be used on the fire. How many pumpers are required and what is the spacing between pumpers if a single 2½-inch hose is used in the relay?

ANSWER

1. First, determine the maximum distance between the pumpers by using the formulas previously introduced:

$$\text{Maximum distance between pumpers} = \frac{165}{FL + BP} \times 100$$

where

$$FL = CQ^2L$$

$$C = 2$$
$$Q = 260/100 = 2.6$$
$$L = 100/100 = 1$$

Now,

$$FL = (2)(2.6)(2.6)(1)$$
$$= 13.52 \text{ or } 14 \text{ psi}$$
$$\text{Total } BP = 210 \times .5$$
$$= 105 \text{ psi}$$
$$BP \text{ per } 100 \text{ feet} = 105/37$$
$$= 2.84 \text{ or } 3 \text{ psi}$$

Then

$$\text{Maximum distance between pumpers} = \frac{165}{14 + 3} \times 100$$
$$= \frac{165}{17} \times 100$$
$$= 971 \text{ feet}$$

Thus the maximum distance in practice is 950 feet.

2. Divide the total distance of the layout by the maximum distance between pumpers. The result plus 1 equals the total number of pumpers required for the relay:

$$\text{Total pumpers required} = \frac{3700}{950} + 1$$
$$= 3.9 + 1 = 4.9, \text{ or } 5$$

3. Multiply the maximum distance between pumpers by the total number of pumpers minus 2:

$$5 - 2 = 3$$
$$3 \times 950 = 2850 \text{ feet}$$

4. Subtract the result from the total length of the hose in the relay, which is 3700 feet:

$$3700 - 2850 = 850 \text{ feet}$$

5. The result is the distance between the last relay pumper and the attack pumper. The relay will then appear as shown in Figure 10.6.   ■

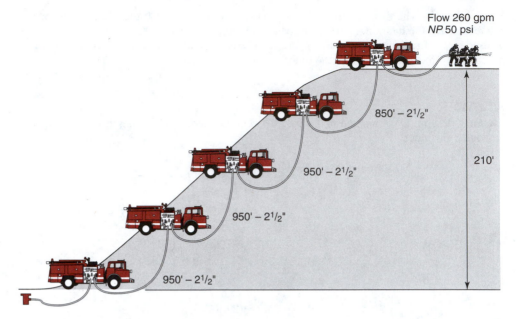

Flow 260 gpm
NP 50 psi

850' – 2½"

950' – 2½"

950' – 2½"

950' – 2½"

210'

FIGURE 10.6 ◆

QUESTION   The distance from a lake where water is being taken to a fire is 4600 feet. The fire is located 190 feet below the level of the pumper. How many pumpers are required and what is the maximum distance between pumpers if a 1-inch tip at 50 psi nozzle pressure is being used on the fire and a single 2½-inch hose is used throughout the relay?

ANSWER

1. First, determine the maximum distance between pumpers by using the formulas previously introduced:

$$\text{Maximum distance between pumpers} = \frac{165}{FL - FP} \times 100$$

Table 10.1 indicates that the flow from a 1-inch tip at 50 psi nozzle pressure is 210 gpm and the friction loss per 100 feet of a single 2½-inch hose at this flow is 9 psi. The total forward pressure is $190 \times .5 = 95$ psi. The forward pressure per 100 feet of hose is $95/46 = 2$ psi. Then,

$$\text{Maximum distance between pumpers} = \frac{165}{9 - 2} \times 100$$

$$= \frac{165}{7} \times 100$$

$$= 2358, \text{ or } 2350 \text{ feet}$$

2. Divide the total distance of the layout by the maximum distance between pumpers. The result plus 1 equals the total number of pumpers required for the relay:

$$\frac{4600}{2350} + 1 = 1.96 + 1 = 2.96, \text{ or } 3$$

3. Multiply the maximum distance between pumpers by the total number of pumpers minus 2:

$$3 - 2 = 1$$

$$1 \times 2350 = 2350 \, \text{feet}$$

4. Subtract the result from the total length of the hose in the relay:

$$4600 - 2350 = 2250 \, \text{feet}$$

5. The result is the distance between the last relay pumper and the attack pumper. The completed relay will appear as shown in Figure 10.7. ■

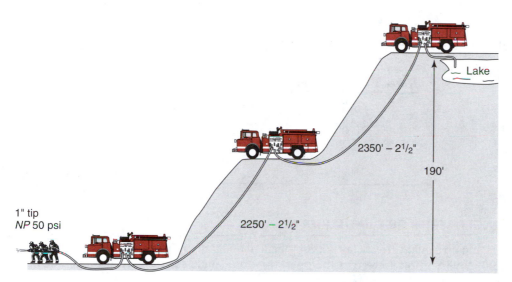

1" tip
NP 50 psi

Lake

2350' − 2½"

190'

2250' − 2½"

FIGURE 10.7 ◆

◆ **LARGE-QUANTITY RELAYS**

Large-quantity relays are those requiring a flow of 500 gpm or more. Such relays are most often required whenever a heavy-stream appliance is needed at a fire or a number of handheld lines requiring a total flow of 500 gpm or more will be used. Some large-quantity relays may be set up with a single series of pumpers, whereas others will require two series of pumpers, the discharge of both terminating in a common appliance.

The factors that determine whether a **single-series** or a **dual-series relay** will be required are the size of the smallest pumper in the relay and the amount of water to be used at the fire. As an example, the largest tip that can be supplied if the smallest pumper in the relay is a 750-gpm pumper is a 500-gpm fog nozzle. If the required flow exceeds 560 gpm, then dual relays are required (see Table 10.2).

A single-series relay designed to supply a 1½-inch tip or a 750-gpm fog nozzle requires that the smallest pumper have a rated capacity of at least 1000 gpm, whereas an appliance with a 1¾-inch tip or a 900-gpm fog nozzle requires that all pumpers have a rated capacity of at least 1250 gpm.

PLACEMENT OF THE ATTACK PUMPER AT THE FIRE

Development of large-quantity relays should start with the placement of the attack pumper that is to supply the heavy-stream appliance. After the pumper is placed, the next decision is whether a single-series relay will suffice or a dual-series relay will be required. Once this decision is made, the total number of pumpers required and their spacing can be determined.

The attack pumper supplying the appliance should be placed close enough to the appliance so that the discharge pressure need not exceed 195 psi. As long as the minimum inlet pressure of 20 psi is maintained, this means that the net pump pressure of the attack pumper will not exceed 175 psi. At this discharge pressure, the pumper should be capable of supplying at least 80 percent of its rated capacity. The dependable flow that can be expected when the pumper is pumping at a capacity of 175 psi net pumper pressure is given as follows:

Rated Capacity (gpm)	Dependable Flow (gpm)
750	600
1000	800
1250	1000
1500	1200

SINGLE-SERIES OR DUAL-SERIES RELAYS

Setting up a dual-series relay is a complicated assignment. The officer in charge should make every effort to restrict the operation to a single-series relay; however, if there are no alternatives, dual relays should be set up.

One of the most significant factors affecting the need for a dual-series relay is the size tip or the selected flow on a fog nozzle that will be used on the appliance. A thorough evaluation should be made to ensure that the tip size or fog nozzle suggested is actually required. For example, perhaps the size of the fire indicates that a 2-inch tip or a 1000-gpm fog nozzle should be used to effect a quick knockdown of the fire; however, most of the pumpers available for use in a relay only have a rated capacity of 1000 gpm. To supply a 2-inch tip or 1000-gpm fog nozzle would require a dual-series relay, whereas a 1¾-inch tip or 750-gpm fog nozzle could be supplied by a single-series relay. The time gained by setting up a single-series rather than a dual-series relay would probably far offset the difference in efficiency between a 1¾-inch tip and a 2-inch tip.

DETERMINING THE NUMBER OF PUMPERS REQUIRED

After deciding whether to use a single-series or dual-series relay, the officer in charge must determine the number of pumpers that will be required between the source of water supply and the attack pumper supplying the appliance. The solution to the number of pumpers required and the maximum distance between pumpers is the same for large-quantity relays as for small-quantity relays. The major change is that the first thought should be given to the use of LDH. Use of LDH normally requires fewer pumpers than the use of 2½-inch or 3½-inch hose.

QUESTION The officer in charge at a fire has a number of 1250-gpm pumpers at his disposal together with an adequate supply of 5-inch hose. The distance from the source of the water to the location where the attack pumper will be placed is 5000 feet over level ground. The decision has been made to attack the fire with a 1000-gpm fog nozzle attached to a portable monitor. Determine how the relay should be set up.

ANSWER

1. First, determine the maximum distance between pumpers by using the formulas previously introduced. There is no back pressure or forward pressure involved in the problem. The formula to use is

$$\text{Maximum distance between pumpers} = \frac{165}{FL} \times 100$$

The friction loss *FL* can be determined by using the formula

$$FL = CQ^2L$$

where
$$C = .08$$
$$Q = 1000/100 = 10$$
$$L = 100/100 = 1$$

Then
$$FL = (.08)(10)(10)(1)$$
$$= 8 \text{ psi}$$

Now
$$\frac{165}{8} \times 100 = 2063$$

Since 5-inch hose is only available in 100-foot sections, the maximum distance in practice is 2000 feet.

2. Divide the total distance of the layout by the maximum distance between pumpers. The result plus 1 equals the total number of pumpers required for the relay:

$$\frac{5000}{2000} + 1 = 2.5 + 1 = 3.5, \text{ or } 4 \text{ pumpers}$$

3. Multiply the maximum distance between pumpers by the total number of pumpers minus 2:

$$4 - 2 = 2$$
$$2 \times 2000 = 4000 \text{ feet}$$

4. Subtract the result from the total length of the hose in the relay:

$$5000 - 4000 = 1000 \text{ feet}$$

5. The result is the distance between the last pumper and the attack pumper. The final layout will appear as shown in Figure 10.8. ◼

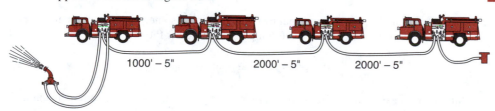

1000' – 5"   2000' – 5"   2000' – 5"

Flow 1000 gpm

FIGURE 10.8 ◆

QUESTION   It is 3300 feet from the hydrant to where an attack pumper will have to be placed to supply an elevated platform. The platform appliance is equipped with a 1200-gpm fog nozzle providing a nozzle pressure of 100 psi. The officer in charge has an adequate supply of 2½-inch hose on the apparatus assigned to her, but only 3500 feet of 5-inch hose. All of the pumpers are rated at 1000 gpm. What type of setup will be required to supply the elevated platform?

ANSWER In the analysis of the problem, the officer in charge is aware that the appliance cannot be supplied by a single relay. The decision is made to divide the required discharge in half and feed the appliance from two separate methods. First, an estimate is made as to whether the appliance can be supplied by the 5-inch hose without resorting to a relay. The formula used for the analysis is

$$FL = CQ^2L$$

where
$$C = .08$$
$$Q = 600/100 = 6$$
$$L = 3300/100 = 33$$

Then
$$FL = (.08)(6)(6)(33)$$
$$= 95.04 \text{ psi}$$

The result indicates that a relay operation is not necessary. At this point the decision is made to provide the additional 600 gpm through a relay using 2½-inch lines. The closest hydrant for use is 3500 feet from where the second attack pumper is to be placed. An analysis is then made of the relay that would be required.

1. First, determine the maximum distance between pumpers by using the formulas previously introduced:

$$\text{Maximum distance between pumpers} = \frac{165}{FL} \times 100$$

The friction loss in 100 feet of two 2½-inch lines can be determined by using the formula

$$FL = CQ^2L$$

where
$$C = .5$$
$$Q = 600/100 = 6$$
$$L = 100/100 = 1$$

Then,
$$FL = (.5)(6)(6)(1)$$
$$= 18 \text{ psi per 100 feet}$$

Then

$$\text{Maximum distance between pumpers} = \frac{165}{18} \times 100$$
$$= 917, \text{ or } 900 \text{ feet}$$

2. Divide the total distance of the layout by the maximum distance between pumpers. The result plus 1 equals the total number of pumpers required for the relay:

$$\frac{3500}{900} + 1 = 3.89 + 1 = 4.89, \text{ or } 5 \text{ pumpers}$$

3. Multiply the maximum distance between pumpers by the total number of pumpers minus 2:

$$5 - 2 = 3$$
$$3 \times 900 = 2700 \text{ feet}$$

4. Subtract the result from the total length of the hose in the relay:

$$3500 - 2700 = 800 \text{ feet}$$

5. The result is the distance between the last pumper and the attack pumper. The entire operation appears as shown in Figure 10.9. ■

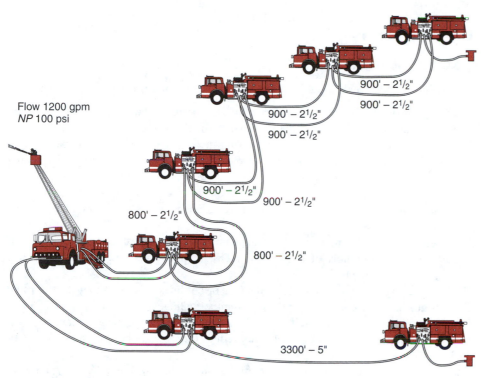

Flow 1200 gpm
NP 100 psi

900' – 2¹/₂"
900' – 2¹/₂"
900' – 2¹/₂"
900' – 2¹/₂"
900' – 2¹/₂"
900' – 2¹/₂"
800' – 2¹/₂"
800' – 2¹/₂"
3300' – 5"

FIGURE 10.9 ◆

◆ **ADJUSTING THE PLANNED RELAY**

It should be remembered that making plans for a relay operation by using recommended formulas and principles is a plan and not an operation that is set in concrete. At times the theoretical outline of the relay does not fit what is needed in practice. For example, the plan as arrived at by formulas and principles may be one where the maximum distance between pumpers is 850 feet, but the distance between the last pumper and the attack pumper projects out to be only 200 feet. When this happens, it is time to make an adjustment. The reason that a result similar to this can be suggested is that it is based on spacing pumpers at the maximum distance as determined by the formula. The maximum distance is as stated—a maximum distance and not an absolute distance that pumpers should be spaced. Adjustments should be made prior to placing the plan in operation. It is difficult to make changes once lines are laid and pumpers placed in operation.

◆ **OPERATIONAL CONSIDERATIONS**

In addition to the items already dealt with, there are a number of additional factors to consider regarding relay operations. Some of these are the following:

1. Setting up the relay.
2. Determining ruptured lines.
3. Shutting down lines.

4. Dealing with unaccountable pressure increases.
5. Maintaining communication.

SETTING UP THE RELAY

The following procedures should be followed when setting up relay operations:

1. The pumper to be used at the water source should be spotted and water taken into the pump as soon as possible. A four-way valve, if available, should be used if water is taken from a hydrant. The use of a four-way valve will permit the replacement of the pumper without shutting down the supply line in the event of mechanical failure. The four-way valve should be used on all pumpers if available.
2. The operator of the first pumper should open the discharge gates slowly and start building up pressure as soon as he or she is informed that connections to the second pumper have been made.
3. The operators of the second and succeeding pumpers should open their drain cocks to vent the air and reduce the shock of arriving water. Drains should be closed as soon as water is flowing freely from them.
4. Each succeeding pump operator should **load** his or her lines as soon as they are informed that hookup has been completed to the next pumper in line.
5. Discharge pressure should be built up to and maintained at 185 psi on all pumpers except the attack pumper. The operators should not reduce the pressure below 185 psi until notified to do so.
6. The operator of the attack pumper should proceed in the same way as if the apparatus were connected to a hydrant. The operator should determine the discharge pressure required to provide good working streams and set the pump accordingly. If the relay has been properly organized, the operator should be able to provide good working streams without reducing the intake pressure to less than 20 psi.
7. All operators must remain continuously alert once the relay has been initiated. They must remember that they are members of a team, and that their operations must be coordinated with all other pumping units in the relay. The successful discharge of water at the end of the line depends upon the efficiency of every operator in the relay.

DETERMINING RUPTURED LINES

Ruptured hose lines must be detected immediately and steps taken for their replacement. A ruptured hose line can first be detected by the operator who is supplying the line. The rupture will cause a rapid drop in the discharge pressure. If two or more lines are laid between pumping units, the discharge gate supplying the ruptured line should be slowly closed. There will be a substantial increase in the discharge pressure over the original pressure if the proper gate has been closed. The discharge pressure on the good lines should be increased to 185 psi if the previous operating pressure was below this limit.

SHUTTING DOWN LINES

If possible, all pump operators in the relay should be notified prior to the closing that a nozzle is to be **shut down**. If time does not permit such notification, then the nozzle should be closed slowly, to allow operators time to observe the change and make necessary adjustments. Sudden closing of a nozzle or a discharge gate anywhere in the relay could result in severe water hammer, with resultant burst lines and possible damage to the pumpers.

DEALING WITH UNACCOUNTABLE PRESSURE INCREASES

Unaccountable increases in pressure may indicate flow-restricting debris in the hose, the closing of a discharge gate on a pumper near the nozzle, or the closing of a nozzle.

MAINTAINING COMMUNICATION

The best means of communication between operators in a relay is by radio. Radio communication is fast and accurate, and allows minor adjustments to be made that result in overall smoothness of operations.

Despite the fact that radios are available on all apparatus and that radio communications are the most effective in relay operations, each fire department should establish standard hand signals that can be used in relay operations. It is very possible that full use of radio communications cannot be made at all relays because of overloaded airways or communications blocks caused by difficult terrain or multistory structures.

◆ SIMPLIFIED RELAY OPERATIONS

At many emergencies, time does not permit the officer in charge to complete the calculations necessary to determine the proper establishment of relay operations. Relays must be started immediately in order to get water on the fire as quickly as possible. Some general procedures can be used that will produce good streams on the fire except where extraordinary conditions exist. These simplified procedures are based on the following assumptions:

1. The officer in charge of the emergency selects a 1-inch tip or a constant-flow fog nozzle of 200 gpm or less, or a combination of smaller nozzles where the total flow does not exceed 200 gpm for use on handheld lines.
2. The largest size of tip used for heavy-stream appliances is a 1½-inch tip or a fog nozzle of 600 gpm or less.
3. The apparatus used in the relay have split beds. The beds carry 2½-inch hoses with a minimum of 600 feet in each bed, or one bed of 2½-inch hose and one bed of LDH.

Based upon these assumptions, relay operations can be set up relatively quickly at fires on level ground or where terrain presents no particular problem.

There is a difference of opinion among authorities regarding what action is to be taken by the first-arriving pumper. Some authorities prefer that the first pumper respond directly to the fire, making an immediate attack on it by using the water available from the apparatus tank while relay operations are being set up. With a tank capacity of 400 gallons, a single 1½-inch line can be used on the fire for approximately 5 minutes without depleting the water in the tank. This type of attack can be advantageous in some situations.

Other authorities prefer that the first-arriving pumper attach to the last available hydrant or water source and make connections as soon as possible. The second-arriving pumper lays out its entire bed of hose, either with a single line or dual lines, depending on whether handheld lines or a heavy stream will be used at the fire. The pumper then clears the road and connects the lines that have been laid to the suction side of the pump. As soon as connections are completed, the operator of the first pumper is notified to start the water flowing. Each succeeding arriving pumper follows the same procedure as the second pumper until sufficient hose is laid to reach the fire. A pumper is spotted near the fire and the working lines taken off it.

Whether the first pumper is to proceed directly to the fire or connect to the hydrant can be decided by the officer in charge at the time of the incident, or departmental policy regarding this type of operation can be established prior to the actual incident. Regardless of the method used, it seems wise to give thought to the problem before the situation occurs.

Summary of Chapter Formula

Maximum Distance Between Pumpers in a Relay

$$\frac{165}{FL + BP - FP} \times 100$$

Lines Laid at Ground Level

$$\frac{165}{FL} \times 100$$

Lines Laid Uphill

$$\frac{165}{FL + BP} \times 100$$

Lines Laid Downhill

$$\frac{165}{FL - FP} \times 100$$

where FL = friction loss per 100 feet of hose layout

BP = back pressure per 100 feet of hose layout

FP = forward pressure per 100 feet of hose layout

Review Questions

1. What are relay operations?
2. When are relay operations employed?
3. Where might relay operations be required in high-density metropolitan areas?
4. What are the factors that determine the time required to set up relay operations?
5. What are the flows from the following tips when used in relay operations: 1, 1⅛, 1¼, 1½, and 2 inches?
6. What factors determine the amount of water that can be moved in a relay operation?
7. Give the maximum dependable flows from the following rated pumpers in a relay operation: 500, 750, 1000, 1500 gpm.
8. Why is it best to place the lowest capacity pumper at the hydrant in a relay operation?
9. Give the maximum efficient carrying capacity of the following sizes of lines: 2½, 3 (2½-in. couplings), 3½, 4, 5, and 6 inches.
10. When evaluating the problem of back pressure, what assumption should be made regarding the grade if the grade is not known?
11. What is considered the maximum pressure with which pumpers in a relay should discharge water?
12. At what minimum pressure should water be provided to the next pumper in a relay operation?
13. Why is it recommended that four-way valves be used on every pumper in a relay?
14. What suction inlet should be used by pump operators in relay operations?

15. What is the primary difference between a small-quantity relay and a large-quantity relay?
16. What are small-quantity relays?
17. What are large-quantity relays?
18. What does the maximum distance that pumpers should be placed apart in a relay depend upon?
19. What formula is used to determine the maximum distance between pumpers in a small-quantity relay?
20. How is the number of pumpers required in a small-quantity relay determined?
21. What are the determining factors regarding the necessity for a single-series relay or a dual-series relay for large-quantity relay operations?
22. What is the starting point for the development of large-quantity relays?
23. Give the flows that should be expected from the following rated capacity pumpers at 175 psi net pump pressure: a 750-gpm pumper; a 1000-gpm pumper; a 1250-gpm pumper; a 1500-gpm pumper.
24. When should the operator of the first pumper in a relay open his or her discharge gates and start building up pressure in the hose lines?
25. Why should pump operators in a relay open the drain cocks on the apparatus prior to receiving water?
26. At what pressure should pumps be operated in a relay unless operators are notified otherwise?
27. What are the guidelines for the operation of the last pumper in a relay?
28. How are ruptured hose lines detected in a relay operation?
29. What are some possible causes for unaccountable increases in pressure when pumping in a relay operation?
30. Explain what is meant by a simplified relay operation.

Test Ten

1. The pumper at the end of a relay is supplying a 1¼-inch tip through a single 2½-inch line. Nozzle pressure is 50 psi. What is the maximum distance that the last pumper in the relay should be placed from the attack pumper supplying the 1¼-inch tip if a single 2½-inch hose is stretched between the two pumpers and the layout is on level ground?
2. How far apart can the pumpers in Question 1 be placed if it is decided to use a single 3-inch hose with 2½-inch couplings between the pumpers rather than a single 2½-inch hose?
3. A number of small lines are being used off the pumper at the end of a relay. The total flow is 250 gpm. If single 2½-inch hose is being used between pumpers and the layout is up a 10 percent grade, what is the maximum distance between the last pumper in the relay and the attack pumper?
4. Water for a relay is being taken from a pond. The distance from the pond to the fire is 4200 feet. A 1⅛-inch tip at 50 psi nozzle pressure is being used on the fire. The relay is at ground level, with a single 2½-inch line being used throughout the relay. How many pumpers are required in the relay? What is the maximum distance apart that the pumpers should be spaced?
5. The distance from the last hydrant to the fire is 2900 feet. The fire is located 80 feet above the hydrant. A 325-gpm fog nozzle is to be used on the fire. If single 2½-inch hose is to be used in the relay, how many pumpers are required and what is the maximum distance between them?
6. The distance between the water supply and the pumper supplying the appliance is 2700 feet. The layout is on level ground. The officer in charge has decided to use a 600-gpm fixed-gallonage nozzle on an elevated platform and to establish a single-series relay to support the attack pumper. It is decided that a single 3½-inch line will be used between pumpers in the relay. How

many pumpers will be required (not counting the attack pumper)? What if it were decided to use 5-inch hose?

7. The distance from a hydrant that is the source of water to a fire is 4300 feet up a 10 percent grade. The officer in charge of the fire has ten 1000-gpm pumpers at his disposal, all carrying 2½-inch or smaller hose. It is decided that a portable monitor with a 1½-inch tip at 80 psi nozzle pressure will be used on the fire. In addition, hose layouts between the pumpers and to the portable monitor will be two 2½-inch lines due to the ease of laying dual lines as compared with laying three lines. The pumper supplying the portable monitor is to be placed 150 feet from the monitor. The officer in charge wants to use as few pumpers as possible so that some can be kept in reserve should additional lines be required in the relay. How many pumpers will be required to complete the operation?

8. In Question 7, what is the required pump discharge pressure on the attack pumper?

9. In Question 7, what is the maximum spacing between pumpers?

10. In Question 7, what should be the discharge pressure of the relaying pumpers?

Fire Ground Hydraulics

11 CHAPTER

Objectives

Upon completing this chapter, the reader should:

- Understand the effect that the pump discharge pressure has on the nozzle pressure.
- Understand the advisability for a department to establish policies regarding initial pump pressures on the fire ground.
- Be able to explain some of the methods available for determining the length of the line that has been laid at a fire.
- Be able to explain the acceptable standards for handheld lines used on the fire ground.
- Be able to solve problems for handheld lines without the use of pencil and paper.
- Be able to explain the acceptable standards and appliance allowances for master streams.
- Be able to solve problems for master streams without the use of pencil and paper.
- Be able to explain some of the material that should be included in a Help Chart.
- Be able to determine the friction loss in 2½-inch hose, 3-inch hose with 2½-inch couplings, 3½-inch hose, 4-inch hose, and 4½-inch hose at various flows using the hand method.

It all comes together on the fire ground—the principles, the theory, the pencil-and-paper problems, the classroom work, and the endless hours of drilling. Despite his or her knowledge and performance at drills, a pump operator is of little value to a company or the fire department if he or she cannot connect up to a water source and transfer the water to the firefighters at the end of the hose line in such volume and pressure that effective streams are produced. Most pump operators achieve good results under ordinary circumstances; good pump operators achieve the necessary results regardless of the circumstances. The true test of a pump operator is his or her ability to handle the unusual and to make the necessary adjustments when things go wrong. The objective of this chapter is to assist in developing the good operator. Both normal and unusual fire ground situations are illustrated, and suggestions are made for methods that might prove beneficial under emergency conditions.

Earlier chapters of this text provided students with a basic understanding of fire hydraulic principles and the formulas used for solving hydraulic problems. In general, the formulas used involved pencil and paper solutions. Such formulas are not practicable on the fire ground.

The terms **fire ground hydraulics** and **field hydraulics** refer primarily to methods of mentally solving for the required pump discharge pressure at fires or other emergencies. There are several methods and variations in use throughout the fire service, each of which has proven satisfactory for providing adequate fire streams at emergencies and other incidents. Although it is not suggested that any one method is most effective, every department should adopt some method for general use in that department. The particular method chosen should be one that proves most effective for the department's individual operations and fire problems.

Because of the many variables in emergency operations, it is not possible to provide examples of every situation that could occur. Sufficient examples of basic operations are provided so that the information offered can be adapted to both standard and unusual situations.

The generally accepted standards for the nozzle pressures on smooth-bore tips is 50 psi for hand lines and 80 psi for master streams. The accepted nozzle pressure is 100 psi for both handheld and master-stream fog nozzles. For low-pressure fog streams, both 75 psi and 50 psi nozzle pressures are acceptable, depending upon the rating of the nozzle.

◆ THE PUMP DISCHARGE EFFECT ON NOZZLE PRESSURE

If four identical hose lines were laid from a pumper and four Pitot gages were attached to the nozzle tips, it is highly unlikely that identical readings would be obtained on each of the gages. A slight difference in the way a line is stretched out on the ground and minor kinks have effects on the friction loss in the line. However, it would be almost impossible for even a trained observer to notice a difference in the hose streams produced. More precisely, it is very unlikely that a trained observer could recognize the difference between a hose stream with a 45 psi nozzle pressure and one with a 55 psi nozzle pressure. Perhaps more important than this is that, whereas many people might think a 10-psi increase in the pump discharge pressure will produce a 10-psi increase in the nozzle pressure, this is not what happens. For example, what would be the nozzle pressure on each of the layouts shown in Figure 11.1?

To solve this problem, it is necessary to return to a formula that has been retired, but was widely accepted and used for at least 75 years. It proved to be satisfactory for determining the information sought, and was referred to as the Underwriters nozzle pressure formula:

$$NP = \frac{EP}{1.1 + KL}$$

where EP = pump discharge pressure

K = a factor for 2½-inch hose (for 1¼-inch tip, the factor is .248)

L = length of 2½-inch hose divided by 50

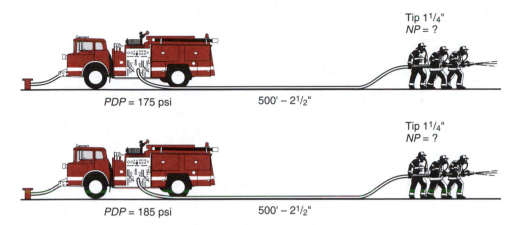

Tip 1¼"
NP = ?

PDP = 175 psi   500' – 2½"

Tip 1¼"
NP = ?

PDP = 185 psi   500' – 2½"

FIGURE 11.1 ◆

Using this formula, consider the following question:

QUESTION What is the nozzle pressure when the layout is 500 feet of 2½-inch hose equipped with a 1¼-inch tip? The pump discharge pressure (*PDP*) is 175 psi (Figure 11.2).

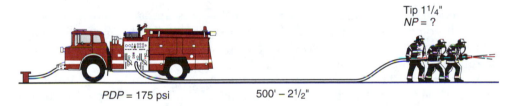

Tip 1¼"
NP = ?

PDP = 175 psi   500' – 2½"

FIGURE 11.2 ◆

ANSWER

$$NP = \frac{PDP}{1.1 + KL}$$

where

$$PDP = 175 \text{ psi}$$
$$K = .248$$
$$L = 500/50 = 10$$

Then

$$NP = \frac{175}{1.1 + (.248)(10)}$$

$$= \frac{175}{1.1 + 2.48}$$

$$= \frac{175}{3.58}$$

$$= 48.88 \text{ psi}$$

QUESTION What is the nozzle pressure if the layout is the same as in the previous problem, but the pump discharge pressure is 185 psi (Figure 11.3)?

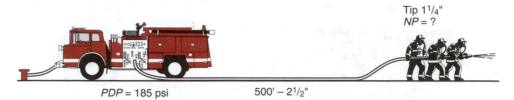

Tip 1¼"
NP = ?

PDP = 185 psi   500' − 2½"

FIGURE 11.3 ◆

ANSWER

$$NP = \frac{PDP}{1.1 + KL}$$

where

$PDP = 185$ psi

$K = .248$

$L = 500/50 = 10$

Then

$$NP = \frac{185}{1.1 + (.248)(10)}$$

$$= \frac{185}{1.1 + 2.48}$$

$$= \frac{185}{3.58}$$

$$= 51.68 \text{ psi}$$

The effect of an increase in the pump discharge pressure on the nozzle pressure depends to some extent upon the length of the hose line. A slightly greater difference occurs in long lays than in short lays. This can be seen by comparing the examples in Figure 11.4 with those in Figure 11.5.

More important than this is the variance that can occur with the pump discharge pressure while still producing an effective fire stream. Today the accepted standardized nozzle pressure for smooth-bore tips is 50 psi for hand lines. A few years ago, there was substantial agreement in the fire service that a satisfactory stream could be

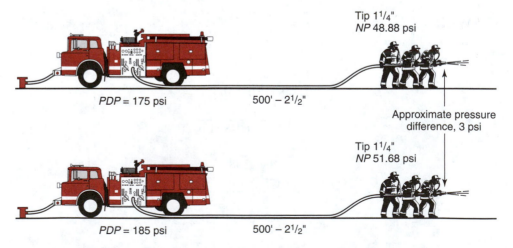

Tip 1¼"
NP 48.88 psi

PDP = 175 psi   500' − 2½"

Approximate pressure difference, 3 psi

Tip 1¼"
NP 51.68 psi

PDP = 185 psi   500' − 2½"

FIGURE 11.4 ◆ The effect of pump discharge pressure on nozzle pressure in long layouts.

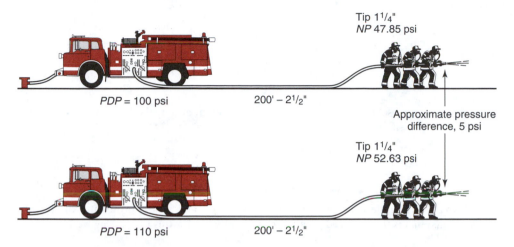

Tip 1¼"
NP 47.85 psi

PDP = 100 psi   200' – 2½"

Approximate pressure
difference, 5 psi

Tip 1¼"
NP 52.63 psi

PDP = 110 psi   200' – 2½"

FIGURE 11.5 ◆ The effect of pump discharge pressure on nozzle pressure in short layouts.

produced with the nozzle pressure between 40 psi and 60 psi. In essence, this is almost the same as establishing a standard of 50 psi. However, it is worth examining the differences in the pump discharge pressures that will produce an effective hose stream with nozzle pressures of 40 psi and 60 psi.

QUESTION What is the required pump discharge pressure for a layout of 500 feet of single 2½-inch hose when the tip is 1¼-inch in size and the nozzle pressure is 40 psi (Figure 11.6)?

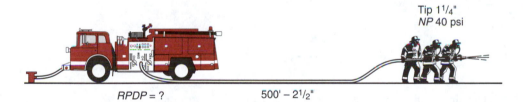

Tip 1¼"
NP 40 psi

RPDP = ?   500' – 2½"

FIGURE 11.6 ◆

ANSWER
$$RPDP = FL + NP$$

First, determine the discharge from the 1¼-inch tip at 40 psi nozzle pressure:

$$\text{Discharge} = 29.7D^2\sqrt{P}$$

where
$$D = 1.25$$
$$P = 40 \text{ psi}$$
$$\sqrt{P} = 6.32$$

Then
$$\text{Discharge} = (29.7)(1.25)(1.25)(6.32)$$
$$= 293.29 \text{ gpm}$$

The friction loss can be determined by using the formula

$$FL = CQ^2L$$

where
$$C = 2$$
$$Q = 293.29/100 = 2.93$$
$$L = 500/100 = 5$$
Then
$$FL = (2)(2.93)(2.93)(5)$$
$$= 85.85 \text{ psi}$$

The required pump discharge pressure can be determined by adding the friction loss in the hose and the nozzle pressure:

$$RPDP = 85.85 + 40$$
$$= 125.85 \text{ psi}$$ ■

QUESTION A single 2½-inch line with a 1¼-inch tip and a nozzle pressure of 60 psi is 500 feet long. What is the required pump discharge pressure for this layout (Figure 11.7)?

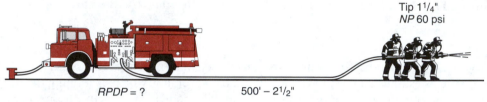

Tip 1¼"
NP 60 psi

RPDP = ?   500' – 2½"

FIGURE 11.7 ◆

ANSWER First, determine the discharge from the 1¼-inch tip at 60 psi nozzle pressure:
$$\text{Discharge} = 29.7D^2\sqrt{P}$$
where
$$D = 1.25$$
$$P = 60 \text{ psi}$$
$$\sqrt{P} = 7.75$$
Then
$$\text{Discharge} = (29.7)(1.25)(1.25)(7.75)$$
$$= 359.65 \text{ gpm}$$

The friction loss in the hose can be determined by using the formula
$$FL = CQ^2L$$
where
$$C = 2$$
$$Q = 359.65/100 = 3.6$$
$$L = 500/100 = 5$$
Then
$$FL = (2)(3.6)(3.6)(5)$$
$$= 129.6 \text{ psi}$$

The required pump discharge pressure can be determined by adding the friction loss in the hose and the nozzle pressure:

$$RPDP = 129.6 + 60$$
$$= 189.6 \text{ psi}$$ ■

Figure 11.8 illustrates the wide variance of pump discharge pressures that can produce effective fire streams. According to the problems just completed and the illustration in Figure 11.8, with a 500-foot layout of a single 2½-inch line equipped with a 1¼-inch tip, the pump discharge pressure could vary between 126 psi and 190 psi and

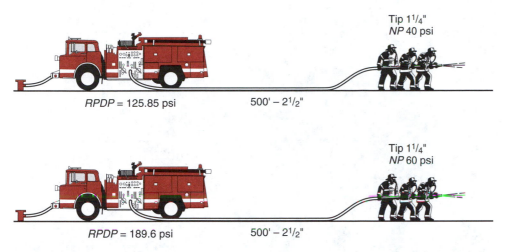

Tip 1¹/₄"
NP 40 psi

RPDP = 125.85 psi   500' – 2¹/₂"

Tip 1¹/₄"
NP 60 psi

RPDP = 189.6 psi   500' – 2¹/₂"

FIGURE 11.8 ◆

still produce an effective fire stream. While all layouts do not produce such a wide variance, the principle does illustrate that the pump discharge is not a firm figure required to produce an effective fire stream. The methods used for determining the pump discharge pressure in this chapter usually result in answers within 5 psi and seldom more than 20 psi of those obtained using the empirical formula. This variance is well within acceptable limits for producing satisfactory hose streams.

◆ **INITIAL PRESSURES**

The first few minutes after arrival at a fire are extremely demanding on the pump operator. Lines must be laid, the pumper must be connected up to a hydrant or other adequate source of water, and lines must be charged as soon as possible so that water can be supplied to the members on the hose line. Any delay in supplying water to the members at the fire could result in considerable fire loss.

It is advisable for departments to establish policies regarding initial pressures. If at all possible, initial pressures should provide nozzle pressures that fall within the acceptable pressure range. Initial pressures should be considered for all sizes of tips carried on the apparatus. The establishment of policies regarding initial pressures takes some of the burden off the pump operator and gives him or her time to make the necessary calculations for determining the required pump discharge pressure.

◆ **THE TIP IN USE**

One of the determining factors for calculating the required pump discharge pressure is the size of tip being used at the fire (Figure 11.9). There might be a considerable difference in pressure requirements for a 1-inch tip and for a 300-gpm constant-flow fog nozzle, particularly when the standard nozzle pressure for a smooth-bore tip is 50 psi and the standard nozzle pressure for a fog nozzle is 100 psi, or in some cases 75 psi or 50 psi for low-pressure fog nozzles. In order to calculate the pressure requirement for a hose layout, it is important that the pump operator be aware of the size of tip on the end of the line.

FIGURE 11.9 ◆ A fog nozzle in use. *Courtesy of Akron Brass Company*

Sometimes the pump operator will know what tip has been placed on the line by the direction of the hose lay. One size of tip may be used on forward lays and a different size on reverse lays. As standard practice, some departments use only one size of tip on a given size of hose line unless the pump operator is notified that a different size of tip has been placed on the line. More and more departments use fog nozzles for most of their operations, which could further complicate the problem for the pump operator. Regardless of the practices and procedures of a department, it is wise for the pump operator to consider the problem before lines are laid at a fire. The standardization of tip usage and the planning of tip selection prior to a fire help simplify operations at the emergency.

◆ THE LENGTH OF THE LINE

The second controlling factor in the pressure requirement for effective streams is the length of the line. Many times it is difficult for the pump operator to determine the exact number of sections of hose that have been laid out. The problem is particularly acute with two-piece companies. A line is laid to a hydrant, and then the hose wagon proceeds down the street with hose flacking out behind it. The wagon stops in front of the fire. The members stretch out a number of sections of hose, place a nozzle on the end, and lead into the fire.

The pump operator spots his or her apparatus at the hydrant, makes the connection from the hydrant to the pump, and then connects the end of the line to the discharge gate. After letting water into the pump, the operator is ready to supply the line. Then comes the big question. How much hose has been laid out?

It helps if the hose wagon is within view of the pump operator. The operator generally knows how much hose is carried in each bed, and he or she can see whether a half bed, a full bed, or perhaps a bed and a half have been laid out. This gives the operator a base from which to estimate—300 feet, 600 feet, 900 feet, and so forth.

However, perhaps the wagon is too far away to see how much hose remains in the bed, or perhaps the wagon has disappeared around a corner or behind an in-line object. The operator then must use his or her judgment of the distance from the pumper to the wagon and how much hose was used to lead in. It is not too difficult to judge distance with a little practice. Good golfers do it every day when estimating the distance to the pin in order to select the proper club to use. Sometimes physical objects can be used as aids. For example, the telephone poles in a given city may be erected according to a standardized spacing. If the distance of the hose layout equals the spacing of one and one-half telephone poles, then it is simple to determine the length of the layout, provided the pump operator is aware of the standardized spacing of the poles. Of course, radios can be used to tell the pump operator how much hose has been laid.

When the line is laid from the fire to the hydrant using hose off the pumper, the pump operator merely has to glance at the amount of hose remaining in the bed to know how much has been laid out. Sometimes the determination can be made to the exact number of sections.

◆ STANDARDS AND ALLOWANCES FOR HANDHELD LINES

Calculations for pressure requirements on the fire ground should be made as simple as possible. The objective is not to test the ability of a pump operator to add and subtract, but rather to achieve effective fire streams. Any tools to help simplify the calculation task should be provided. One help for the operator would be the standardization of the various components that must be considered. The primary concern should not be whether a standard is a little high or a little low, but that the overall results are satisfactory.

A number of different standards have been proposed by various experts in the field of hydraulics for the various components that must be considered. Most of the standards result in satisfactory operations; however, in order to provide a method of producing effective streams, only one set of standards should be used by any department. Too many variables would do nothing but confuse the pump operator. The following appear to be acceptable standards for handheld lines. Other standards for master streams are presented later in the chapter.

1. The standard nozzle pressure for smooth-bore tips is 50 psi.
2. The standard nozzle pressure for most fog nozzles is 100 psi. However, 75 or 50 psi is sometimes used for low-pressure fog nozzles.
3. The friction loss in a standpipe is 10 psi, provided the total flow is 350 gpm or more. If the flow is less than 350 psi, no friction loss is considered for the standpipe.
4. The back pressure in multistory buildings is 5 psi per floor. The number of floors above street level is normally one less than the floor on which the line is working. However, most multistory buildings do not have a thirteenth floor. With these buildings, when the line is working above the twelfth floor, the number of floors above the street is generally two fewer than the designated floor. Lines working on the roof of buildings twelve stories or fewer in height are the same distance above the ground as the number of floors designated in the building. In buildings fourteen stories and higher, this number is normally one fewer than the designated stories in the building.

5. Allow 5 psi for each 10 feet of elevation when pumping up or down hills. In essence, this means that the back pressure or forward pressure is one-half of the elevation (for example, elevation is 30 feet, back pressure is 15 psi).
6. Pump to the nearest 5 psi. It is difficult to read the gage closer than in 5-pound increments.
7. When pumping uphill or downhill, assume a grade of 10 percent if the grade is not given.

◆ FOG NOZZLES

Increasingly more departments use fog nozzles in lieu of smooth-bore tips (Figure 11.10). In fact, some departments have changed over to using fog nozzles for all operations except those in which a particular situation dictates otherwise. It is understandable why such a change would be made. The use of fog nozzles provides much more versatility than the use of smooth-bore tips, yet they are capable of producing effective straight streams when required (Figure 11.11).

FIGURE 11.10 ◆ Many departments use fog nozzles for handheld lines. *Courtesy of Task Force Tips*

FIGURE 11.11 ◆ Fog nozzles can produce effective straight lines. *Courtesy of Akron Brass Company*

There are many varieties of fog nozzles in general use. Some allow the nozzle operator to select patterns that provide a constant flow, some provide various flows at a constant pressure, and others vary the flow and pressure as determined by the nozzle operator. Because of the many variations of fog nozzles in common use, it is impossible to illustrate every conceivable pattern of flow; however, some generalities can be developed that will provide effective steams on the fire ground for most fog nozzles. The generalities are based on the use of average flows from fog nozzles used on different sizes of hose lines. Of course, when constant-flow fog nozzles are used, the friction loss resulting from the constant flow should be used rather than the average flow as indicated in the chart. The use of constant-pressure fog nozzles lessens the problem, as it is unnecessary for the pump operator to calculate nozzle pressure requirements.

The flows from fog nozzles vary according to the hose on which they are used. Following are some examples of the various flows for different sizes of hose. The flows are not exact, but provide some idea of the range within which flows are available:

1-inch hose	20, 30, 40, 50, 60 gpm
1½-inch hose	40, 60, 80, 95, 125, 150 gpm
1¾-inch hose	95, 125, 150, 175, 200 gpm
2-inch hose	100, 125, 150, 175, 200 gpm
2½-inch hose	175, 200, 225, 250, 275, 300, 350 gpm

The following average flows are used for the problems in this chapter.

Hose Size (in.)	Average Flow (gpm)	Friction Loss per 100 Feet (psi)
1	40	24
1½	100	24
1¾	150	35
2	150	18
2½	300	18

◆ SOLVING PROBLEMS FOR HANDHELD LINES

IFSTA defines a handline nozzle as "any nozzle that one to three firefighters can safely handle and that flows less than 350 gpm." This book defines a handheld line as any line equipped with a handline nozzle as defined by IFSTA.

The required pump discharge pressure for handheld lines can be determined by the use of the following formulas:

Lines laid at ground level:

$$RPDP = FL + NP$$

Lines laid uphill:

$$RPDP = FL + NP + BP$$

Lines laid downhill:

$$RPDP = FL + NP - FP$$

Lines laid into building standpipes:

$$RPDP = FL + NP + AFL + BP$$

where $RPDP$ = required pump discharge pressure

          FL = friction loss in the hose

          NP = nozzle pressure

          AFL = friction loss in the standpipe if total flow is 350 gpm or more; if less than 350 gpm, no loss is applied

          BP = back pressure

          FP = forward pressure

Although other size handlines are in use in the fire service, the most commonly used are 1, 1½, 1¾, 2, and 2½ inches. These lines are used as examples in this text. Similar types of tactics can be developed by departments using hose other than those represented here.

ONE-INCH LINES

Table 11.1 gives the flow in gpm and the friction loss for three smooth-bore tips and one fog nozzle that are used on 1-inch lines. Flow and friction losses are based upon the standard nozzle pressure of 50 psi for smooth-bore tips and 100 psi for the fog nozzle.

TABLE 11.1 ◆ Friction Loss in 1-inch Hose			
Nozzle Size (in.)	*Nozzle Pressure (psi)*	*Flow (gpm)*	*Friction Loss per 100 Feet of 1-inch Hose (psi)*
¼	50	13	3
⁵⁄₁₆	50	20	6
⅜	50	29	13
Fog nozzle (average)	100	40	24

Ground-Level Problems

To determine the required pump discharge pressure when lines are laid at ground level, find the friction loss in the hose and add it to the nozzle pressure.

QUESTION A 400-foot length of 1-inch hose equipped with a 1¼-inch tip at 50 psi nozzle pressure has been laid from a pumper. What is the required pump discharge pressure (Figure 11.12)?

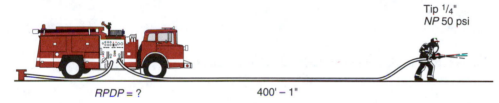

Tip ¹/₄"
NP 50 psi

RPDP = ?   400' – 1"

FIGURE 11.12 ◆

ANSWER The friction loss in the line is (4)(3) = 12. Then

$$RPDP = 12 + 50$$
$$= 62 \text{ psi (pump at 60 psi)}$$

QUESTION What is the required pump discharge pressure for a layout of 300 feet of 1-inch hose equipped with a fog nozzle at 100 psi nozzle pressure (Figure 11.13)?

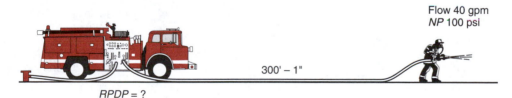

Flow 40 gpm
NP 100 psi

RPDP = ?   300' – 1"

FIGURE 11.13 ◆

ANSWER The friction loss in the hose is (3)(24) = 72 psi. Then

$$RPDP = 72 + 100$$
$$= 172 \text{ psi (pump at 170 psi)}$$

Lines Laid Uphill

To determine the required pump discharge pressure when lines are laid uphill, find the friction loss in the hose and add it to the nozzle pressure and the back pressure.

QUESTION Engine 3 is pumping through 500 feet of 1-inch hose equipped with a $\frac{5}{16}$-inch tip at 50 psi nozzle pressure. The line is laid up a 12 percent grade and is working on a rubbish fire. What is the required pump discharge pressure (Figure 11.14)?

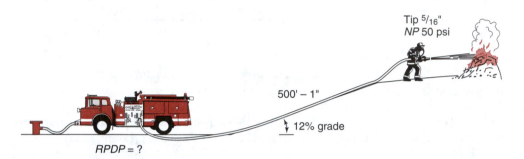

Tip $\frac{5}{16}$"
NP 50 psi

500' – 1"

12% grade

RPDP = ?

FIGURE 11.14 ◆

ANSWER The friction loss in the hose is (5)(6) = 30 psi. Then

$$H = (5)(12) = 60 \text{ feet}$$
$$BP = \frac{1}{2} \text{ of } 60 = 30 \text{ psi}$$

and

$$RPDP = FL + NP + BP$$
$$= 30 + 50 + 30$$
$$= 110 \text{ psi}$$

QUESTION A 150-foot length of 1-inch hose equipped with a fog nozzle with a flow of 40 gpm at a nozzle pressure of 100 psi has been laid uphill to a point 40 feet above the level of the pump. What is the required pump discharge pressure (Figure 11.15)?

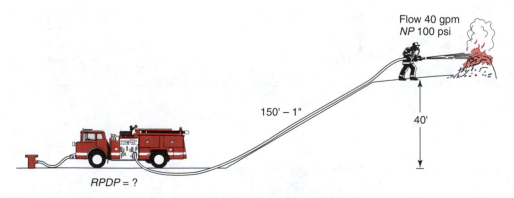

Flow 40 gpm
NP 100 psi

150' – 1"

40'

RPDP = ?

FIGURE 11.15 ◆

ANSWER The friction loss in the hose is (1.5)(24) = 36 psi. The back pressure is one-half of 40, or 20 psi. Then

$$RPDP = FL + NP + BP$$
$$= 36 + 100 + 20$$
$$= 156 \text{ psi (pump at 155 psi)}$$

∎

ONE AND THREE-QUARTER-INCH LINES

One and one-half inch hose, 1¾-inch hose, and 2-inch hose are the lines most often used for inside work at fires of any size. These lines can be advanced rapidly, provide a fair amount of water, and can be handled by a maximum of two firefighters. The amount of water discharged is generally sufficient for most dwelling fires and fires in small commercial occupancies.

Many fire departments carry preconnected, wyed 1½-inch or 1¾-inch assemblies. The preconnected assemblies are usually supplied by a single 2½-inch line. The wyed lines may be 100, 150, or 200 feet in length, depending upon the needs and operational procedures of the department.

To simplify field hydraulics where preconnected assemblies are in common use, it is best to establish a pressure requirement for the entire assembly, which includes the wye, the hose, and the nozzles. To determine the required pump discharge pressure when the assembly is used at ground level, it is only necessary to determine the friction loss in the 2½-inch line and add the pressure required for the assembly. If elevation is involved, the back pressure or forward pressure must be considered.

For the purpose of illustration, an assembly is designed for use in this chapter. It may be one that is used somewhere in the country or it may be one strictly from the imagination. However, it is sufficient for demonstrating the procedure involved. The assembly consists of two 150-foot lengths of 1¾-inch hose, each equipped with a fog nozzle with a flow of 150 gpm at 100 psi. The friction loss in the 1¾-inch hose used in the assembly can be determined from Table 11.2.

Figure 11.16 illustrates a single 1¾-inch line 150 feet in length, with a fog nozzle having a flow of 150 gpm at 100 psi nozzle pressure. The friction loss in the hose in this layout is (1.5)(35) = 52.5, or 53 psi. The required pump discharge pressure is

$$RPDP = FL + NP$$
$$= 53 + 100$$
$$= 153 \text{ psi}$$

TABLE 11.2 ◆ Friction Loss in 1¾-inch Hose

Nozzle Size (in.)	Nozzle Pressure (psi)	Flow*(gpm)	Friction Loss per 100 Feet of 1¾-inch Hose (psi)
½	50	50	4
⅝	50	80	10
¾	50	120	22
Fog nozzle (average)	100	150	35

*Flows are rounded off.

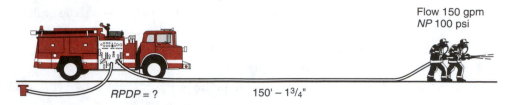

Flow 150 gpm
NP 100 psi

RPDP = ?   150' – 1³/₄"

FIGURE 11.16 ◆

Figure 11.17 illustrates two of these lines taken off a wye to complete a wyed assembly. The pressure requirement at the wye is also 153 psi. Therefore, the pressure requirement for the entire assembly shown in Figure 11.17 is 153 psi. For simplification in field problems, 155 psi is used. This is the assembly designed for illustration purposes in this chapter.

The normal procedure is to lay a single 2½-inch line to supply this assembly. The friction loss per 100 feet of single 2½-inch line in which is flowing the combined discharge from the two fog nozzles (300 gpm) can be determined by the formula

$$FL = CQ^2L$$

where

$$C = 2$$
$$Q = 300/100 = 3$$
$$L = 100/100 = 1$$

Then

$$FL = (2)(3)(3)(1)$$
$$= 18 \text{ psi}$$

The problem of determining the required pump discharge pressure on the fire ground has now become a simple matter. Whenever lines are laid at ground level, add 18 psi for every 100 feet of single 2½-inch hose used to supply the assembly to the 155-psi pressure requirement for the assembly. No loss will be factored in for the wye appliance, as the total flow from the two nozzles is less than 350 gpm. Similar standards can be developed by department officials for preconnected assemblies carried on their apparatus.

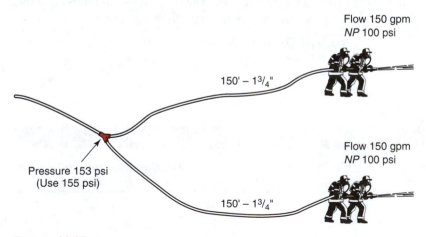

Flow 150 gpm
NP 100 psi

150' – 1³/₄"

Pressure 153 psi
(Use 155 psi)

Flow 150 gpm
NP 100 psi

150' – 1³/₄"

FIGURE 11.17 ◆

QUESTION A single 2½-inch line 300 feet in length has been laid into a standardized 1¾-inch wyed assembly. What is the required pump discharge pressure (Figure 11.18)?

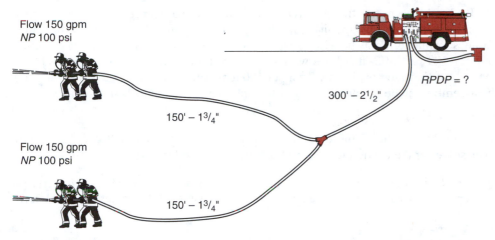

Flow 150 gpm
NP 100 psi

Flow 150 gpm
NP 100 psi

RPDP = ?

300' – 2¹⁄₂"

150' – 1³⁄₄"

150' – 1³⁄₄"

FIGURE 11.18 ◆

ANSWER To solve this problem, determine the friction loss in the single 2½-inch line and add the result to the assembly pressure requirement of 155 psi. The friction loss in the 2½-inch line is (3)(18) = 54 psi. Then

$$RPDP = 54 + 155$$
$$= 209 \text{ psi (pump at 210 psi)}$$

■

QUESTION A single 2½-inch line 400 feet in length has been laid into a standardized 1¾-inch wyed assembly. What is the required pump discharge pressure (Figure 11.19)?

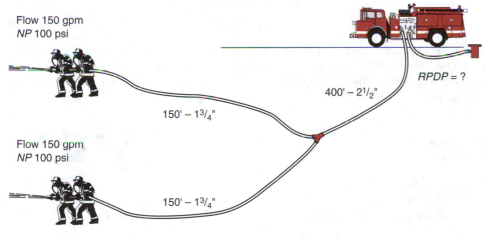

Flow 150 gpm
NP 100 psi

Flow 150 gpm
NP 100 psi

RPDP = ?

400' – 2¹⁄₂"

150' – 1³⁄₄"

150' – 1³⁄₄"

FIGURE 11.19 ◆

ANSWER To solve this problem, determine the friction loss in the single 2½-inch line and add the result to the assembly pressure requirement of 155 psi. The friction loss = (4)(18) = 54 psi. Then

$$RPDP = 72 + 155$$
$$= 227 \text{ psi (pump at 210)}$$

■

It should be noted that the required pump discharge pressure when using a 1¾-inch assembly builds up fairly rapidly when it is supplied by a single 2½-inch line. If one is available, it may be wise to supply the assembly with 3-inch hose. The friction loss per 100 feet of 3-inch hose with 2½-inch couplings and a flow of the 300 gpm needed for the assembly is only 7 psi per 100 feet. This means that the supply line for the assembly can be greater than twice as long as when a single 2½-inch line is used.

Ground-Level Problems

To determine the required pump discharge pressure when lines are laid at ground level, solve for the friction loss in the hose and add it to the nozzle pressure.

QUESTION The layout is 550 feet of single 1¾-inch line equipped with a ⅝-inch tip at 50 psi nozzle pressure. What is the required pump discharge pressure for this layout (Figure 11.20)?

Tip ⅝"
NP 50 psi

550' – 1¾"   RPDP = ?

FIGURE 11.20 ◆

ANSWER The friction loss in the hose is (5.5)(10) = 55 psi (Table 11.2).
The required pump discharge pressure is

$$RPDP = FL + NP$$
$$= 55 + 50$$
$$= 105 \text{ psi}$$ ■

QUESTION A 400-foot length of 3-inch hose with 2½-inch couplings has been laid from Engine 4. The 3-inch hose has been wyed into two 1¾-inch lines, each 150 feet in length, equipped with fog nozzles with a flow of 150 gpm at 100 psi nozzle pressure. What is the required pump discharge pressure (Figure 11.21)?

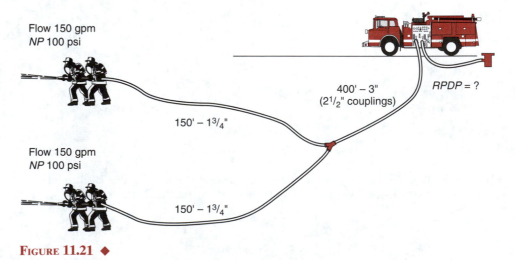

Flow 150 gpm
NP 100 psi

400' – 3"   RPDP = ?
(2½" couplings)

150' – 1¾"

Flow 150 gpm
NP 100 psi

150' – 1¾"

FIGURE 11.21 ◆

ANSWER The friction loss in the 3-inch hose is (4)(7) = 28 psi.
 The required pump discharge pressure is

$$RPDP = FL + \text{assembly requirement}$$
$$= 28 + 155$$
$$= 183 \text{ psi (pump at 185 psi)}$$ ■

Lines Laid Uphill

 To determine the required pump discharge pressure when lines are laid uphill, de-termine the friction loss in the hose and add it to the nozzle pressure or assembly pres-sure requirement and the back pressure.

QUESTION Engine 5 is pumping uphill through 400 feet of single 1¾-inch hose equipped with a ½-inch nozzle at 50 psi nozzle pressure. Determine the required pump discharge pressure for this layout (Figure 11.22).

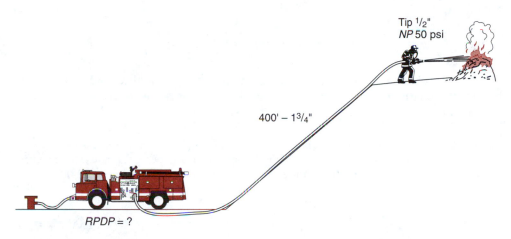

Tip ¹/₂"
NP 50 psi

400' – 1³/₄"

RPDP = ?

FIGURE 11.22 ◆

ANSWER The friction loss in the hose is (4)(4) = 16 psi. The head is 40 feet, (4)(10), and the back pressure is one-half of the head, or (½)(40) = 20 psi.
 The required pump discharge pressure is

$$RPDP = FL + NP + BP$$
$$= 20 + 50 + 20 \text{ psi}$$
$$= 90 \text{ psi}$$ ■

QUESTION A 1¾-inch line 200 feet in length has been laid uphill to a point where the noz-zle is being used on a fire that is 10 feet above the level of the pump. The line is equipped with a fog nozzle with a flow of 150 gpm at 100 psi nozzle pressure. At what pressure should the pump operator discharge water (Figure 11.23)?

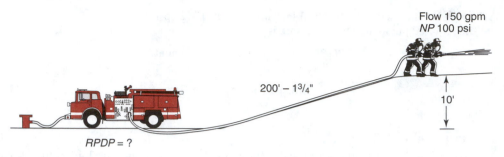

Flow 150 gpm
NP 100 psi

200' – 1³/₄"

10'

RPDP = ?

FIGURE 11.23 ◆

ANSWER The friction loss in the hose is (2)(35) = 70 psi. The back pressure is one-half of 10, or 5 psi.

The required pump discharge pressure is

$$RPDP = FL + NP + BP$$
$$= 70 + 100 + 5$$
$$= 175 \text{ psi}$$

■

Lines Laid Downhill

To determine the required pump discharge pressure when lines are laid downhill, determine the friction loss in the hose and add it to the nozzle pressure or pressure requirement and then subtract the forward pressure.

QUESTION A 1¾-inch line is equipped with a ¾-inch tip with a nozzle pressure of 50 psi. The line supplying the tip is 700 feet in length. It is laid down a 12 percent grade, where it is being used on a shed fire. What is the required pump discharge pressure (Figure 11.24)?

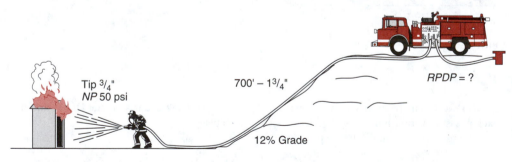

Tip ³/₄"
NP 50 psi

700' – 1³/₄"

RPDP = ?

12% Grade

FIGURE 11.24 ◆

ANSWER The friction loss in the hose is (7)(22) = 154 psi. The head is (7)(12) = 84 feet. The forward pressure is one-half the head, or (½)(84) = 42 psi.

The required pump discharge pressure is

$$RPDP = FL + NP - FP$$
$$= 154 + 50 - 42$$
$$= 162 \text{ psi (pump at 160 psi)}$$

■

QUESTION A single 2½-inch line has been laid to a dwelling fire from a hydrant that is located 50 feet above the fire. The line is 400 feet in length and is attached to a standardized 1¾-inch assembly. What is the required pump discharge pressure (Figure 11.25)?

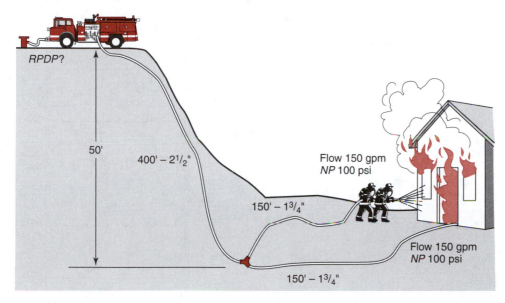

RPDP?

50'

400' – 2½"

Flow 150 gpm
NP 100 psi

150' – 1¾"

Flow 150 gpm
NP 100 psi

150' – 1¾"

FIGURE 11.25 ◆

ANSWER The friction loss in the hose is (4)(18) = 72 psi. The forward pressure is (½)(50) = 25 psi.

The required pump discharge pressure is

$$RPDP = FL + \text{assembly requirement} - FP$$
$$= 72 + 155 - 25$$
$$= 202 \text{ psi (pump at 200 psi)} \qquad \blacksquare$$

Lines Laid into a Standpipe

To determine the required pump discharge when lines are laid into a standpipe system, add the friction loss in the hose to the nozzle pressure and the back pressure. No loss is applied for the friction loss in the standpipe, as the total flow is less than 350 gpm.

QUESTION A single 2½-inch line 500 feet in length has been laid into the standpipe of an eight-story building. A 1¾-inch line 250 feet in length has been taken off the standpipe and is working on the fifth floor. The line is equipped with a ⅝-inch tip at 50 psi nozzle pressure. What is the required pump discharge pressure (Figure 11.26)?

ANSWER The friction loss in a 100-foot length of a single 2½-inch line with a flow of 80 gpm is 1 psi. Then, the friction loss in 500 feet is 5 psi.

The friction loss in the 1¾-inch line is (2.5)(10) = 25 psi. The total friction loss in the hose is 5 + 25 = 30 psi. The back pressure is (5)(4) = 20 psi.

The required pump discharge pressure is

$$RPDP = FL + NP + BP$$
$$= 30 + 50 + 20$$
$$= 100 \text{ psi} \qquad \blacksquare$$

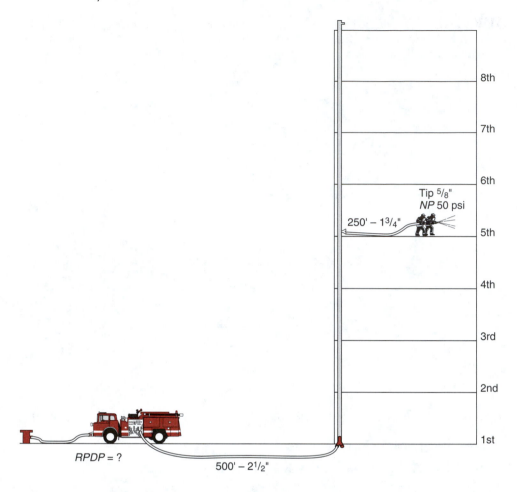

8th
7th
6th

Tip ⁵/₈"
NP 50 psi

250' – 1³/₄"

5th
4th
3rd
2nd

RPDP = ?

500' – 2¹/₂"

1st

Figure 11.26 ◆

QUESTION A single 2½-inch line 300 feet in length has been laid downhill into the stand-pipe of a ten-story building. The inlets of the standpipe are 60 feet below the pump. A single 1¾-inch line 150 feet in length has been taken off the standpipe on the seventh floor and is working on a fire on that floor. Water is flowing at 150 gpm with a nozzle pressure of 100 psi. What is the required pump discharge pressure for this layout (Figure 11.27)?

ANSWER The friction loss in 100 feet of single 2½-inch hose with a flow of 150 gpm is 5 psi. Then the friction loss in 300 feet is (3)(5) = 15 psi.

The friction loss in the 1¾-inch hose is (1.5)(35) = 52.5, or 53 psi.

The total friction loss in the hose is 15 + 53 = 68 psi.

The back pressure in the building is (5)(6) = 30 psi.

The forward pressure is (½)(60) = 30 psi.

The required pump discharge pressure is

$$RPDP = FL + NP + BP - FP$$
$$= 68 + 100 + 30 - 30$$
$$= 168 \text{ psi (pump at 170 psi)}$$

■

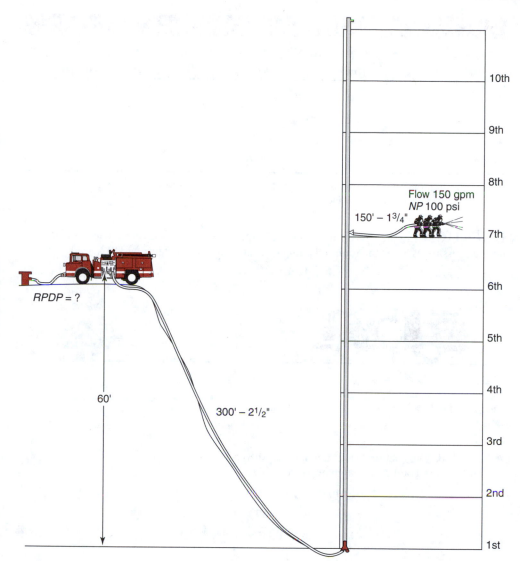

Flow 150 gpm
NP 100 psi

150' – 1³/₄"

RPDP = ?

60'

300' – 2¹/₂"

10th
9th
8th
7th
6th
5th
4th
3rd
2nd
1st

FIGURE 11.27 ◆

TWO AND ONE-HALF-INCH LINES

The smooth-bore tips most often used on handheld 2½-inch lines are the 1-, 1⅛-, and 1¼-inch sizes. For illustration, these tips and a 300-gpm fog nozzle with a nozzle pressure of 100 psi are used. The friction loss in 100 feet of 2½-inch hose for these nozzles is shown in Table 11.3. Each department should develop standards for tips carried on its own apparatus.

Ground-Level Problems

The required pump discharge pressure for lines of 2½-inch hose laid at ground level can be determined by finding the friction loss in the hose and adding the nozzle pressure.

TABLE 11.3 ◆ Friction Loss in 2½-inch Hose

Nozzle Size (in.)	Nozzle Pressure (psi)	Flow* (gpm)	Friction Loss per 100 Feet of 2½-inch Hose (psi)
1	50	200	8
1⅛	50	250	13
1¼	50	325	21
Fog nozzle	100	300	18

*Flows shown for the smooth-bore tips are rounded off. The actual flows at 50 psi are as follows: 1 inch, 210 gpm; 1⅛ inch, 265 gpm; and 1¼ inch, 328 gpm.

QUESTION Engine 8 has laid 600 feet of 2½-inch hose equipped with a 1-inch tip at 50 psi nozzle pressure. What is the required pump discharge pressure (Figure 11.28)?

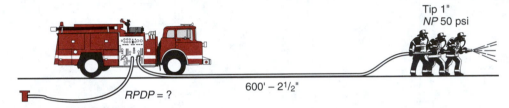

FIGURE 11.28 ◆

ANSWER The friction loss in the hose is (6)(8) = 48 psi.
The required pump discharge pressure is

$$RPDP = FL + NP$$
$$= 48 + 50$$
$$= 98 \text{ psi (pump at 100 psi)}$$

QUESTION A layout is 500 feet of single 2½-inch hose equipped with a 300-gpm, constant-flow fog nozzle at 100 psi nozzle pressure. What is the required pump discharge pressure (Figure 11.29)?

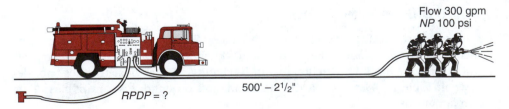

FIGURE 11.29 ◆

ANSWER The friction loss in the hose is (5)(18) = 90 psi.
The required pump discharge pressure is

$$RPDP = FL + NP$$

$$= 90 + 100$$

$$= 190 \text{ psi}$$

■

Lines Laid Uphill

The required pump discharge pressure for lines laid uphill can be determined by adding the friction loss in the hose and the back pressure to the nozzle pressure.

QUESTION   A handheld 2½-inch line is equipped with a 1¼-inch tip at 50 psi nozzle pressure. The 400-foot line is being used on a fire in a commercial building located 80 feet above the pump. The flow from the tip is 325 gpm. What is the required pump discharge pressure for this layout (Figure 11.30)?

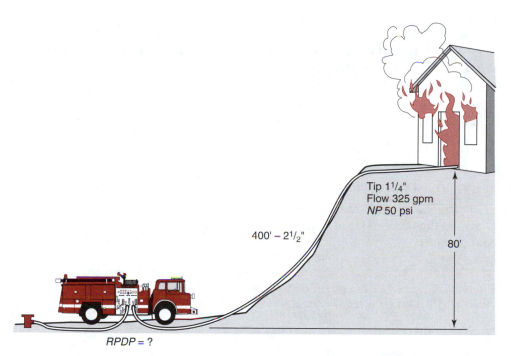

Tip 1¼"
Flow 325 gpm
NP 50 psi

400' – 2½"

80'

RPDP = ?

FIGURE 11.30 ◆

ANSWER   The friction loss in the hose is (4)(21) = 84 psi. The back pressure is (½)(80) = 40 psi.

The required pump discharge pressure is

$$RPDP = FL + NP + BP$$

$$= 84 + 50 + 40$$

$$= 174 \text{ psi (pump at 175 psi)}$$

■

QUESTION   A 300-gpm fog nozzle is attached to 400 feet of 2½-inch hose. This line is laid up a 15 percent grade. The nozzle pressure is 100 psi. What is the pump discharge pressure (Figure 11.31)?

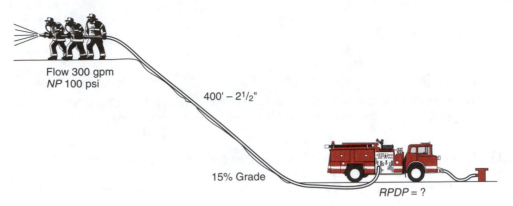

Flow 300 gpm
NP 100 psi

400' – 2½"

15% Grade

RPDP = ?

FIGURE 11.31 ◆

ANSWER The friction loss in the hose is (4)(18) = 72 psi. The head is (4)(15) = 60 feet. The back pressure is (½)(60) = 30 psi.

The required pump discharge pressure is

$$RPDP = FL + NP + BP$$
$$= 72 + 100 + 30$$
$$= 202 \text{ psi (pump at 200 psi)}$$

Lines Laid Downhill

The required pump discharge pressure can be determined for lines laid downhill by adding the friction loss in the hose to the nozzle pressure and subtracting the forward pressure.

QUESTION A 2½-inch line equipped with a 300-gpm fog nozzle at 100 psi nozzle pressure is working on a structure fire that is located 70 feet below the level of the pump. The line is 550 feet in length. What is the required pump discharge pressure (Figure 11.32)?

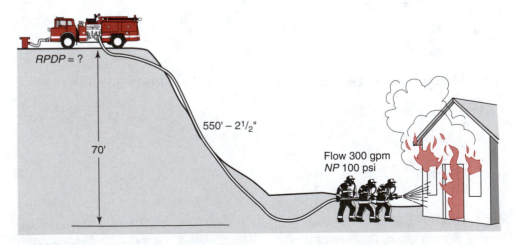

RPDP = ?

70'

550' – 2½"

Flow 300 gpm
NP 100 psi

FIGURE 11.32 ◆

ANSWER The friction loss in the hose line is (5.5)(18) = 99 psi. The forward pressure is (½)(70) = 35 psi.

The required pump discharge pressure is

$$RPDP = FL + NP - FP$$
$$= 99 + 100 - 35$$
$$= 164 \text{ psi (pump at 165 psi)}$$

■

QUESTION A layout is 650 feet of 2½-inch hose equipped with a 1⅛-inch tip at 50 psi nozzle pressure. The line is laid down a 12 percent grade. What is the required pump discharge pressure (Figure 11.33)?

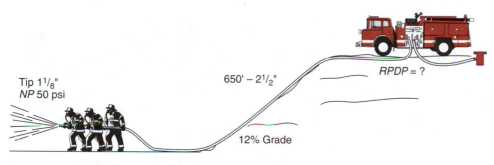

Tip 1⅛"
NP 50 psi

650' – 2½"

RPDP = ?

12% Grade

FIGURE 11.33 ◆

ANSWER The friction loss in the hose is (6.5)(13) = 84.5 psi, or 85 psi. The head is (6.5)(12) = 78 feet. The forward pressure is (½)(78) = 39 psi.

The required pump discharge pressure is

$$RPDP = FL + NP - FP$$
$$= 85 + 50 - 39$$
$$= 96 \text{ psi (pump at 95 psi)}$$

■

Lines Laid into a Standpipe

The required pump discharge pressure for lines laid into a standpipe system can be determined by adding the friction loss in the hose and the back pressure to the nozzle pressure. No appliance friction loss is added if the total flow is less than 350 gpm.

QUESTION A single 2½-inch line 200 feet in length has been laid from Engine 7 into the standpipe of a twelve-story building. A 2½-inch line 150 feet in length and equipped with a 1⅛-inch tip at 50 psi nozzle pressure has been taken off the standpipe and is working on the fifth floor. What is the required pump discharge pressure (Figure 11.34)?

ANSWER The friction loss in the hose is (3.5)(13) = 45.5 psi (use 45 psi, which is easier to add than 46). The back pressure is (5)(4) = 20 psi. There is no appliance friction loss in the standpipe, as the flow is less than 350 gpm.

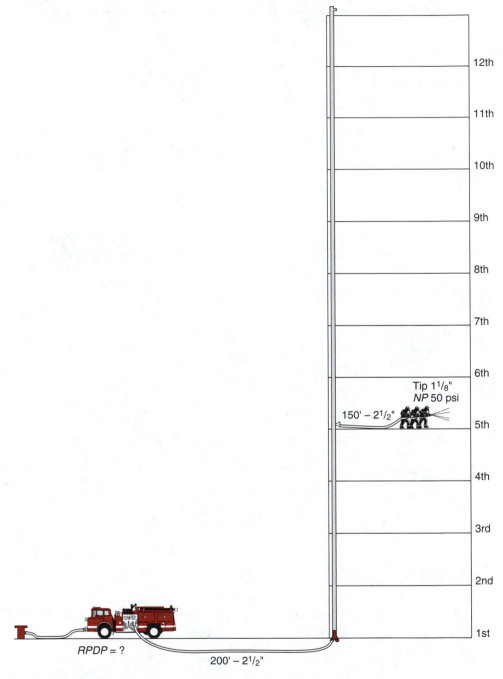

FIGURE 11.34 ◆

The required pump discharge pressure is

$$RPDP = FL + NP + BP$$
$$= 45 + 50 + 20$$
$$= 115 \text{ psi}$$

The terms **master streams** and **heavy streams** refer to hose streams that are discharged from appliances and use 1½-inch or larger smooth-bore tips or fog nozzles with flows of 500 gpm or greater. These streams are the big guns of a fire department.

Master streams are provided by appliances such as portable monitors, deck guns, and aerial equipment such as ladder pipes, elevated platforms, and water towers (Figures 11.35 through 11.37).

STANDARDS AND APPLIANCE ALLOWANCES FOR MASTER STREAMS

The following **appliance allowances** are used in this book only for the purpose of illustrating the principles involved. They should not be considered as standards for use in normal field operations. Departments having standards different from those outlined here should use their own standards. It is recommended that departments that do not have any such standards should develop them. The allowances used in this book are as follows:

1. The standard nozzle pressure for smooth-bore tips is 80 psi.
2. The standard nozzle pressure for fog nozzles is 100 psi.
3. When pumping uphill or downhill, assume a grade of 10 percent, unless the grade is given or known. Allow 5 psi for each 10 feet of elevation. This can best be determined by taking one-half of the elevation.
4. Pump to the nearest 5 psi.
5. The appliance allowance for portable monitors, deck guns, and prepiped elevated equipment refers to the pressure requirement at the appliance inlet. When a separate hose line is required to the appliance (detachable ladder pipes and some deck guns), the pressure

FIGURE 11.35 ◆ An unmounted deck gun showing mounting plate. *Courtesy of Elkhart Brass Mfg Company*

FIGURE 11.36 ◆ An elevated platform discharging water. It takes many lines to supply these appliances. *Courtesy of Pierce Manufacturing, Inc.*

FIGURE 11.37 ◆ A low-level portable monitor requires a minimum of labor. *Courtesy of Task Force Tips*

requirement refers to the pressure required at the inlet of the feeder hose. A siamese assembly could be attached to the feeder hose.

6. The appliance allowance for portable monitors using smooth-bore tips is 105 psi. This allowance includes the nozzle pressure (80 psi) and the friction loss in the monitor (25 psi).

7. The appliance allowance for portable monitors using fog nozzles is 125 psi. This allowance includes the nozzle pressure (100 psi) and the friction loss in the monitor (25 psi).

8. The appliance allowance for deck guns using smooth-bore tips is 110 psi. This allowance includes the nozzle pressure (80 psi), the friction loss in the deck gun (25 psi), and the back pressure to the deck gun (5 psi).

9. The appliance allowance for deck guns using fog nozzles is 130 psi. This allowance includes the nozzle pressure (100 psi), the friction loss in the deck gun (25), and the back pressure to the deck gun (5 psi).

10. The appliance allowance for ladder pipes and elevated platforms is 130 psi for smooth-bore tips and 150 psi for fog nozzles. The allowance includes the nozzle pressure, the friction loss in the appliance, and a 25-psi allowance for back pressure. This allowance uses a raised height of 50 feet. It may be a little higher or a little lower than this, but use of 50 feet makes the solution to the problem simpler, while the difference is not significant in the overall solution to the problem.

Departments having elevated units in service should develop appliance allowances to fit their individual needs. The appliance allowance should include the nozzle pressure, the friction loss in the appliance, plus the friction loss in the prepiping or hose used to feed the appliance. The friction loss in the prepiping and the friction loss in the appliance are generally available from the manufacturers of the apparatus and the appliances. If not, it is a good practice for departments to do their own testing to determine these losses.

SOLVING PROBLEMS FOR MASTER STREAMS

Once an appliance allowance is established, the required pump discharge pressure for master streams can be determined by the use of the following formulas:

Determining the Required Pump Discharge Pressure for Master Streams When Lines Are Laid at Ground Level

$$RPDP = FL + AA$$

where

$RPDP$ = required pump discharge pressure

FL = friction loss in the hose

AA = appliance allowance

Determining the Required Pump Discharge Pressure for Master Streams When Lines Are Laid Uphill

$$RPDP = FL + AA + BP$$

where

$RPDP$ = required pump discharge pressure

FL = friction loss in the hose

AA = appliance allowance

BP = back pressure

> 🖩 **Determining the Required Pump Discharge Pressure for Master Streams When Lines Are Laid Downhill**
>
> $$RPDP = FL + AA - FP$$
>
> where
> $RPDP$ = required pump discharge pressure
> FL = friction loss in the hose
> AA = appliance allowance
> FP = forward pressure

DETERMINING THE FRICTION LOSS IN THE SUPPLY LINES FOR MASTER STREAMS

The friction loss in the hose when large quantities of water are transferred can become excessive. To reduce the loss to a minimum, the attack pumper should be placed as close as possible to the master stream appliance. If the company carries 5-inch hose, one of the best means of accomplishing this is to lay a 5-inch line from a strong hydrant to where it is intended to place the attack pumper. This is the next best thing to having a hydrant next to the attack pumper's placement. The friction loss in a 5-inch line carrying 1000 gpm is only 8 psi per 100 feet. However, if there is any doubt concerning the hydrant's capability of supplying the required pressure, it is wise to attach a pumper to the hydrant to supply the 5-inch hose line.

The lines from the attack pumper to the appliance should be as short as possible. Again, one of the first choices should be large-diameter hose. However, there are a number of different hose lays that can be used for the purpose of supplying the appliance. What is used depends upon what is immediately available.

Table 11.4 lists the friction loss per 100 feet of some possible layouts for smooth-bore tips from 1½ to 3 inches and fog nozzles with flows of 500, 750, and 1000 gpm. The layouts listed are siamesed lines of equal length. It should be remembered that the full capacity of a pumper cannot be obtained unless the net pump pressure is kept at 150 psi or less. From the table it can be seen that 2½-inch and 3-inch smooth-bore tips require two pumpers to provide the necessary water.

To determine the friction loss in a layout, divide the length of the layout by 100 and multiply the result by the friction loss for that layout and nozzle.

QUESTION What is the friction loss in the hose of a layout of two 2½-inch lines 300 feet in length that is supplying a 1¾-inch smooth-bore tip (Figure 11.38)?

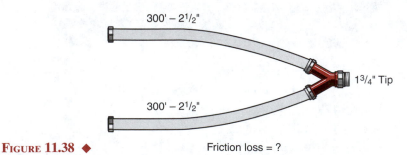

300' – 2¹/₂"

1³/₄" Tip

300' – 2¹/₂"

FIGURE 11.38 ◆

Friction loss = ?

TABLE 11.4 ◆ Master Streams*

	Friction Loss per 100 Feet for Given Nozzle								
Layout	1½", 600 gpm	1¾", 800 gpm	2", 1000 gpm	2¼", 1300 gpm	2½", 1700 gpm	3", 2400 gpm	Fog, 500 gpm	Fog, 750 gpm	Fog, 1000 gpm
One 3" (2½" couplings)	28						20		
One 3" (3" couplings)	24						17		
One 3½"	12	22	34				9	19	34
Two 2½"	18	32					13	28	
Three 2½"	8	14	22				6	12	22
Two 3" (2½" couplings)	7	13	20				5	11	20
One 3" (2½" couplings), one 2½"	10	19	30				8	17	30
One 3" (3" couplings), one 2½"	10	17	27				7	15	27
Two 2½", one 3" (2½" couplings)	6	10	16	27			4	9	16
Two 3" (2½" couplings), one 2½"	4	8	12	20			3	7	12
One 4"	7	13	20	34			5	11	20
One 4½"	4	6	10	17			3	6	10
One 5"	3	5	8	14	23		2	5	8
One 6"	2	3	5	8	14	29	1	3	5

*Smooth-bore tips have 80 psi nozzle pressure. Fog nozzles have 100 psi nozzle pressure. Friction losses are rounded off. The formula used for their determination is $FL = CQ^2$. Discharge is rounded off. Actual discharge (in gpm) is: 1½" tip, 598; 1¾" tip, 814; 2" tip, 1063; 2¼" tip, 1345; 2½" tip, 1660; 3" tip, 2392.

ANSWER From Table 11.4, the friction loss in two 2½-inch lines supplying a 1¾-inch smooth-bore tip is 32 psi for each 100 feet. To determine the total friction loss in the layout, use the formula

$$TFL = (300/100) \times 32$$
$$= 3 \times 32$$
$$= 96 \, psi$$

Note: This is excessive for most heavy-stream appliances. Using three 2½-inch lines instead of two reduces the friction loss by 54 psi. For three 2½-inch lines

$$TFL = (300/100) \times 14$$
$$= 3 \times 14$$
$$= 42 \, psi$$

Thus the difference between the two layouts is

$$96 - 42 = 54 \, psi$$

■

QUESTION A heavy-stream appliance is equipped with a 1000-gpm fog nozzle. The distance from the attack pumper to the appliance is 400 feet. Consider that the following hoses are available: which would produce the smallest amount of friction loss (Figure 11.39)?

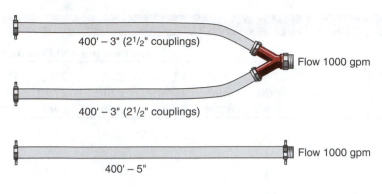

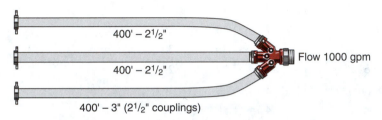

FIGURE 11.39 ◆ Which layout has the least amount of friction loss?

- Two 3-inch lines (2½-inch couplings)
- One 5-inch line
- Two 2½-inch lines and one 3-inch line (2½-inch couplings)

ANSWER The one 5-inch line is by far the better one to use. The comparison is as follows:

- Two 3-inch lines

$$TFL = (4)(20) = 80\,\text{psi}$$

- One 5-inch line

$$TFL = (4)(8) = 32\,\text{psi}$$

- Two 2½-inch lines and one 3-inch line

$$TFL = (4)(16) = 64\,\text{psi}$$

DETERMINING THE REQUIRED PUMP DISCHARGE PRESSURE FOR MASTER STREAMS

When lines are laid at ground level, the required pump discharge pressure for master streams can be determined by finding the friction loss in the hose and adding it to the appliance allowance.

QUESTION The layout from Engine 9 is two 2½-inch lines into the inlets of a portable monitor. The 2½-inch lines are each 400 feet in length. What is the required pump discharge pressure if the monitor is equipped with a 1½-inch tip with a nozzle pressure of 80 psi (Figure 11.40)?

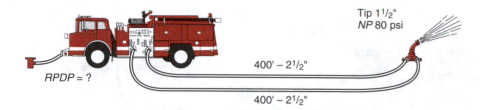

FIGURE 11.40 ◆

ANSWER The friction loss in the hose is (4)(18) = 72 psi. The appliance allowance is 105 psi. The required pump discharge pressure is

$$RPDP = FL + AA$$
$$= 72 + 105$$
$$= 177 \text{ psi (pump at 175 psi)}$$ ■

QUESTION An elevated platform has been raised to 50 feet. The heavy stream appliance on the platform is equipped with a 750-gpm fog nozzle. Engine 4 has laid 500 feet of single 5-inch hose into the inlet of the platform. What is the required pump discharge pressure for this setup (Figure 11.41)?

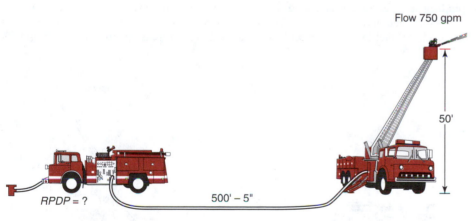

FIGURE 11.41 ◆

ANSWER The friction loss in the hose is (5)(5) = 25 psi. The appliance allowance is 150 psi. The required pump discharge pressure is

$$RPDP = FL + AA$$
$$= 25 + 150$$
$$= 175 \text{ psi}$$ ■

QUESTION Engine 10 has laid three 2½-inch lines into a deck gun mounted on Engine 5. The lines are 300 feet in length. The deck gun is equipped with a 2-inch tip. What is the required pump discharge pressure for Engine 10 (Figure 11.42)?

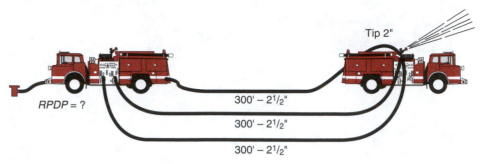

FIGURE 11.42 ◆

ANSWER The friction loss in the hose is (3)(22) = 66 psi. The appliance allowance is 110 psi. The required pump discharge pressure is

$$RPDP = FL + AA$$
$$= 66 + 110$$
$$= 176 \text{ psi (pump at 175 psi)}$$

■

QUESTION A 100-foot aerial ladder is equipped with a ladder pipe that has attached a 500-gpm fog nozzle. The ladder pipe is raised to 50 feet. A single 3-inch line 100 feet in length is attached to the ladder pipe. The other end of the 3-inch line has attached a three-into-one siamese. A chart carried on the apparatus indicates that the friction loss in the 3-inch line with 500 gpm flowing is 20 psi. The standard friction loss in the siamese is 10 psi. Engine 5 has laid two 3-inch lines equipped with 2½-inch couplings and one 2½-inch line into the siamese. The lines are each 200 feet in length. What is the required pump discharge pressure for Engine 5 (Figure 11.43)?

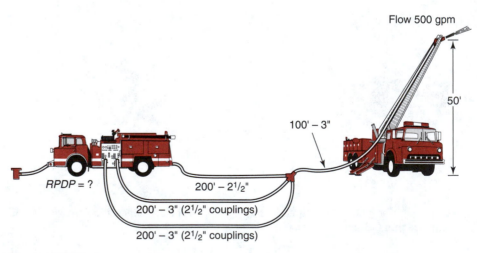

FIGURE 11.43 ◆

ANSWER The friction loss in the layout is (2)(3) = 6 psi. The friction loss in the 3-inch line is 20 psi. The friction loss in the siamese is 10 psi.

The required pump discharge pressure is

$$RPDP = FL(\text{layout}) + AA + FL(3 \text{ in.}) + AFL(\text{siamese})$$

$$= 6 + 150 + 20 + 10$$
$$= 186 \, \text{psi (pump at 185 psi)}$$ ■

LINES LAID UPHILL

The required pump discharge pressure for lines to master streams laid uphill can be determined by adding the friction loss in the hose and the back pressure to the appliance allowance.

QUESTION   A single 4-inch line 500 feet in length has been laid uphill and into a portable monitor that is located 40 feet above the level of the pump. The monitor is equipped with a 1½-inch tip at 80 psi nozzle pressure. What is the required pump discharge pressure for this setup (Figure 11.44)?

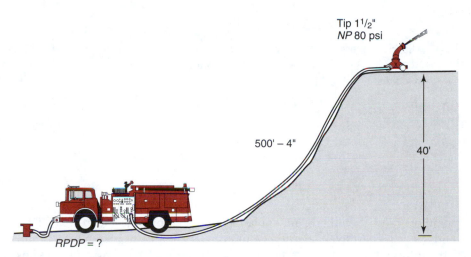

Tip 1½"
NP 80 psi

500' – 4"

40'

RPDP = ?

FIGURE 11.44 ◆

ANSWER   The friction loss in the hose is $(5)(7) = 35$ psi. The back pressure is $(\frac{1}{2})(40) = 20$ psi. The appliance allowance is 105 psi.
    The required pump discharge pressure is

$$RPDP = FL + NP + BP$$
$$= 35 + 105 + 20$$
$$= 160 \, \text{psi}$$ ■

LINES LAID DOWNHILL

The required pump discharge pressure for lines laid downhill into master-stream appliances can be determined by adding the friction loss to the appliance allowance and subtracting the forward pressure.

QUESTION   A single 4½-inch line 300 feet in length has been laid downhill and into the inlets of an elevated platform. The platform is equipped with a 2-inch tip at 80 psi nozzle pressure. What is the required pump discharge pressure for this layout (Figure 11.45)?

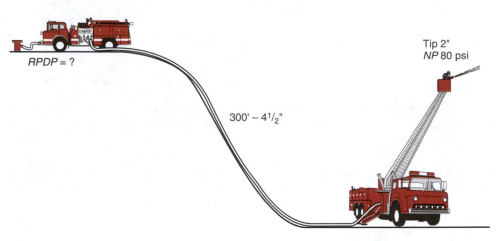

FIGURE 11.45 ◆

ANSWER The friction loss in the hose is (3)(10) = 30 psi. The head is (3)(10) = 30 feet. The forward pressure is (½)(30) = 15 psi. The appliance allowance is 130 psi.

The required pump discharge pressure is

$$RPDP = FL + AA - FP$$
$$= 30 + 130 - 15$$
$$= 145 \text{ psi}$$

◆ **A HELP CHART**

Every pump operator should prepare and attach to the apparatus a help chart. There is no standard help chart available that would prove valuable to all pump operators because the information that should be on the chart varies depending upon the hose and other equipment carried on the apparatus, and on other apparatus in the community that may need to be supplied occasionally by the pump operator.

The material in the help chart should be confined to what is necessary. Superficial material should be eliminated. Some of the information that should be included on the chart is

1. The friction loss per 100 feet for every size of hose carried on the apparatus based on the flow to the various nozzles carried on the apparatus.
2. The pressure required for all preconnected hose layouts.
3. The pressure requirement for all preconnected wyed lines that need to be supplied.
4. The appliance allowance for all master-stream appliances carried on the apparatus or those available within a community to which a pump operator may be required to supply.
5. The friction loss per 100 feet for hose carried on other companies that may be attached to the pumper during fire operations. The friction loss should be based upon the flows for the standard-size nozzles normally attached to the hose.
6. Information from all companies from other jurisdictions that might be supplied during mutual aid responses.

Once a help chart is developed, it should be placed in a plastic or other container that is moisture-proof and attached to an inside door compartment next to the pump panel or, if it is small enough, to the pump panel itself.

◆ THE HAND METHOD FOR DETERMINING FRICTION LOSS IN HOSE

The following basic information for determining the friction loss in hose by the hand method was provided through the courtesy of the Maryland Fire and Rescue Institute, University of Maryland, College Park, Maryland. Some modifications have been made to the basic information for the purpose of increasing the number of hose lines for which this system can be used.

Despite all the planning done by a pump operator and the designing of a help chart to be carried on the apparatus, the occasion may arise where it becomes necessary to determine the friction loss for certain sizes of hose lines on the fire ground. A portion of the system described here has been used by a number of departments for many years. The results obtained appear to have been satisfactory for use on the fire ground.

THE BASIC AND THE EXTENDED HAND

The design of the basic hand starts with the left hand palm up. Numbers from 1 to 5 are assigned respectively to the end of each finger, starting with the thumb and continuing clockwise to the little finger. The gaps between the fingers are assigned 1.5, 2.5, 3.5, etc.

At the base of each finger, a number is placed that corresponds to the number at the tip of the finger: 200 for finger 2, 300 for finger 3, etc. In the spaces between the fingers are written the numbers 250, 350, 450, etc. The completed basic hand is shown in Figure 11.46.

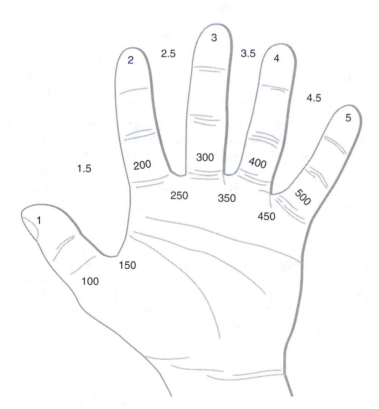

FIGURE 11.46 ◆ A basic hand.

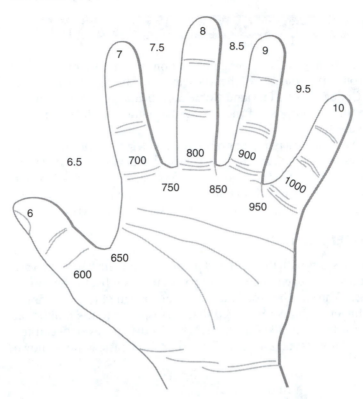

FIGURE 11.47 ◆ An extended hand.

The extended hand is constructed in the same manner, but with the numbers 6 through 10. Gaps between the fingers are assigned 6.5, 7.5, etc. The same system as before is used for placing numbers at the bottom of and between the fingers. An extended hand is shown in Figure 11.47.

The numbers on the tips of the fingers and between the tips of the fingers are used in determining the amount of friction loss in a hose layout. The numbers at the base of the fingers are used in determining the amount of water flowing in the line. For example, if the problem is to find the friction loss per 100 feet of a line with a flow of 300 gpm, the finger having the 300 written on it is used for solving the problem. In this case, the 3 on the tip of the finger is used in the calculations described below.

ACCURACY

The accepted pencil-and-paper formula used for determining the friction loss in 100 feet of hose is $FL = CQ^2$, where C is a coefficient and Q is the flow divided by 100. With the exception of single 3½-inch hose, the result obtained by the hand method for the hose presented here is exactly the same as that determined by the accepted pencil-and-paper formula. The result for the 3½-inch hose is sufficiently accurate for use on the fire ground.

SOLVING FOR 2½-INCH HOSE

The system used for determining the friction loss in 100 feet of single 2½-inch hose is as follows. The number to use for solving the problem is at the end of the fingers for flows that end in 00 and between the fingers for flows that end in 50. The first step is to find and square the appropriate number. For example, if it is desired to determine the friction loss in 2½-inch hose when the flow is 300 gpm, the first step is to square the 3 at the tip of that finger. To square a number means to multiply it by itself. To square 3, multiply 3 × 3.

Most people have no problem squaring whole numbers, but have a problem when it comes to squaring a number such as 3.5 or 4.5. Following is a simple procedure for accomplishing this. First, the result will always end in 25. With the numbers used in problems in this portion of the book, the numbers will end in .25. Take the first numeral in the number to be multiplied; for example, in 3.5 take the 3, and increase it by 1. Now multiply the original numeral (3) by the increased numeral (4). In this case 3 × 4 = 12. Add the .25 to make it 12.25. The result is the solution to squaring 3.5; in this case, 3.5 × 3.5 equals 12.25.

Figure 11.48 shows the hand to be used for determining the friction loss in single 2½-inch hose. After squaring the number, multiply the result by 2.

QUESTION What is the friction loss in 100 feet of single 2½-inch hose when the flow is 200 gpm?

ANSWER First, square the 2 at the tip of the finger with the 200 gpm on it: (2)(2) = 4. Now multiply the 4 by 2: 2 × 4 = 8. The result indicates that the friction loss in 100 feet of single 2½-inch hose when the flow is 200 gpm is 8 psi. ■

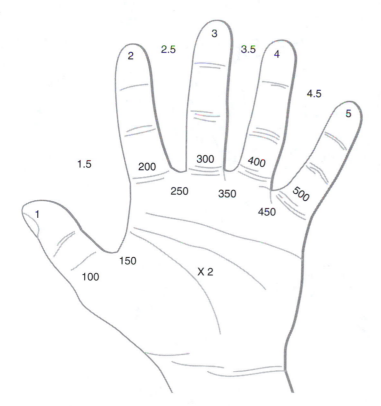

FIGURE 11.48 ◆ Determining the friction loss in single 2½-inch hose.

QUESTION What is the friction loss in 100 feet of single 2½-inch hose when the flow is 350 gpm?

ANSWER First, square the 3.5 that is between the fingers where the 350 gpm is placed. To do this, increase the 3 by 1 to get 4. Then multiply $3 \times 4 = 12$. Add .25 to get 12.25. This is the square of 3.5.

Next, multiply the result by 2: $(12.25)(2) = 24.5$ psi. The friction loss in 100 feet of single 2½-inch hose when 350 gpm is flowing is 24.5 psi. ■

SOLVING FOR 3-INCH HOSE WITH 2½-INCH COUPLINGS

The information on 3-inch hose is shown in Figure 11.49. To solve for the friction loss in 100 feet of single 3-inch hose with 2½-inch couplings, square the number at the tip or between the fingers, multiply the result by 8, then move the decimal point one place to the left.

QUESTION The flow in a 3-inch line with 2½-inch couplings is 400 gpm. What is the friction loss in 100 feet of this line?

ANSWER Square the 4 at the end of the finger: $(4)(4) = 16$. Multiply the result by 8: $(16)(8) = 128$. Move the decimal point one place to the left:

$$128 = 12.8 \text{ psi}$$

■

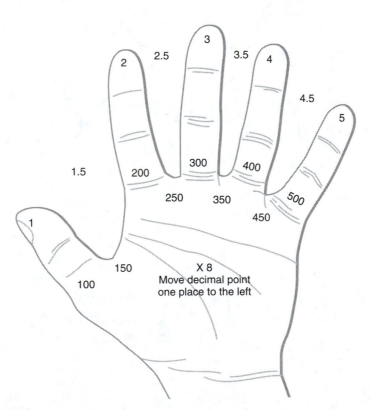

FIGURE 11.49 ◆ Determining the friction loss in single 3-inch hose with 2½-inch couplings.

QUESTION What is the friction loss in 100 feet of single 3-inch hose with 2½-inch couplings if the flow is 350 gpm?

ANSWER Square the 3.5 in the gap between the fingers showing 2 and 3. To get the square, increase 3 by 1, to get 4. Then multiply 3 × 4 = 12 Add .25 to get 12.25. This is the square of 3.5. Multiply 12.25 by 8: (12.25)(8) = 98. Move the decimal point one place to the left.

$$98 = 9.8 \text{ psi}$$

■

SOLVING FOR 3½-INCH HOSE

The information on the 3½-inch hose is shown in Figure 11.50. The answer is not exactly the same as determined by the pencil-and-paper formula. However, the result is sufficient for use on the fire ground.

To solve for the friction loss in single 3½-inch hose, square the number at the end of the finger or between two fingers, depending upon the flow. Divide the result by 3.

QUESTION What is the friction loss in 100 feet of 3½-inch hose if the flow is 300 gpm?

ANSWER Square the 3 at the end of the finger with 300 on it: (3)(3) = 9. Divide the result by 3: 9/3 = 3 psi.

By the hand method, the friction loss in 100 feet of 3½-inch hose when 300 gpm is flowing is 3 psi. By the pencil-and-paper formula, the answer is 3.06 psi.

■

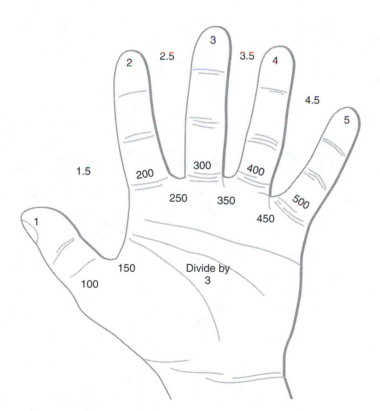

FIGURE 11.50 ◆ Determining the friction loss in single 3½-inch hose.

QUESTION The flow through a single 3½-inch line is 450 gpm. What is the friction loss in 100 feet of this line?

ANSWER Square the 4.5 between the 4 and 5 fingers. To do this, increase the 4 by 1, to get 5. Now multiply 4 × 5 = 20. Add .25 to obtain 20.25. This is the square of 4.5.
　　Divide the result by 3: 20.25 divided by 3 = 6.75 psi. The answer by the pencil-and-paper method is 6.89 psi. ■

SOLVING FOR 4-INCH HOSE

The information for the 4-inch hose is shown on Figure 11.51.
　　To solve for the friction loss in a 100-foot section of single 4-inch hose, square the number on the end of the finger or between two fingers identified by the flow, then multiply the result by two. Next, move the decimal point one place to the left.
　　Note that the solution for 4-inch hose is the same as for 2½-inch hose, with the exception that the decimal point is moved one place to the left.

QUESTION What is the friction loss in 100 feet of single 4-inch hose if the flow is 800 gpm?

ANSWER Square the 8 on the end of the finger identified by the 800 gpm flow: (8)(8) = 64. Multiply the result by 2: (2)(64) = 128. Move the decimal point one place to the left:

$$128 = 12.8 \text{ psi}$$ ■

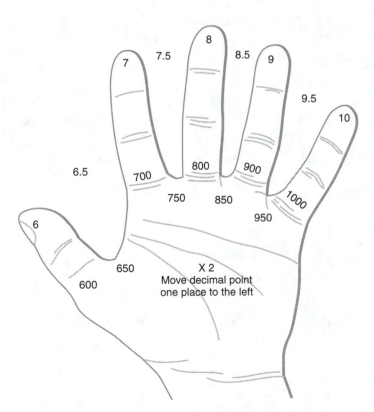

FIGURE 11.51 ◆ Determining the friction loss in single 4-inch hose.

QUESTION The flow in a single line of 4-inch hose is 750 gpm. What is the friction loss in 100 feet of this layout?

ANSWER Square the 7.5 found between the 7 and 8 fingers. To do this, increase the 7 by 1, to get 8. Multiply $7 \times 8 = 56$. Add .25 to obtain 56.25. This is the square of 7.5. Multiply the result by 2: $(2)(56.25) = 112.50$. Move the decimal point one place to the left:

$$112.50 = 11.25 \text{ psi}$$

■

SOLVING FOR 4½-INCH HOSE

The information on 4½-inch hose is shown in Figure 11.52.

To determine the friction loss in 100 feet of single 4½-inch hose, square the number identified by the flow and then move the decimal point one place to the left.

QUESTION What is the friction loss in 100 feet of single 4½-inch hose if the flow is 800 gpm?

ANSWER Square the 8 identified by the 800-gpm flow: $(8)(8) = 64$. Move the decimal point one place to the left:

$$64 = 6.4 \text{ psi}$$

■

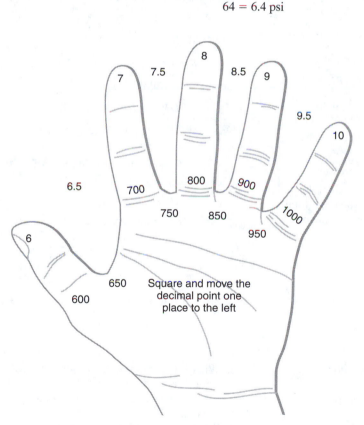

FIGURE 11.52 ◆ Determining the friction loss in single 4½-inch hose.

QUESTION The flow in a single line of 4½-inch hose is 650 gpm. What is the friction loss in 100 feet of this layout?

ANSWER Square the 6.5 identified by the 650-gpm flow. To do this, increase the 6 by 1, to get 7. Multiply $6 \times 7 = 42$. Add .25 to obtain 42.25. This is the square of 6.5. Move the decimal point one place to the left:

$$42.25 = 4.23 \text{ psi}$$ ■

Summary of Chapter Formulas

Handheld Lines

Lines Laid at Ground Level

$$RPDP = FL + NP$$

Lines Laid Uphill

$$RPDP = FL + NP + BP$$

Lines Laid Downhill

$$RPDP = FL + NP - FP$$

Lines Laid into Building Standpipes

$$RPDP = FL + NP + AFL + BP$$

where $RPDP$ = required pump discharge pressure

       FL = friction loss in the hose

       NP = nozzle pressure

     AFL = friction loss in the standpipe if the total flow is 350 gpm or more; if less than 350 gpm, no loss is applied

       BP = back pressure

       FP = forward pressure

Master Streams

Lines Laid at Ground Level

$$RPDP = FL + AA$$

Lines Laid Uphill

$$RPDP = FL + AA + BP$$

Lines Laid Downhill

$$RPDP = FL + AA - FP$$

where $RPDP$ = required pump discharge pressure

       FL = friction loss in the hose

       AA = appliance allowance

       BP = back pressure

       FP = forward pressure

Review Questions

1. To what does the term field hydraulics generally refer?
2. In the past, what was accepted as the range of nozzle pressures for handheld lines?
3. In the past, what was accepted as the range of nozzle pressures for master streams?
4. What is the standard nozzle pressure for smooth-bore tips on handheld lines?
5. What is the standard nozzle pressure for smooth-bore tips on master streams?
6. What is the standard nozzle pressure for fog nozzles?
7. What is allowed for friction loss per floor for back pressure?
8. What is allowed for friction loss for elevation for lines laid uphill or downhill?
9. Give the field hydraulic formula for determining the required pump discharge

pressure for handheld lines under the following conditions: (a) lines laid at ground level, (b) lines laid uphill, (c) lines laid downhill.

10. What are some of the means that a pump operator has to help determine the amount of hose laid out at the fire?

11. What is the standard allowance for the friction loss in a standpipe?

12. How much pressure loss is allowed for back pressure per floor in multistory buildings?

13. What pressure loss is allowed for back pressure for each 10 feet of elevation when pumping uphill?

14. What grade should be assumed on a hill if the grade is not given or known?

15. What formula is used for determining the required pump discharge pressure when handlines are laid at ground level?

16. What formula is used for determining the required pump discharge pressure when handlines are laid uphill?

17. What formula is used for determining the required pump discharge pressure when handlines are laid downhill?

18. What sizes of hose lines are most often used for inside work at fires of any size?

19. To what does the term master streams refer?

20. What is probably the best means to use when it is desired to place the attack pumper close to the master-stream appliance?

21. What is the maximum pressure at which a pumper can pump and still provide the full capacity of the pumper?

22. What sizes of tips attached to master-stream appliances will almost always need two pumps to supply the required amount of water?

23. What information should a pump operator put on a help chart?

24. For the hand method of determining friction loss on the fire ground, what is the construction of the basic hand?

25. How accurate is the hand method compared with the standard formula for determining friction loss?

26. What is the first step in determining the friction loss in 2½-inch hose using the hand method?

27. What is meant by squaring a number?

28. When using the hand method for solving friction loss in 3-inch hose with 2½-inch couplings, what is done after squaring the number at the tip of the finger?

29. What is the next step after squaring the number on the end of the finger when solving for friction loss by the hand method for single 3½-inch hose?

30. With what size hose is it only necessary to square the number indicated by the flow and then move the decimal point one place to the left to determine the friction loss in 100 feet of the hose by the hand method?

■ ■

Test Eleven

Determine the required pump discharge pressure for the following layouts:

1. Ground level: 650 feet of 1-inch hose with a ¼-inch tip.

2. Up a 12% grade: 200 feet of 1-inch hose with a fog nozzle.

3. Downhill: 700 feet of 1-inch hose with a ⅜-inch tip.

4. Ground level: 400 feet of 1¾-inch hose with a ½-inch tip.

5. Ground level: 300 feet of 2½-inch hose wyed into a standard 1¾-inch assembly.

6. Uphill: 350 feet of 1¾-inch hose with a ⅝-inch tip.

7. Down a 15% grade: 400 feet of 1¾-inch hose with a fog nozzle.

8. Ground level: 400 feet of 2½-inch hose into standpipe of an eight-story building. Single 1¾-inch line, 150 feet in length, with a ⅝-inch tip, taken off standpipe and working on the fourth floor.

9. Ground level: 500 feet of 2½-inch hose with a 1¼-inch tip.

10. Up an 8% grade: 350 feet of 2½-inch hose with a 1-inch tip.

11. Downhill to a point 50 feet below the level of the pump: 600 feet of 2½-inch hose with a 300-gpm constant-flow fog nozzle.

12. Ground level: 300 feet of 2½-inch hose into standpipe of a ten-story building. Single 2½-inch line, 200 feet in length, with 1-inch tip, taken off a standpipe and working on the fifth floor.

13. Ground level: two 2½-inch lines, each 300 feet in length, into a portable monitor with a 1¾-inch tip.

14. Ground level: three 2½-inch lines, each 300 feet in length, into ladder pipe with a 500-gpm fog nozzle.

15. Ground level: a single 3-inch line with 2½-inch couplings, 300 feet in length, into a deck gun with a 500-gpm fog nozzle.

16. Ground level: single 3-inch line with 3-inch couplings, 200 feet in length, into a portable monitor with a 1½-inch tip.

17. Ground level: two 3-inch lines with 2½-inch couplings, 250 feet in length, into elevated platform with a 1¾-inch tip.

18. Ground level: a single 5-inch line, 500 feet in length, into an elevated platform with a 1000-gpm fog nozzle.

19. Ground level: single 3½-inch line, 400 feet in length, into a deck gun with a 500-gpm fixed-gallonage fog nozzle.

20. Ground level: one 4½-inch line, 300 feet in length, into an elevated platform with a 1000-gpm fog nozzle.

Solve the following problems using the hand method for determining friction loss.

21. 400 feet of single 2½-inch line with a flow of 300 gpm.

22. 600 feet of single 4½-inch hose with a flow of 500 gpm.

23. 300 feet of single 3½-inch hose with a flow of 400 gpm.

24. 600 feet of single 4-inch hose with a flow of 700 gpm.

25. 400 feet of single 3-inch hose with 2½-inch couplings with a flow of 300 gpm.

Glossary

absolute pressure. Pressure above absolute zero.

absolute zero pressure. A complete absence of pressure, or a perfect vacuum.

air leak. A leakage of air into the intake side of a pump during drafting operations. Air leaks interfere with the proper priming of a pump.

air lock. Air trapped in a pump or the plumbing on the intake side of a pump during drafting, which prevents the complete priming of the pump.

angle of discharge. The angle at which water is being discharged from a nozzle tip as measured from level ground.

appliance allowance. A pressure allowance for the friction loss in an appliance.

appliance friction loss. The loss of pressure as the water flows through fittings such as reducers, increasers, manifolds, siameses, wyes, standpipe systems, portable monitors, and aerial apparatus.

atmospheric pressure. The weight of air or atmosphere. At sea level, the atmospheric pressure is approximately 14.7 psi.

attack pumper. A pumper supplying water to the nozzles at a fire.

automatic nozzle. A nozzle that produces a nearly consistent nozzle pressure through a wide variety of volumes. Also referred to as a constant-pressure nozzle.

average daily consumption. The average amount of water used each day in a city. The average daily consumption of cities in the United States is approximately 140 to 150 gallons per capita.

back pressure. The pressure necessary for a pumper to overcome due to head when lines are laid uphill, into standpipes, into deck guns and elevated platforms, or whenever the discharge of water is above the level of the pump.

bleeder valve. A valve on a pumper, used to relieve pressure on a hose line.

broken connection. A situation where water is taken from a hydrant or other positive-pressure source, discharged into an open container, and then taken from the container by drafting operations.

Btu (British thermal unit). The amount of heat required to raise the temperature of 1 pound of water 1°F.

calorie. The amount of heat required to raise the temperature of 1 gram of water 1°C.

cavitation. A condition within a pump when the pump operator attempts to discharge more water than the pump is capable of producing.

centrifugal pump. A pump that uses impellers to impart velocity to water by centrifugal force.

cistern. A large water-storage tank, normally located beneath the street of a municipality.

clapper valve. An automatic valve installed in hydraulic systems that permits the flow of water in one direction only.

coefficient of discharge. The percentage of water actually discharged from a nozzle tip or hydrant outlet compared with the theoretical amount of water that should be discharged from the tip.

compound gage. A gage connected to the intake side of a pump, which registers both positive pressure and negative pressure.

constant-flow nozzle. A nozzle designed to provide the same flow at a specific nozzle pressure regardless of the setting of the flow pattern.

dead end main. A main supplying water from one direction only.

deck gun. A heavy-stream appliance mounted on a hose wagon or pumper.

density. The weight per unit volume of a substance.

direct attack. An attack where water is discharged directly onto the materials involved with fire.

discharge. The amount of water issuing from an opening.

distribution mains. Mains used to supply water directly to fire hydrants and to various occupancies within a community for domestic purposes.

drafting operations. The taking of water from a source other than a hydrant or from a source other than where water is delivered to the intake of a pump under pressure.

dry-barrel hydrant. A hydrant in which the valve controlling the water is located below the frost line; consequently, the barrel of the hydrant is dry except when in use.

dry hydrant. A hydrant that is not connected to a positive-pressure source of water.

dual-series relay. A relay that requires two series of pumpers, the discharge of both terminating in a common appliance.

effective fire stream. A fire stream that accomplishes its objective with the least amount of effort and use of water.

elevated platform. A mechanically raised platform mounted on a fire apparatus. In addition to its use for rescue and firefighting service, the platform is used as a mount for a heavy-stream appliance.

elevation pressure. Refers to a pressure loss or gain whenever a nozzle or other discharge opening is located above or below the pressure source.

equivalent-nozzle-diameter. The size of a nozzle tip having the same discharge capacity as two or more other nozzle tips combined.

fail safe principle. Using more water or personnel than appears to be needed in order to make sure that the objective is accomplished.

field hydraulics. Primarily, methods of mentally solving for the pump discharge pressure at fires or other emergencies.

fire department hydraulics. The portion of general hydraulics pertaining to water and its use in firefighting and fire protection.

fire stream. A stream of water from the time it leaves a nozzle until it reaches the point of intended use, or until it reaches its projection limit, whichever occurs first.

firefighting tactics. The distribution of firefighters and equipment at the scene of a fire for the purpose of extinguishing the fire with a minimum of loss of life and property.

fixed-gallonage nozzle. A nozzle designed to provide the same flow at a specific nozzle pressure regardless of the setting of the flow pattern.

flow gage. An instrument that measures the flow (gpm) from a specific outlet.

flow hydrants. Those hydrants used to determine the amount of water available from a water system when testing the system.

flow pressure. The pressure of water after it has been placed in motion.

flush-type hydrant. A hydrant in which the outlets and the control valves are located below ground level.

fog stream. A stream of water projected from a fog nozzle and discharged in the form of a fine mist.

force. The amount of energy being applied at a given point.

forward pressure. The pressure aid given to a pump, hydrant, or other source of water pressure when the discharge of water is below the level of the pressure source.

four-way valve. A valve that permits changing from a hydrant stream to a pumper stream without shutting off the flow of water.

friction loss. The loss of energy, in the form of pressure, due to friction.

friction loss coefficients. Constants for various hose layouts used in the friction loss formula.

gage pressure. Pressure above atmospheric pressure.

getting away. A condition where the nozzle operator loses control of the hose line.

grade. The rise or decline of elevation per 100 feet in the horizontal direction.

gravity system. A type of water system where water is stored at same point that is elevated above the distribution system and flows into the distribution system by gravity.

gravity tank. A water storage tank that delivers water to a point of use by means of gravity flow.

head. The vertical distance from the surface of the water to the point being considered.

high-pressure fog nozzle. A specially designed nozzle that produces fog only.

high-pressure system. A water distribution system in which high pressures are supplied at hydrant outlets at all times.

horizontal reach. The reach of a fire stream when the angle of discharge is 45 degrees or less.

horizontal stream. A stream that has an angle of discharge of 45 degrees or less.

hydraulics. The branch of physics dealing with the mechanical properties of water and other liquids and with the application of these properties to engineering.

inches of mercury. A reading below zero on the compound gage.

indirect attack. A method of fire attack in which water is applied to the heated atmosphere rather than to the burning material itself.

knock down. To reduce the flames and heat at a fire to the point at which the fire is considered under control.

ladder pipe. A heavy-stream appliance attached to the end of an aerial ladder and used for directing large streams of water from an above-ground position.

large-quantity relays. Relays requiring a flow of 500 gpm or more.

latent heat of fusion. The amount of heat absorbed or released by a substance as it passes between the solid and liquid phases. Latent heat is measured in Btu's or calories per unit weight.

latent heat of vaporization. The amount of heat absorbed or given off as a substance passes between the liquid and gaseous phases.

layout. The configuration of hose lines from a pressure source to the point of use.

lift. The vertical distance from the surface of the water to the center of the pump when a pumper is drafting.

load a line. Fill a hose line with water.

master stream. Any stream that is too large to be handheld or a stream providing 350 gpm or more.

maximum daily consumption. The maximum amount of water used in a city during any 24-hour period in a 3-year period.

maximum lift. The maximum height to which a pumper can draft water.

momentary nozzle reaction. The nozzle reaction that occurs when a nozzle is first opened.

multipurpose nozzle. A nozzle designed to provide fire streams that include a solid-bore stream and a fog nozzle at the same time or independently.

needed fire flow. The amount of water needed to confine a major fire to the buildings within a block or other major complex.

net pump pressure. The amount of pressure actually produced by a pump.

nonsolid streams. Water discharged from a fog nozzle in the form of a spray or fog mist.

nozzle pressure. The discharge pressure of water from a nozzle.

nozzle reaction. A force moving in the opposite direction of water leaving a nozzle.

officer in charge. The ranking officer in charge of an emergency.

open butt. A hose line without a nozzle attached, or a hydrant outlet from which water is flowing.

partial vacuum. A reduction of pressure below atmospheric pressure.

peak hourly consumption. The maximum amount of water that can be expected to be used in any given hour of a day.

pitot tube. A device that is inserted into a stream of water from a nozzle or hydrant opening and used to determine the discharge pressure.

portable monitor. A heavy-stream appliance that can be moved to various locations for discharging master streams.

positive-displacement pump. A pump capable of pumping air as well as water.

pressure. Force per unit area.

pressure gage. The gage connected to the discharge side of a pump.

pressure hydrant. The hydrant used to measure the static and residual pressure on a water system during testing operations.

primary feeders. Large pipes used for moving water from the source of supply or storage area to the secondary feeders.

priming pump. A positive-displacement pump used to remove the air from the main pump of a pumper when drafting.

priming valve. A valve used to open a line between the main pump and the priming pump.

pump drain. A valve used to drain the water from a pump.

pumper. A fire apparatus equipped with a pump and used to move water from a supply source to the working lines.

pumping system. A water system that depends on the use of pumps to provide the pressures necessary to overcome friction loss in the system and to provide the required pressures for fire protection.

reach. The distance that a fire stream can be effectively projected from a nozzle and still be classified as a good stream.

relative pressure. Pressure relative to atmospheric pressure. Pressure above atmospheric pressure is referred to as positive pressure, and pressure below atmospheric pressure is referred to as negative pressure.

relay operations. Operations that involve the movement of water and require the use of two or more pumpers in series so that the water discharged from one pumper passes into the intake of another pumper.

relay pumper. A pumper in line in a relay operation between the supply pumper and the attack pumper.

relief valve. A valve designed to sense an increase of water flow and open a valve to dump some of the water in order to maintain the pressure for which it is set.

required pump discharge pressure. The pressure required to overcome the friction loss in a hose, the back pressure, and the appliance friction loss and to provide the desired nozzle pressure.

residual pressure. The pressure remaining at a hydrant outlet after water is flowing.

riser. A vertical pipe in a building or appliance, used to move water from a lower level to an upper level.

secondary feeders. Water mains used to tie the grid system to the primary feeders so as to aid in the concentration of the needed fire flow at any point within the grid network.

selectable-gallonage nozzle. A nozzle designed to enable the nozzle operator to select a discharge rate that he or she deems best suitable for the task at hand.

shut down. The closing of a nozzle tip or the closing of all discharge gates on a pumper.

siamesed lines. A combination of two or more lines to form a single line.

similar right triangles. Two different size right triangles which have identical angles.

single-series relay. A relay that requires a single line of pumpers in series to supply the working line.

small-quantity relays. Relays in which a single $2\frac{1}{2}$-inch hose can be used between relaying pumpers.

smooth-bore tip. A nozzle tip designed to project a solid stream.

solid stream. A straight stream that has not broken into showers of spray.

specific heat. The amount of heat required to raise the temperature of 1 pound of a substance 1°F, or the number of calories required to raise the temperature of 1 gram of a substance 1°C. (From a fire protection standpoint, the specific heat of a substance should be thought of as its thermal capacity, or its ability to absorb heat.)

spray stream. A stream of water projected from a fog nozzle and discharged in the form of water droplets.

sprinkler head. A device for automatically distributing water on a fire or holding a fire in check.

standpipe. A pipe riser used to transfer water from the street level to outlets on upper floors of multistory buildings.

static pressure. The pressure of water when it is not in motion.

straight stream. A stream of water projected from a fog nozzle, if selected.

stream penetration. The penetration of water from a hose stream into a building.

suction gage. A gage that measures the intake pressure of a pump.

suction hose. Hard or flexible hose that is reinforced against collapsing due to negative pressure inside the hose. Used in drafting operations.

suction inlet. The intake side of a pump.

suction strainer. The strainer used at the end of a suction hose.

supply pumper. A pumper receiving water from a pressure or static source and transferring it to a relay or attack pumper.

system adequacy. The condition of a water system being able to deliver the needed fire flow for the required duration of hours while the domestic consumption is at the maximum daily rate.

system reliability. The condition of a water system being able to supply the needed

fire flow for the required duration of hours while the domestic consumption is at the maximum daily rate, under certain emergency or unusual conditions.

two and one-half-inch equivalent. The length of 2½-inch hose that has the same friction loss as a layout of hose other than single 2½-inch hose when the same amount of water is flowing in the other layout as in the single 2½-inch line.

vacuum. Pressure below atmospheric pressure. It is also referred to as negative pressure.

velocity flow. The speed with which water passes a given point.

vertical reach. The reach of a fire stream when the angle of discharge is greater than 45 degrees.

vertical stream. A fire stream where the angle of discharge is greater than 45 degrees.

volume. The amount of space included within the bounding surfaces of rectangular or cylindrical containers.

water hammer. The shock caused by the sudden stoppage of flowing water.

water triangle. A diagram used to explain the three limits to a pumper setup.

wet-barrel hydrant. A hydrant in which the barrel is filled with water at all times.

wet-standpipe system. A fire protection system that is filled with water under pressure and to which hose lines are attached for the purpose of firefighting operations.

whirlpool. A whirling motion of water, causing a vacuum at the center.

wyed lines. A single line divided into two or more lines.

Abbreviations

A	area; also, one side of a right triangle
AA	appliance allowance; a pressure allowance for heavy-stream appliances that includes the friction loss in the appliance, the nozzle pressure, and, with ladder pipes and elevated platforms, the back pressure
AFL	appliance friction loss; a pressure allowance for the friction loss of an appliance
AWWA	American Water Works Association
BP	back pressure
C	friction loss coefficient
C2	constant used in the formula for determining the equivalent length of 2½-inch hose for a given layout
D	diameter of a circle
DIS	discharge; measured in gallons per minute (gpm)
F	force
FL	by the formula, the friction loss in 100 feet of the hose considered
FP	forward pressure
fpm	feet per minute; a measurement of velocity
fps	feet per second; a measurement of velocity
G	percent of grade (elevation)
gpm	gallons per minute
H	head; the vertical distance of the surface of a body of water above the point being considered
h	height
HF	horizontal factor; a factor used in the horizontal reach formula
Hg	chemical symbol for mercury
Hg	used in formulas to indicate "inches of mercury"
IFSTA	International Fire Service Training Association
ipm	inches per minute; a measurement of velocity
ips	inches per second; a measurement of velocity
L	length of a hose line or hose length in hundreds of feet
LDH	large-diameter hose; hose 3½ inches and larger
MNR	momentary nozzle reaction
N	number of smaller tips required to have the same discharge capacity as a larger tip
NFPA	National Fire Protection Association
NP	nozzle pressure
NR	nozzle reaction
P	pressure; the measurement of the energy in water, usually expressed in pounds per square inch (psi)
PDP	pump discharge pressure
psi	pounds per square inch

psia	pounds per square inch absolute
psig	pounds per square inch gage
Q	flow in gpm, divided by 100
RPDP	required pump discharge pressure
S	distance or side
TFL	total friction loss
V	velocity flow, usually measured in feet per second (fps); volume
VF	vertical factor; a factor used in the vertical reach formula
w	width

Summary of Chapter Formulas

Hydraulic problems are solved by selecting the proper formula, substituting numbers for the letters in the formula, and then completing the necessary mathematical computations. Many times, separate computations are required before the proper numbers can be inserted in the formula. In many instances, such as in field hydraulics, the formulas are used as guidelines and checklists because the actual calculations are done mentally. The luxury of referring to a written formula is normally not available to fire service personnel either on the fire ground or when competing in promotional examinations; therefore, it is imperative that formulas be memorized, together with all factors and constants required for the solution of problems.

Following is a list of formulas presented in this book. They are listed as presented, chapter by chapter.

CHAPTER ONE

$$P = .434H$$

where
$$P = \text{pressure}$$
$$H = \text{head}$$

$$H = 2.304P$$

where
$$H = \text{head}$$
$$P = \text{pressure}$$

$$BP = .434H$$

where
$$BP = \text{back pressure}$$
$$H = \text{head}$$

$$FP = .434H$$

where
$$FP = \text{forward pressure}$$
$$H = \text{head}$$

$$H = GL$$

where
$$H = \text{resultant head}$$
$$G = \text{percent of grade}$$
$$L = \frac{\text{length of line}}{100}$$

$$\text{Force } (F) = P \times A \text{ (pressure} \times \text{area)}$$

$$\frac{\text{Pressure required to}}{\text{open a clapper valve}} = \frac{\text{force on one side}}{\text{area of the other side}}$$

CHAPTER TWO

Area of a Square

$$\text{Area} = S^2$$

where S = side

Area of a Rectangle

$$\text{Area} = lw$$

where l = length
w = width

Area of a Circle

$$\text{Area} = .7854D^2$$

where D = diameter

Volume of a Rectangular Container

$$\text{Volume} = lwh$$

where l = length
w = width
h = height

Volume of a Cylindrical Container

$$\text{Volume} = .7854D^2H$$

where D = diameter
H = height

Gallon Capacity of Containers

$$\text{Gallon capacity} = 7.48V$$

where V = volume in cubic feet

Weight Capacity of Containers

$$\text{Weight capacity} = 62.5V$$

where V = volume in cubic feet

$$\text{Gallon capacity of hose} = \frac{\text{volume (cu. in.)}}{231}$$

$$\text{Weight capacity of hose} = \text{gallons} \times 8.35$$

CHAPTER THREE

No formulas.

CHAPTER FOUR

Horizontal Reach

$$S = \sqrt{(HF)(P)}$$

where
- S = distance
- HF = horizontal factor
- P = nozzle pressure

or

$$S = \frac{1}{2}P + 26$$

where
- S = distance
- P = nozzle pressure

Vertical Reach

$$S = \sqrt{(VF)(P)}$$

where
- S = distance
- VF = vertical factor
- P = nozzle pressure

Stream Penetration

$$\frac{A}{B} = \frac{C}{D}$$

where
- A = height of ceiling above window sill
- B = distance from outer wall to point of contact
- C = height of window sill above ground level
- D = distance from nozzle to the building

$$C^2 = A^2 + B^2$$

where
- C = hypotenuse of a right triangle
- A = one side of the right triangle
- B = the other side of the right triangle

Velocity Flow

$$V = 8\sqrt{H}$$

where
- V = velocity flow
- H = head

and

$$V = 12.14\sqrt{P}$$

where
- V = velocity flow
- P = discharge pressure

Comparing the Velocity Flow in a Hose with the Velocity Flow at the Nozzle Tip

$$VD^2 = vd^2$$

where
V = velocity flow through the hose
v = velocity flow from the nozzle tip
D = diameter of the hose
d = diameter of the nozzle tip

Nozzle Reaction

$$NR = 1.57D^2P$$

where
NR = nozzle reaction
D = nozzle diameter
P = nozzle pressure

Momentary Nozzle Reaction

$$MNR = 1.88D^2P$$

where
MNR = momentary nozzle reaction
D = nozzle diameter
P = nozzle pressure

Nozzle Reaction on a Fog Nozzle

$$NR = (.0505)(\text{gpm})(\sqrt{P})$$

where
NR = nozzle reaction
P = nozzle pressure

Formula for Fog Nozzles Rated at 100 psi

$$NR = \frac{\text{gpm}}{2}$$

CHAPTER FIVE

Discharge

$$\text{Discharge} = 29.7D^2\sqrt{P}$$

where
D = diameter of opening
P = discharge pressure

Open-Butt Discharge

$$\text{Discharge} = 27D^2\sqrt{P}$$

where
D = diameter of opening
P = discharge pressure

Comparing Discharge When Nozzle Pressures Are Identical

$$\text{Increase of discharge} = \left(\frac{D}{d}\right)^2$$

where
D = diameter of larger tip
d = diameter of smaller tip

Comparing Discharge When Tip Sizes Are Identical

$$\text{Increase in discharge} = \sqrt{\frac{P}{p}}$$

where
P = larger pressure
p = smaller pressure

Sprinkler Discharge

$$\text{Discharge} = \frac{1}{2}P + 15$$

where
P = discharge pressure

Equivalent Nozzle Diameter

$$END = \frac{\sqrt{D1^2 + D2^2 + D3^2 + \text{etc.}}}{8}$$

where
END = equivalent nozzle diameter
D_1 = number of eighths in diameter of first nozzle tip (in.)
D_2 = number of eighths in diameter of second nozzle tip (in.)
D_3 = number of eighths in diameter of third nozzle tip (in.)

Several Tips Replacing One Tip

$$N = \left(\frac{D}{d}\right)^2$$

where
N = number of smaller tips required
D = diameter of larger tip given in eighths (in.)
d = diameter of smaller tip given in eighths (in.)

CHAPTER SIX

Friction Loss in Relationship to the Pump Discharge Pressure

$$FL = PDP - NP$$

where
FL = friction loss in hose
PDP = pump discharge pressure
NP = nozzle pressure

Total Friction Loss in the Hose

$$TFL = (FL)(L)$$

where
TFL = total friction loss in hose
FL = friction loss in 100 feet of hose
L = total length of line/100

Rate of Friction Loss Increase Caused by an Increase in the Velocity Flow

$$\text{Rate of increase} = \left(\frac{V}{v}\right)^2$$

where
$$V = \text{new velocity flow}$$
$$v = \text{old velocity flow}$$

New Friction Loss Resulting from an Increase in the Velocity Flow

$$FL = \left(\frac{V}{v}\right)^2 (fl)$$

where
$$FL = \text{new friction loss}$$
$$fl = \text{old friction loss}$$
$$V = \text{new velocity flow}$$
$$v = \text{old velocity flow}$$

Friction Loss Formula

$$FL = CQ^2L$$

where
$$FL = \text{friction loss in a layout}$$
$$C = \text{friction loss coefficient (from Table 6.1)}$$
$$Q = \text{flow in hundreds of gpm (flow/100)}$$
$$L = \text{hose length in hundreds of}$$
$$\text{feet (length/100)}$$

Relationship Between Different-Size Hose Lines

$$\text{Equivalent length} = \frac{C_1}{C_2} \times L$$

where
$$C_1 = \text{coefficient of smaller line}$$
$$C_2 = \text{coefficient of larger line}$$
$$L = \text{length of smaller line}$$

CHAPTER SEVEN

Terms used in the following formulas have the following meanings:

$$PDP = \text{pump discharge pressure}$$
$$RPDP = \text{required pump discharge pressure}$$
$$NP = \text{nozzle pressure}$$
$$FL = \text{friction loss in the hose}$$
$$BP = \text{back pressure}$$
$$FP = \text{forward pressure}$$
$$AFL = \text{appliance friction loss}$$
$$C = \text{friction loss constant}$$
$$Q = \text{flow/100}$$
$$L = \text{length of line/100}$$

Single Lines Laid at Ground Level (no appliances involved)

$$RPDP = NP + FL$$

A Single Line Wyed into Two Lines

$$RPDP = NP + FL + AFL$$

Siamesed Lines into a Single Line

$$RPDP = NP + FL + AFL$$

Lines Laid Uphill (no appliances involved)

$$RPDP = NP + FL + BP$$

Lines Laid Downhill (no appliances involved)

$$RPDP = NP + FL - FP$$

Lines Laid into Standpipe Systems

$$RPDP = NP + FL + BP + AFL$$

Lines Laid into Portable Monitors

$$RPDP = NP + FL + AFL$$

Lines Laid into Deck Guns

$$RPDP = NP + FL + BP + AFL$$

Lines Laid into Ladder Pipes

$$RPDP = NP + FL + BP + AFL$$

Lines Laid into Aerial Platforms

$$RPDP = NP + FL + BP + AFL$$

CHAPTER EIGHT

To change hose other than 2½-inch hose into 2½-inch hose equivalent, use

$$E\ 2\frac{1}{2}\text{-inch} = L/C$$

where

$E\ 2\frac{1}{2}$-inch = equivalent length of 2½-inch hose for the line being considered

L = length of line being considered

C = coefficient from Table 8.1

An additional formula for the same purpose is

$$E\ 2\frac{1}{2}\text{-inch} = (L/100) \times C2$$

where

L = length of hose to be changed

$C2$ = equivalent length from Table 8.2

CHAPTER NINE

Determining the Water Available from a Pump

$$\text{Water available} = \frac{\text{pound-gallons}}{\text{pressure}}$$

where

Pound-gallons = rated capacity of pump times the initial rated pressure of the pump

Pressure = net pump pressure (when greater than rated net pump pressure)

Theoretical Height to Which Water Will Rise

$$h = 1.13Hg$$

where

h = height in feet

Hg = inches of mercury

Pressure Reduction Required to Lift Water a Given Amount

$$Hg = .885h$$

where

Hg = inches of mercury

h = height

CHAPTER TEN

Determining the Maximum Distance Between Pumpers

$$\text{Maximum distance between pumpers} = \frac{165}{FL + BP - FP} \times 100$$

where

FL = friction loss per 100 feet of hose layout

BP = back pressure per 100 feet of hose layout

FP = forward pressure per 100 feet of hose layout

CHAPTER ELEVEN

Handheld Lines

Lines laid at ground level: $RPDP = FL + NP$

Lines laid uphill: $RPDP = FL + NP + BP$

Lines laid downhill: $RPDP = FL + NP - FP$

Lines laid into building standpipes: $RPDP = FL + NP + AFL + BP$

where

$RPDP$ = required pump discharge pressure

FL = friction loss in the hose

NP = nozzle pressure

AFL = friction loss in the standpipe if total flow is 350 gpm or more; if less than 350 gpm, no loss is applied

BP = back pressure

FP = forward pressure

Master Streams

Lines laid at ground level: $RPDP = FL + AA$

Lines laid uphill: $RPDP = FL + AA + BP$

Lines laid downhill: $RPDP = FL + AA - FP$

where

$RPDP$ = required pump discharge pressure

FL = friction loss in the hose

AA = appliance allowance

BP = back pressure

FP = forward pressure

Square Roots Table

Square Roots of Numbers (1 to 250)

n	$\sqrt{n}$	n	$\sqrt{n}$	n	$\sqrt{n}$	n	$\sqrt{n}$	n	$\sqrt{n}$	n	$\sqrt{n}$
1	1.	51	7.1414	101	10.0499	151	12.2882	201	14.1774		
2	1.414	52	7.2111	102	10.0995	152	12.3288	202	14.2127		
3	1.732	53	7.2801	103	10.1489	153	12.3693	203	14.2478		
4	2.000	54	7.3485	104	10.1980	154	12.4097	204	14.2829		
5	2.236	55	7.4163	105	10.2470	155	12.4499	205	14.3178		
6	2.449	56	7.4833	106	10.2956	156	12.4900	206	14.3527		
7	2.646	57	7.5498	107	10.3441	157	12.5300	207	14.3875		
8	2.828	58	7.6158	108	10.3923	158	12.5698	208	14.4222		
9	3.000	59	7.6811	109	10.4403	159	12.6095	209	14.4568		
10	3.162	60	7.7460	110	10.4881	160	12.6491	210	14.4914		
11	3.3166	61	7.8102	111	10.5357	161	12.6886	211	14.5258		
12	3.4641	62	7.8740	112	10.5830	162	12.7279	212	14.5602		
13	3.6056	63	7.9373	113	10.6301	163	12.7671	213	14.5945		
14	3.7417	64	8.0000	114	10.6771	164	12.8062	214	14.6287		
15	3.8730	65	8.0623	115	10.7238	165	12.8452	215	14.6629		
16	4.0000	66	8.1240	116	10.7703	166	12.8841	216	14.6969		
17	4.1231	67	8.1854	117	10.8167	167	12.9228	217	14.7309		
18	4.2426	68	8.2462	118	10.8628	168	12.9615	218	14.7648		
19	4.3589	69	8.3066	119	10.9087	169	13.0000	219	14.7986		
20	4.4721	70	8.3666	120	10.9545	170	13.0384	220	14.8324		
21	4.5826	71	8.4261	121	11.0000	171	13.0767	221	14.8661		
22	4.6904	72	8.4853	122	11.0454	172	13.1149	222	14.8997		
23	4.7958	73	8.5440	123	11.0905	173	13.1529	223	14.9332		
24	4.8990	74	8.6023	124	11.1355	174	13.1909	224	14.9666		
25	5.0000	75	8.6603	125	11.1803	175	13.2288	225	15.0000		
26	5.0990	76	8.7178	126	11.2250	176	13.2665	226	15.0333		
27	5.1962	77	8.7750	127	11.2694	177	13.3041	227	15.0665		
28	5.2915	78	8.8318	128	11.3137	178	13.3417	228	15.0997		
29	5.3852	79	8.8882	129	11.3578	179	13.3791	229	15.1327		
30	5.4772	80	8.9443	130	11.4018	180	13.4164	230	15.1658		
31	5.5678	81	9.0000	131	11.4455	181	13.4536	231	15.1987		
32	5.6569	82	9.0554	132	11.4891	182	13.4907	232	15.2315		
33	5.7446	83	9.1104	133	11.5326	183	13.5277	233	15.2643		
34	5.8310	84	9.1652	134	11.5758	184	13.5647	234	15.2971		
35	5.9161	85	9.2195	135	11.6190	185	13.6015	235	15.3297		
36	6.0000	86	9.2736	136	11.6619	186	13.6382	236	15.3623		
37	6.0828	87	9.3274	137	11.7047	187	13.6748	237	15.3948		
38	6.1644	88	9.3808	138	11.7473	188	13.7113	238	15.4272		
39	6.2450	89	9.4340	139	11.7898	189	13.7477	239	15.4596		
40	6.3246	90	9.4868	140	11.8322	190	13.7840	240	15.4919		
41	6.4031	91	9.5394	141	11.8743	191	13.8203	241	15.5242		
42	6.4807	92	9.5917	142	11.9164	192	13.8564	242	15.5563		
43	6.5574	93	9.6437	143	11.9583	193	13.8924	243	15.5885		
44	6.6332	94	9.6954	144	12.0000	194	13.9284	244	15.6205		
45	6.7082	95	9.7468	145	12.0416	195	13.9642	245	15.6525		
46	6.7823	96	9.7980	146	12.0830	196	14.0000	246	15.6844		
47	6.8557	97	9.8489	147	12.1244	197	14.0357	247	15.7162		
48	6.9282	98	9.8995	148	12.1655	198	14.0712	248	15.7480		
49	7.0000	99	9.9499	149	12.2066	199	14.1067	249	15.7797		
50	7.0711	100	10.0000	150	12.2474	200	14.1421	250	15.8114		

Answers to Tests

CHAPTER ONE

1. 55.55 psi
2. 45.14 psi
3. 195.84 psi
4. 34.2 feet
5. 48.61 psi
6. 62.93 psi
7. 42.32 psi
8. 16.93 *BP*
9. 1692.6 pounds
10. 61.11 psi

CHAPTER TWO

1. 126 square feet
2. 2574 cubic feet
3. 132.73 square feet
4. 3887.73 cubic feet
5. 19,836.96 gallons
6. 23,029.18 gallons
7. 742,500 pounds
8. 2,650,725 pounds
9. 10,404.1 pounds
10. 3394.73 gallons
11. 68.14 pounds
12. 65.28 gallons
13. 153.31 pounds
14. 25 gallons
15. 6.25 gallons

CHAPTER FOUR

1. 79 feet
2. 107.7 feet
3. 15 feet
4. 55.56 feet
5. 6 feet
6. 62 fps
7. 105.13 fps
8. 31.25 fps
9. 144 fps
10. 119.22 pounds
11. 176.25 pounds
12. 2.56 times greater

CHAPTER FIVE

1. 348.5 gpm
2. 129.47 gpm
3. 293.29 gpm
4. 1127.4 gpm
5. 3615.8 gpm
6. 27,412.56 gallons
7. Two
8. Approximately 375 gpm
9. 1.77 times greater
10. 470.25 gpm
11. 1.25 times greater
12. 156.25 gpm
13. 15,750 gallons
14. 1½ inch
15. 2 inches
16. Six

Note: Answers obtained may be slightly different, depending on how the numbers are rounded off.

CHAPTER SIX

1. 35 psi
2. 1.96 times greater
3. 85.5 psi
4. 12.99 times greater
5. 32 times
6. 171.88 psi
7. 43.2 psi
8. 55.96 psi
9. 77.25 psi
10. 96.75 psi
11. 57.29 psi
12. 50.7 psi
13. 91.26 psi
14. 38.4 psi
15. 138.24 psi
16. 87.19 psi
17. 61.25 psi
18. The second line requires 110.25 psi
19. 52.1 psi
20. 80.94 psi

CHAPTER SEVEN

1. 166 psi
2. 70.25 psi
3. 199 psi
4. 111.27 psi
5. 142.03 psi
6. 230.5 psi
7. 161.12 psi
8. 72.5 psi
9. 63.39 psi
10. 163.38 psi
11. 174.5 psi
12. 205 psi
13. 130 psi
14. 182.88 psi
15. 219.14 psi

Note: In Problem 15, the pressure on 5-inch lines is limited to 185 psi. Therefore, the answer to the problem is too great. A different setup would be required.

CHAPTER EIGHT

1. 154.15 psi
2. 159.47 psi
3. 192.3 psi
4. 186.51 psi
5. 116.73 psi
6. 204.4 psi
7. 80 feet
8. 2308 feet
9. 160 feet
10. 103 feet
11. 1533 feet
12. 46 feet
13. 176.83 psi
14. 196.7 psi

CHAPTER NINE

1. 6.86 psi
2. 18.08 feet
3. 15.93 inches of Hg
4. 19.09 feet
5. 135 psi

CHAPTER TEN

1. 750 feet
2. 1800 feet
3. 900 feet
4. Five pumpers; 1150 feet
5. 6 pumpers; 700 feet
6. Two pumpers; a relay is not needed
7. Seven pumpers
8. 139.5 psi (pump at 140)
9. Five pumpers at 700 feet, one pumper at 650 feet
10. 165 psi

CHAPTER ELEVEN

1. 69.5 psi (pump at 70)
2. 160 psi
3. 106 psi (pump at 105)
4. 66 psi (pump at 65)
5. 209 psi (pump at 210)
6. 102.5 psi (pump at 105)
7. 210 psi
8. 85 psi
9. 155 psi
10. 92 psi (pump at 90)
11. 183 psi (pump at 185)
12. 110 psi
13. 201 psi (pump at 200)
14. 168 psi (pump at 170)
15. 190 psi
16. 153 psi (pump at 155)
17. 163 psi (pump at 165)
18. 190 psi
19. 166 psi (pump at 165)
20. 180 psi
21. 72 psi
22. 15 psi
23. 16 psi
24. 58.8 or 59 psi
25. 28.8 or 29 psi

Index